T0139959

Applied Mathematical Sciences

Volume 188

For further volumes:
http://www.springer.com/series/34

Applied Mathematical Sciences

Volume 188

Founding Editors
F. John, Joseph LaSalle and Lawrence Sirovich

Editors
S.S. Antman
ssa@math.umd.edu

P. Holmes
pholmes@math.princeton.edu

K. Sreenivasan
katepalli.sreenivasan@nyu.edu

Advisors
L. Greengard
L. Kadanoff
R.V. Kohn
B. Matkowsky
R. Pego
C. Peskin
A. Singer
A. Stevens
A. Stuart

For further volumes:
http://www.springer.com/series/34

Fioralba Cakoni • David Colton

A Qualitative Approach
to Inverse Scattering Theory

 Springer

Fioralba Cakoni
Department of Mathematical Sciences
University of Delaware
Newark, DE, USA

David Colton
Department of Mathematical Sciences
University of Delaware
Newark, DE, USA

ISSN 0066-5452   ISSN 2196-968X (electronic)
ISBN 978-1-4899-7960-5   ISBN 978-1-4614-8827-9 (eBook)
DOI 10.1007/978-1-4614-8827-9
Springer New York Heidelberg Dordrecht London

Mathematics Subject Classification (2010): 35P25, 35R25, 35R30, 65M30, 65R30, 78A45

Printed on acid-free paper

Springer is part of Springer Science+Business Media (www.springer.com)

To the Memory of
Natasha Colton
1975–2013

Preface

The field of inverse scattering theory has been a particularly active field in applied mathematics for the past 25 years. The aim of research in this field has been to not only detect but also to identify unknown objects through the use of acoustic, electromagnetic, or elastic waves. Although the success of such techniques as ultrasound and x-ray tomography in medical imaging has been truly spectacular, progress has lagged in other areas of application, which are forced to rely on different modalities using limited data in complex environments. Indeed, as pointed out in [88] concerning the problem of locating unexploded ordinance, "Target identification is the great unsolved problem. We detect almost everything, we identify nothing."

Until a few years ago, essentially all existing algorithms for target identification were based on either a weak scattering approximation or on the use of nonlinear optimization techniques. A survey of the state of the art for acoustic and electromagnetic waves as of 1998 can be found in [54]. However, as the demands of imaging increased, it became clear that incorrect model assumptions inherent in weak scattering approximations imposed severe limitations on when reliable reconstructions were possible. On the other hand, it was also realized that for many practical applications nonlinear optimization techniques require a priori information that is in general not available. Hence, in recent years, alternative methods for imaging have been developed that avoid incorrect model assumptions but, as opposed to nonlinear optimization techniques, only seek limited information about the scattering object. Such methods come under the general title of *qualitative methods in inverse scattering theory*. Examples of such an approach are the linear sampling method, [54, 107], the factorization method [98, 107], the method of singular sources [138, 139], the probe method [91, 92], and the use of convex scattering supports [74, 116], all of which seek to determine an approximation to the shape of the scattering obstacle but in general provide only limited information about the material properties of the scatterer.

This book is designed to be an introduction to qualitative methods in inverse scattering theory, focusing on the basic ideas of the linear sampling method and its close relative, the factorization method. The obvious question is: an introduction for whom? One of the problems in making these new ideas in inverse scattering theory available to the wider scientific and engineering community is that the research papers in this area make use of mathematics that may be beyond the training of a reader who is not a professional mathematician. This book represents an effort to overcome this problem and to write a monograph that is accessible to anyone having a mathematical background only in advanced calculus and linear algebra. In particular, the necessary material on functional analysis, Sobolev spaces, and the theory of ill-posed problems will be given in the first two chapters. Of course, to do this in a short book such as this one, some proofs will not be given, nor will all theorems be proven in complete generality. In particular, we will use the mapping and discontinuity properties of double- and single-layer potentials with densities in the Sobolev spaces $H^{1/2}(\partial D)$ and $H^{-1/2}(\partial D)$, respectively, but will not prove any of these results, referring for their proofs to the monographs [111] and [127]. We will furthermore restrict ourselves to a simple model problem, the scattering of time-harmonic electromagnetic waves by an infinite cylinder. This choice means that we can avoid the technical difficulties of three-dimensional inverse scattering theory for different modalities and instead restrict our attention to the simpler case of two-dimensional problems governed by the Helmholtz equation. For a glimpse of the problems arising in the three-dimensional "real world," we refer the reader to [26].

Although, for the foregoing reasons we do not discuss the qualitative approach to the inverse scattering problem for modalities other than electromagnetic waves, the reader should not assume that such approaches do not exist! Indeed, having mastered the material in this book, the reader will be fully prepared to understand the literature on qualitative methods for inverse scattering problems arising in other areas of application, such as acoustics and elasticity. In particular, for qualitative methods in the inverse scattering problem for acoustic waves and underwater sound see [12, 133, 158, 159, 160], whereas for elasticity we refer the reader to [5, 37, 38, 73, 132, 135, 150].

We would like to acknowledge the scientific and financial support of the Air Force Office of Scientific Research and in particular Dr. Arje Nachman of AFOSR and Dr. Richard Albanese of Brooks Air Force Base. Finally, a special thanks to our colleague Peter Monk, who has been a participant with us in developing the qualitative approach to inverse scattering theory and whose advice and insights have been indispensable to our research efforts.

In closing, we note that this book is an updated and expanded version of an earlier book by the authors that originally appeared in the Springer Series on Interactions of Mechanics and Mathematics entitled *Qualitative Methods in Inverse Scattering Theory.*

Newark, Delaware   Fioralba Cakoni, David Colton

Contents

1   **Functional Analysis and Sobolev Spaces**   1
    1.1   Normed Spaces   1
    1.2   Bounded Linear Operators   6
    1.3   Adjoint Operator   14
    1.4   Sobolev Space $H^p[0, 2\pi]$ 17
    1.5   Sobolev Space $H^p(\partial D)$ 23

2   **Ill-Posed Problems**   27
    2.1   Regularization Methods   28
    2.2   Singular Value Decomposition   30
    2.3   Tikhonov Regularization   36

3   **Scattering by Imperfect Conductors**   45
    3.1   Maxwell's Equations   45
    3.2   Bessel Functions   47
    3.3   Direct Scattering Problem   51

4   **Inverse Scattering Problems for Imperfect Conductors**   63
    4.1   Far-Field Patterns   64
    4.2   Uniqueness Theorems for Inverse Problem   67
    4.3   Linear Sampling Method   72
    4.4   Determination of Surface Impedance   78
    4.5   Limited Aperture Data   81
    4.6   Near-Field Data   83

5   **Scattering by Orthotropic Media**   85
    5.1   Maxwell Equations for an Orthotropic Medium   85
    5.2   Mathematical Formulation of Direct Scattering Problem   89
    5.3   Variational Methods   94
    5.4   Solution of Direct Scattering Problem 105

6   Inverse Scattering Problems for Orthotropic Media 111
   6.1   Formulation of Inverse Problem 112
   6.2   Interior Transmission Problem 114
   6.3   Transmission Eigenvalue Problem 122
         6.3.1   The Case $n = 1$ 123
         6.3.2   The Case $n \neq 1$ 136
         6.3.3   Discreteness of Transmission Eigenvalues 136
         6.3.4   Existence of Transmission Eigenvalues for $n \neq 1$ 140
   6.4   Uniqueness 147
   6.5   Linear Sampling Method 151
   6.6   Determination of Transmission Eigenvalues from Far-Field
         Data 160

7   Factorization Methods 165
   7.1   Factorization Method for Obstacle Scattering 166
         7.1.1   Preliminary Results 166
         7.1.2   Properties of Far-Field Operator 176
         7.1.3   Factorization Method 180
   7.2   Factorization Method for an Inhomogeneous Medium 186
         7.2.1   Preliminary Results 186
         7.2.2   Properties of Far-Field Operator 191
         7.2.3   Factorization Method 193
   7.3   Justification of Linear Sampling Method 199
   7.4   Closing Remarks 202

8   Mixed Boundary Value Problems 203
   8.1   Scattering by a Partially Coated Perfect Conductor 204
   8.2   Inverse Scattering Problem for Partially Coated Perfect
         Conductor 211
   8.3   Numerical Examples 216
   8.4   Scattering by Partially Coated Dielectric 221
   8.5   Inverse Scattering Problem for Partially Coated Dielectric 231
   8.6   Numerical Examples 237
   8.7   Scattering by Cracks 240
   8.8   Inverse Scattering Problem for Cracks 251
   8.9   Numerical Examples 258

9   Inverse Spectral Problems for Transmission Eigenvalues ... 263
   9.1   Entire Functions 263
   9.2   Transformation Operators 267
   9.3   Transmission Eigenvalues 269
   9.4   An Inverse Spectral Theorem 274

10 A Glimpse at Maxwell's Equations 279

References 287

Index 295

1

Functional Analysis and Sobolev Spaces

Much of the recent work on inverse scattering theory is based on the use of special topics in functional analysis and the theory of Sobolev spaces. The results that we plan to present in this book are no exception. Hence we begin our book by providing a short introduction to the basic ideas of functional analysis and Sobolev spaces that will be needed to understand the material that follows. Since these two topics are the subject matter of numerous books at various levels of difficulty, we can only hope to present the bare rudiments of each of these fields. Nevertheless, armed with the material presented in this chapter, the reader will be well prepared to follow the arguments presented in subsequent chapters of this book.

We begin our presentation with the definition and basic properties of normed spaces and in particular Hilbert spaces. This is followed by a short introduction to the elementary properties of bounded linear operators and in particular compact operators. Included here is a proof of the Riesz theorem for compact operators on a normed space and the spectral properties of compact operators. We then proceed to a discussion of the adjoint operator in a Hilbert space and a proof of the Hilbert–Schmidt theorem. We conclude our chapter with an elementary introduction to Sobolev spaces. Here, following [111], we base our presentation on Fourier series rather than the Fourier transform and prove special cases of Rellich's theorem, the Sobolev embedding theorem, and the trace theorem.

1.1 Normed Spaces

We begin with the basic definition of a normed space X. We will always assume that $X \neq \{0\}$.

Definition 1.1. Let X be a vector space over the field $\mathbb{C}$ of complex numbers. A function $||\cdot|| : X \to \mathbb{R}$ such that

F. Cakoni and D. Colton, *A Qualitative Approach to Inverse Scattering Theory*,
Applied Mathematical Sciences 188, DOI 10.1007/978-1-4614-8827-9_1,
© Springer Science+Business Media New York 2014

1. $\|\varphi\| \geq 0$,
2. $\|\varphi\| = 0$ if and only if $\varphi = 0$,
3. $\|\alpha\varphi\| = |\alpha| \, \|\varphi\|$ for all $\alpha \in \mathbb{C}$,
4. $\|\varphi + \psi\| \leq \|\varphi\| + \|\psi\|$

for all $\varphi, \psi \in X$ is called a *norm* on X. A vector space X equipped with a norm is called a *normed space*.

Example 1.2. The vector space $\mathbb{C}^n$ of ordered n-tuples of complex numbers $(\xi_1, \xi_2, \cdots, \xi_n)$, with the usual definitions of addition and scalar multiplication, is a normed space with norm

$$\|x\| := \left(\sum_1^n |\xi_i|^2 \right)^{\frac{1}{2}},$$

where $x = (\xi_1, \xi_2, \cdots, \xi_n)$. Note that the *triangle inequality* $\|x + y\| \leq \|x\| + \|y\|$ is simply a restatement of Minkowski's inequality for sums [115].

Example 1.3. Consider the vector space X of continuous complex valued functions defined on the interval $[a, b]$ with the obvious definitions of addition and scalar multiplication. Then

$$\|\varphi\| := \max_{a \leq x \leq b} |\varphi(x)|$$

defines a norm on X, and we refer to the resulting normed space as $C\,[a, b]$.

Example 1.4. Let X be the vector space of square integrable functions on $[a, b]$ in the sense of Lebesgue. Then it is easily seen that

$$\|\varphi\| := \left[\int_a^b |\varphi(x)|^2 \, dx \right]^{\frac{1}{2}}$$

defines a norm on X. We refer to the resulting normed space as $\mathrm{L}^2[a, b]$.

Given a normed space X, we now introduce a topological structure on X. A sequence $\{\varphi_n\}, \varphi_n \in X$, *converges* to $\varphi \in X$ if $\|\varphi_n - \varphi\| \to 0$ as $n \to \infty$, and we write $\varphi_n \to \varphi$. If Y is another normed space, a function $A : X \to Y$ is *continuous* at $\varphi \in X$ if $\varphi_n \to \varphi$ implies that $A\varphi_n \to A\varphi$. In particular, it is an easy exercise to show that $\|\cdot\|$ is continuous. A subset $U \subset X$ is *closed* if it contains all limits of convergent sequences of U. The *closure* $\overline{U}$ of U is the set of all limits of convergent sequences of U. A set U is called *dense* in X if $\overline{U} = X$.

In applications we are usually only interested in normed spaces that have the property of *completeness*. To define this property, we first note that a sequence $\{\varphi_n\}, \varphi_n \in X$, is called a *Cauchy sequence* if for every $\epsilon > 0$ there exists an integer $N = N(\epsilon)$ such that $\|\varphi_n - \varphi_m\| < \epsilon$ for all $m, n \geq N$. We then call a subset U of X *complete* if every Cauchy sequence in U converges to an element of U.

Definition 1.5. A complete normed space X is called a *Banach space*.

It can be shown that for each normed space X there exists a Banach space $\hat{X}$ such that X is isomorphic and isometric to a dense subspace of $\hat{X}$, i.e., there is a linear bijective mapping I from X onto a dense subspace of $\hat{X}$ such that $||I\varphi||_{\hat{X}} = ||\varphi||_X$ for all $\varphi \in X$ [115]. $\hat{X}$ is said to be the *completion* of X. For example, $[a, b]$ with the norm $||x|| = |x|$ for $x \in [a, b]$ is the completion of the set of rational numbers in $[a, b]$ with respect to this norm. It can be shown that the completion of the space of continuous complex valued functions on the interval $[a, b]$ with respect to the norm $||\cdot||$ defined by

$$||\varphi|| := \left[\int_a^b |\varphi(x)|^2 \, dx \right]^{\frac{1}{2}}$$

is the space $L^2[a, b]$ defined earlier.

We now introduce vector spaces that have an *inner product* defined on them.

Definition 1.6. Let X be a vector space over the field $\mathbb{C}$ of complex numbers. A function $(\cdot, \cdot) : X \times X \to \mathbb{C}$ such that

1. $(\varphi, \varphi) \geq 0$,
2. $(\varphi, \varphi) = 0$ if and only if $\varphi = 0$,
3. $(\varphi, \psi) = \overline{(\psi, \varphi)}$,
4. $(\alpha\varphi + \beta\psi, \chi) = \alpha(\varphi, \chi) + \beta(\psi, \chi)$ for all $\alpha, \beta \in \mathbb{C}$

for all $\varphi, \psi, \chi \in X$ is called an *inner product* on X.

Example 1.7. For $x = (\xi_1, \xi_2, \cdots, \xi_n), y = (\eta_1, \eta_2, \cdots, \eta_n)$ in $\mathbb{C}^n$,

$$(x, y) := \sum_1^n \xi_i \overline{\eta_i}$$

is an inner product on $\mathbb{C}^n$.

Example 1.8. An inner product on $L^2[a, b]$ is given by

$$(\varphi, \psi) := \int_a^b \varphi\overline{\psi} \, dx.$$

Theorem 1.9. *An inner product satisfies the* Cauchy–Schwarz *inequality*

$$|(\varphi, \psi)|^2 \leq (\varphi, \varphi)(\psi, \psi)$$

for all $\varphi, \psi \in X$ with equality if and only if φ and ψ are linearly dependent.

Proof. The inequality is trivial for $\varphi = 0$. For $\varphi \neq 0$ and

$$\alpha = -\frac{\overline{(\varphi, \psi)}}{(\varphi, \varphi)^{1/2}} \quad , \quad \beta = (\varphi, \varphi)^{1/2}$$

we have that

$$0 \leq (\alpha\varphi + \beta\psi, \alpha\varphi + \beta\psi) = |\alpha|^2(\varphi, \varphi) + 2\mathrm{Re}\left\{\alpha\overline{\beta}(\varphi, \psi)\right\} + |\beta|^2(\psi, \psi)$$
$$= (\varphi, \varphi)(\psi, \psi) - |(\varphi, \psi)|^2,$$

from which the inequality of the theorem follows. Equality holds if and only if $\alpha\varphi + \beta\psi = 0$, which implies that φ and ψ are linearly dependent since $\beta \neq 0$. $\square$

A vector space with an inner product defined on it is called an *inner product space*. If X is an inner product space, then $||\varphi|| := (\varphi, \varphi)^{\frac{1}{2}}$ defines a norm on X. If X is complete with respect to this norm, then X is called a *Hilbert space*. A *subspace* U of an inner product space X is a vector subspace of X taken with the inner product on X restricted to $U \times U$.

Example 1.10. With the inner product of the previous example, $\mathrm{L}^2[a, b]$ is a Hilbert space.

Two elements φ and ψ of a Hilbert space are called *orthogonal* if $(\varphi, \psi) = 0$, and we write $\varphi \perp \psi$. A subset $U \subset X$ is called an *orthogonal system* if $(\varphi, \psi) = 0$ for all $\varphi, \psi \in U$ with $\varphi \neq \psi$. An orthogonal system U is called an *orthonormal system* if $||\varphi|| = 1$ for every $\varphi \in U$. The set

$$U^{\perp} := \{\psi \in X : \psi \perp U\}$$

is called the *orthogonal complement* of the subset U.

Now let $U \subset X$ be a subset of a normed space X, and let $\varphi \in X$. An element $v \in U$ is called a *best approximation* to φ with respect to U if

$$||\varphi - v|| = \inf_{u \in U} ||\varphi - u||.$$

Theorem 1.11. *Let U be a subspace of a Hilbert space X. Then v is a best approximation to $\varphi \in X$ with respect to U if and only if $\varphi - v \perp U$. For each $\varphi \in X$ there exists at most one best approximation with respect to U.*

Proof. The theorem follows from

$$||(\varphi - v) + \alpha u||^2 = ||\varphi - v||^2 + 2\alpha\mathrm{Re}(\varphi - v, u) + \alpha^2||u||^2, \qquad (1.1)$$

which is valid for all $v, u \in U$ and $\alpha \in \mathbb{R}$. In particular, if $u \neq 0$, then the minimum of the right-hand side of (1.1) occurs when

$$\alpha = -\frac{\text{Re}(\varphi - v, u)}{||u||^2},$$

and hence $||(\varphi - v) + \alpha u||^2 > ||\varphi - v||^2$, unless $\varphi - v \perp U$. On the other hand, if $\varphi - v \perp U$, then $||(\varphi - v) + \alpha u||^2 \geq ||\varphi - v||^2$ for all α and u, which implies that v is a best approximation to φ. Finally, if there are two best approximations v_1 and v_2, then $(\varphi - v_1, u) = (\varphi - v_2, u) = 0$, and hence $(\varphi, u) = (v_1, u) = (v_2, u)$ for every $u \in U$. Thus $(v_1 - v_2, u) = 0$ for every $u \in U$ and, setting $u = v_1 - v_2$, we see that $v_1 = v_2$. □

Theorem 1.12. *Let U be a complete subspace of a Hilbert space X. Then for every element of X there exists a unique best approximation with respect to U.*

Proof. Let $\varphi \in X$ and choose $\{u_n\}, u_n \in U$, such that

$$||\varphi - u_n||^2 \leq d^2 + \frac{1}{n}, \tag{1.2}$$

where $d := \inf_{u \in U} ||\varphi - u||$. Then, from the easily verifiable *parallelogram equality*

$$||\varphi + \psi||^2 + ||\varphi - \psi||^2 = 2 \left(||\varphi||^2 + ||\psi||^2 \right),$$

we have that

$$||(\varphi - u_n) + (\varphi - u_m)||^2 + ||u_n - u_m||^2 = 2 ||\varphi - u_n||^2 + 2 ||\varphi - u_m||^2$$

$$\leq 4d^2 + \frac{2}{n} + \frac{2}{m},$$

and, since $\frac{1}{2}(u_n + u_m) \in U$, we have that

$$||u_n - u_m||^2 \leq 4d^2 + \frac{2}{n} + \frac{2}{m} - 4 \left\| \varphi - \frac{1}{2}(u_n + u_m) \right\|^2$$

$$\leq \frac{2}{n} + \frac{2}{m}.$$

Hence $\{u_n\}$ is a Cauchy sequence and, since U is complete, u_n converges to an element $v \in U$. Passing to the limit in (1.2) implies that v is a best approximation to φ with respect to U. Uniqueness follows from Theorem 1.11. □

We note that if U is a closed (and hence complete) subspace of a Hilbert space X, then we can write $\varphi = v + \varphi - v$, where $\varphi - v \perp U$, i.e., X is the *direct sum* of U and its orthogonal complement, which we write as

$$X = U \oplus U^\perp.$$

If U is a subset of a vector space X, then the set spanned by all finite linear combinations of elements of U is denoted by $\text{span}\, U$. A set $\{\varphi_n\}$ in a Hilbert space X such that $\text{span}\{\varphi_n\}$ is dense in X is called a *complete set*.

Theorem 1.13. *Let $\{\varphi_n\}_1^\infty$ be an orthonormal system in a Hilbert space X. Then the following are equivalent:*

a. $\{\varphi_n\}_1^\infty$ is complete.
b. Each $\varphi \in X$ can be expanded in a Fourier series

$$\varphi = \sum_1^\infty (\varphi, \varphi_n)\, \varphi_n.$$

c. For every $\varphi \in X$ we have Parseval's equality

$$\|\varphi\|^2 = \sum_1^\infty |(\varphi, \varphi_n)|^2.$$

d. $\varphi = 0$ is the only element in X with $(\varphi, \varphi_n) = 0$ for every integer n.

Proof. a $\Rightarrow$ b: Theorems 1.11 and 1.12 imply that

$$u_n = \sum_1^n (\varphi, \varphi_k)\, \varphi_k$$

is the best approximation to φ with respect to $\mathrm{span}\{\varphi_1, \varphi_2, \cdots, \varphi_n\}$. Since $\{\varphi_n\}_1^\infty$ is complete, there exists $\hat{u}_n \in \mathrm{span}\{\varphi_1, \varphi_2, \cdots, \varphi_n\}$ such that $\|\hat{u}_n - \varphi\|$ $\to 0$ as $n \to \infty$, and since $\|\hat{u}_n - \varphi\| \geq \|u_n - \varphi\|$, we have that $u_n \to \varphi$ as $n \to \infty$.
b $\Rightarrow$ c: we have that

$$\|u_n\|^2 = (u_n, u_n) = \sum_1^n |(\varphi, \varphi_k)|^2.$$

Now let $n \to \infty$ and use the continuity of $\|\cdot\|$.
c $\Rightarrow$ d: this is trivial.
d $\Rightarrow$ a: set $U := \overline{\mathrm{span}\{\varphi_n\}}$, and assume $X \neq U$. Then there exists $\varphi \in X$ with $\varphi \notin U$. Since U is a closed subspace of X, U is complete. Hence, by Theorem 1.12, the best approximation v to φ with respect to U exists and satisfies $(v - \varphi, \varphi_n) = 0$ for every integer n. By assumption this implies $v = \varphi$, which is a contradiction. Hence $X = U$. $\square$

As a consequence of part b of the preceding theorem, a complete orthonormal system in a Hilbert space X is called an *orthonormal basis* for X. It can be shown that every Hilbert space has a (possibly uncountable) orthonormal basis [115].

1.2 Bounded Linear Operators

An operator $A : X \to Y$ mapping a vector space X into a vector space Y is called *linear* if

$$A\left(\alpha\varphi + \beta\psi\right) = \alpha A\varphi + \beta A\psi$$

for all $\varphi, \psi \in X$ and $\alpha, \beta \in \mathbb{C}$.

Theorem 1.14. *Let X and Y be normed spaces and $A : X \to Y$ a linear operator. Then A is continuous if it is continuous at one point.*

Proof. Suppose A is continuous at $\varphi_0 \in X$. Then for every $\varphi \in X$ and $\varphi_n \to \varphi$ we have that

$$A\varphi_n = A\left(\varphi_n - \varphi + \varphi_0\right) + A\left(\varphi - \varphi_0\right) \to A\varphi_0 + A\left(\varphi - \varphi_0\right) = A\varphi$$

since $\varphi_n - \varphi + \varphi_0 \to \varphi_0$. □

A linear operator $A : X \to Y$ from a normed space X into a normed space Y is called *bounded* if there exists a positive constant C such that

$$\|A\varphi\| \leq C \|\varphi\|$$

for every $\varphi \in X$. The *norm* of A is the smallest such C, i.e., (dividing by $\|\varphi\|$ and using the linearity of A)

$$\|A\| := \sup_{\|\varphi\|=1} \|A\varphi\| \quad , \quad \varphi \in X.$$

If $Y = \mathbb{C}$, then A is called a bounded linear *functional*. The space X^* of bounded linear functionals on a normed space X is called the *dual space* of X.

Theorem 1.15. *Let X and Y be normed spaces and $A : X \to Y$ a linear operator. Then A is continuous if and only if it is bounded.*

Proof. Let $A : X \to Y$ be bounded, and let $\{\varphi_n\}$ be a sequence in X such that $\varphi_n \to 0$ as $n \to \infty$. Then $\|A\varphi_n\| \leq C \|\varphi_n\|$ implies that $A\varphi_n \to 0$ as $n \to \infty$, i.e., A is continuous at $\varphi = 0$. By Theorem 1.14 A is continuous for all $\varphi \in X$.

Conversely, let A be continuous, and assume that there is no C such that $\|A\varphi\| \leq C \|\varphi\|$ for all $\varphi \in X$. Then there exists a sequence $\{\varphi_n\}$ with $\|\varphi_n\| = 1$ such that $\|A\varphi_n\| \geq n$. Let $\psi_n := \|A\varphi_n\|^{-1} \varphi_n$. Then $\psi_n \to 0$ as $n \to \infty$, and hence by the continuity of A we have that $A\psi_n \to A0 = 0$, which is a contradiction since $\|A\psi_n\| = 1$ for every integer n. Hence A must be bounded. □

Example 1.16. Let $K(x, y)$ be continuous on $[a, b] \times [a, b]$, and define $A : L^2[a, b] \to L^2[a, b]$ by

$$(A\varphi)(x) := \int_a^b K(x, y)\varphi(y)\, dy.$$

Then

$$\|A\varphi\|^2 = \int_a^b |(A\varphi)(x)|^2 \, dx$$

$$= \int_a^b \left| \int_a^b K(x,y)\varphi(y) \, dy \right|^2 dx$$

$$\leq \int_a^b \int_a^b |K(x,y)|^2 \, dy \int_a^b |\varphi(y)|^2 \, dy \, dx$$

$$= \|\varphi\|^2 \int_a^b \int_a^b |K(x,y)|^2 \, dx \, dy.$$

Hence A is bounded and

$$\|A\| \leq \left[\int_a^b \int_a^b |K(x,y)|^2 \, dx \, dy \right]^{\frac{1}{2}}.$$

Let X be a Hilbert space and $U \subset X$ a nontrivial subspace. A bounded linear operator $P : X \to U$ with the property that $P\varphi = \varphi$ for every $\varphi \in U$ is called a *projection operator* from X onto U. Suppose U is a nontrivial closed subspace of X. Then $X = U \oplus U^\perp$, and we define the *orthogonal projection* $P : X \to U$ by $P\varphi = v$, where v is the best approximation to φ. Then clearly $P\varphi = \varphi$ for $\varphi \in U$ and P is bounded since $\|\varphi\|^2 = \|P\varphi + (\varphi - P\varphi)\|^2 = \|P\varphi\|^2 + \|\varphi - P\varphi\|^2 \geq \|P\varphi\|^2$ by the orthogonality property of v (Theorem 1.11). Since $\|P\varphi\| \leq \|\varphi\|$ and $P\varphi = \varphi$ for $\varphi \in U$, we in fact have that $\|P\| = 1$.

Our next step is to introduce the central idea of *compactness* into our discussion. A subset U of a normed space X is called *compact* if every sequence of elements in U contains a subsequence that converges to an element in U. U is called *relatively compact* if its closure is compact. A linear operator $A : X \to Y$ from a normed space X into a normed space Y is a *compact operator* if it maps each bounded set in X into a relatively compact set in Y. This is equivalent to requiring that for each bounded sequence $\{\varphi_n\}$ in X the sequence $\{A\varphi_n\}$ must have a convergent subsequence in Y. Note that since compact sets are bounded, compact operators are clearly bounded. It is also easy to see that linear combinations of compact operators are compact and the product of a bounded operator and a compact operator is a compact operator.

Theorem 1.17. *Let X be a normed space and Y a Banach space. Suppose $A_n : X \to Y$ is a compact operator for each integer n and there exists a linear operator A such that $\|A - A_n\| \to 0$ as $n \to \infty$. Then A is a compact operator.*

Proof. Let $\{\varphi_m\}$ be a bounded sequence in X. We will use a diagonalization procedure to show that $\{A\varphi_m\}$ has a convergent subsequence in Y. Since A_1 is a compact operator, $\{\varphi_m\}$ has a subsequence $\{\varphi_{1,m}\}$ such that $\{A_1\varphi_{1,m}\}$ is convergent. Similarly, $\{\varphi_{1,m}\}$ has a subsequence $\{\varphi_{2,m}\}$ such that $\{A_2\varphi_{2,m}\}$

is convergent. Continuing in this manner, we see that the diagonal sequence $\{\varphi_{m,m}\}$ is a subsequence of $\{\varphi_m\}$ such that for every fixed positive integer n the sequence $\{A_n\varphi_{m,m}\}$ is convergent. Since $\{\varphi_m\}$ is bounded, say $\|\varphi_m\| \leq C$ for all m, $\|\varphi_{m,m}\| \leq C$ for all m. We now use the fact that $\|A - A_n\| \to 0$ as $n \to \infty$ to conclude that for each $\epsilon > 0$ there exists an integer $n_0 = n_0(\epsilon)$ such that

$$\|A - A_{n_0}\| < \frac{\epsilon}{3C},$$

and since $\{A_{n_0}\varphi_{m,m}\}$ is convergent, there exists an integer $N = N(\epsilon)$ such that

$$\|A_{n_0}\varphi_{j,j} - A_{n_0}\varphi_{k,k}\| < \frac{\epsilon}{3}$$

for $j, k > N$. Hence, for $j, k > N$ we have that

$$\begin{aligned}
\|A\varphi_{j,j} - A\varphi_{k,k}\| &\leq \|A\varphi_{j,j} - A_{n_0}\varphi_{j,j}\| + \|A_{n_0}\varphi_{j,j} - A_{n_0}\varphi_{k,k}\| \\
&\quad + \|A_{n_0}\varphi_{k,k} - A\varphi_{k,k}\| \\
&\leq \|A - A_{n_0}\|\,\|\varphi_{j,j}\| + \frac{\epsilon}{3} + \|A_{n_0} - A\|\,\|\varphi_{k,k}\| \\
&< \epsilon.
\end{aligned}$$

Thus $\{A\varphi_{m,m}\}$ is a Cauchy sequence and therefore convergent in the Banach space Y. □

Example 1.18. Consider the operator $A : \mathrm{L}^2[a,b] \to \mathrm{L}^2[a,b]$ defined as in the previous example by

$$(A\varphi)(x) := \int_a^b K(x,y)\varphi(y)\,dy,$$

where $K(x,y)$ is continuous on $[a,b] \times [a,b]$. Let $\{\varphi_n\}$ be a complete orthonormal set in $\mathrm{L}^2[a,b]$. Then it is easy to show that $\{\varphi_n(x)\varphi_m(y)\}$ is a complete orthonormal set in $\mathrm{L}^2\left([a,b] \times [a,b]\right)$. Hence

$$K(x,y) = \sum_{i,j=1}^{\infty} a_{ij}\varphi_i(x)\varphi_j(y)$$

in the mean square sense and by Parseval's equality

$$\int_a^b \int_a^b |K(x,y)|^2 \, dx\,dy = \sum_{i,j=1}^{\infty} |a_{ij}|^2.$$

Furthermore,

$$\int_a^b \int_a^b \left| K(x,y) - \sum_{i,j=1}^{n} a_{ij}\varphi_j(x)\varphi_j(y) \right|^2 dx\,dy = \sum_{i,j=n+1}^{\infty} |a_{ij}|^2,$$

which can be made as small as we please for n sufficiently large. Hence A can be approximated in norm by A_n, where

$$(A_n\varphi)(x) := \int_a^b \left[\sum_{i,j=1}^n a_{ij}\varphi_i(x)\varphi_j(y) \right] \varphi(y)\, dy.$$

But $A_n : L^2[a,b] \to L^2[a,b]$ has a finite-dimensional range. Hence if $U \subset X$ is bounded, then $A_n(U)$ is a set in a finite-dimensional space $A_n(X)$. By the Bolzano–Weierstrass theorem, $A_n(U)$ is relatively compact, i.e., A_n is a compact operator. Theorem 1.17 now implies that A is a compact operator.

Lemma 1.19 (Riesz's Lemma). *Let X be a normed space, $U \subset X$ a closed subspace such that $U \neq X$, and $\alpha \in (0,1)$. Then there exists $\psi \in X$, $\|\psi\| = 1$, such that $\|\psi - \varphi\| \geq \alpha$ for every $\varphi \in U$.*

Proof. There exists $f \in X$, $f \notin U$, and since U is closed, we have that

$$\beta := \inf_{\varphi \in U} \|f - \varphi\| > 0.$$

Now choose $g \in U$ such that

$$\beta \leq \|f - g\| \leq \frac{\beta}{\alpha}$$

and define

$$\psi := \frac{f - g}{\|f - g\|}.$$

Then $\|\psi\| = 1$ and for every $\varphi \in U$ we have, since $g + \|f - g\|\varphi \in U$, that

$$\|\psi - \varphi\| = \frac{1}{\|f - g\|} \|f - (g + \|f - g\|\varphi)\| \geq \frac{\beta}{\|f - g\|} \geq \alpha.$$

$\square$

Riesz's lemma is the key step in the proof of a series of basic results on compact operators that will be needed in the sequel. The following is the first of these results and will be used in the following chapter on ill-posed problems.

Theorem 1.20. *Let X be a normed space. Then the identity operator $I : X \to X$ is a compact operator if and only if X has finite dimension.*

Proof. Assume that I is a compact operator and X is not finite dimensional. Choose $\varphi_1 \in X$ with $\|\varphi_1\| = 1$. Then $U_1 := \text{span}\{\varphi_1\}$ is a closed subspace of X, and by Riesz's lemma there exists $\varphi_2 \in X$, $\|\varphi_2\| = 1$, with $\|\varphi_2 - \varphi_1\| \geq \frac{1}{2}$. Now let $U_2 := \text{span}\{\varphi_1, \varphi_2\}$. Using Riesz's lemma again, there exists $\varphi_3 \in X$, $\|\varphi_3\| = 1$, and $\|\varphi_3 - \varphi_1\| \geq \frac{1}{2}$, $\|\varphi_3 - \varphi_2\| \geq \frac{1}{2}$. Continuing in this manner, we obtain a sequence $\{\varphi_n\}$ in X such that $\|\varphi_n\| = 1$ and $\|\varphi_n - \varphi_m\| \geq \frac{1}{2}$

for $n \neq m$. Hence $\{\varphi_n\}$ does not contain a convergent subsequence, i.e., $I : X \to X$ is not compact. This contradicts our assumption. Hence if I is a compact operator, then X has finite dimension. Conversely, if X has finite dimension, then $I(X)$ is finite-dimensional, and by the Bolzano–Weierstrass theorem, $I(X)$ is relatively compact, i.e., $I : X \to X$ is a compact operator.

□

The next theorem, due to Riesz [144], is one of the most celebrated theorems in all of mathematics, having its origin in Fredholm's seminal paper of 1903 [69].

Theorem 1.21 (Riesz's Theorem). *Let $A : X \to X$ be a compact operator on a normed space X. Then either (1) the homogeneous equation*

$$\varphi - A\varphi = 0$$

has a nontrivial solution $\varphi \in X$ or (2) for each $f \in X$ the equation

$$\varphi - A\varphi = f$$

has a unique solution $\varphi \in X$. If $I - A$ is injective (and hence bijective), then $(I - A)^{-1} : X \to X$ is bounded.

Proof. The proof will be divided into four steps.
Step 1: let $L := I - A$, and let $N(L) := \{\varphi \in X : L\varphi = 0\}$ be the *null space* of L. We will show that there exists a positive constant C such that

$$\inf_{\chi \in N(L)} \|\varphi - \chi\| \leq C \|L\varphi\|$$

for all $\varphi \in X$. Suppose this is not true. Then there exists a sequence $\{\varphi_n\}$ in X such that $\|L\varphi_n\| = 1$ and $d_n := \inf_{\chi \in N(L)} \|\varphi_n - \chi\| \to \infty$. Choose $\{\chi_n\} \subset N(L)$ such that $d_n \leq \|\varphi_n - \chi_n\| \leq 2d_n$, and set

$$\psi_n := \frac{\varphi_n - \chi_n}{\|\varphi_n - \chi_n\|}.$$

Then $\|\psi_n\| = 1$ and $\|L\psi_n\| \leq d_n^{-1} \to 0$. But since A is compact, by passing to a subsequence if necessary, we may assume that the sequence $\{A\psi_n\}$ converges to an element $\varphi_0 \in X$. Since $\psi_n = (L + A)\psi_n$, we have that $\{\psi_n\}$ converges to φ_0, and hence $\varphi_0 \in N(L)$. But

$$\inf_{\chi \in N(L)} \|\psi_n - \chi\| = \|\varphi_n - \chi_n\|^{-1} \inf_{\chi \in N(L)} \|\varphi_n - \chi_n - \|\varphi_n - \chi_n\| \chi\|$$

$$= \|\varphi_n - \chi_n\|^{-1} \inf_{\chi \in N(L)} \|\varphi_n - \chi\| \geq \frac{1}{2},$$

which contradicts the fact that $\psi_n \to \varphi_0 \in N(L)$.

Step 2: we next show that the range of L is a closed subspace of X. $L(X) :=$ $\{x \in X : x = L\varphi \text{ for some } \varphi \in X\}$ is clearly a subspace. Hence if $\{\varphi_n\}$ is a sequence in X such that $\{L\varphi_n\}$ converges to an element $f \in X$, then we must show that $f = L\varphi$ for some $\varphi \in X$. By the foregoing result the sequence $\{d_n\}$, where $d_n := \inf_{\chi \in N(L)} \|\varphi_n - \chi\|$, is bounded. Choosing $\chi_n \in N(L)$ as above and writing $\tilde{\varphi}_n := \varphi_n - \chi_n$, we have that $\{\tilde{\varphi}_n\}$ is bounded and $L\tilde{\varphi}_n \to f$. Since A is compact, by passing to a subsequence if necessary, we may assume that $\{A\tilde{\varphi}_n\}$ converges to an element $\tilde{\varphi}_0 \in X$. Hence $\tilde{\varphi}_n$ converges to $f + \varphi_0$, and by the continuity of L, we have that $L(f + \varphi_0) = f$. Hence $L(X)$ is closed.

Step 3: the next step is to show that if $N(L) = \{0\}$, then $L(X) = X$, i.e., if case (1) of the theorem does not hold, then case (2) is true. To this end, we note that from our previous result the sets $L^n(X), n = 1, 2, \cdots$, form a nonincreasing sequence of closed subspaces of X. Suppose that no two of these spaces coincide. Then each is a proper subspace of its predecessor. Hence, by Riesz's lemma, there exists a sequence $\{\psi_n\}$ in X such that $\psi_n \in L^n(X)$, $\|\psi_n\| = 1$, and $\|\psi_n - \psi\| \geq \frac{1}{2}$ for all $\psi \in L^{n+1}(X)$. Thus, if $m > n$, then

$$A\psi_n - A\psi_m = \psi_n - (\psi_m + L\psi_n - L\psi_m)$$

and $\psi_m + L\psi_n - L\psi_m \in L^{n+1}(X)$ since

$$\psi_m + L\psi_n - L\psi_m = L^{n+1}(L^{m-n-1}\varphi_m + \varphi_n - L^{m-n}\varphi_m).$$

Hence $\|A\psi_n - A\psi_m\| \geq \frac{1}{2}$, contrary to the compactness of A. Thus we can conclude that there exists an integer n_0 such that $L^n(X) = L^{n_0}(X)$ for all $n \geq n_0$. Now let $\varphi \in X$. Then $L^{n_0}\varphi \in L^{n_0}(X) = L^{n_0+1}(X)$, and so $L^{n_0}\varphi = L^{n_0+1}\psi$ for some $\psi \in X$, i.e., $L^{n_0}(\varphi - L\psi) = 0$. But since $N(L) = \{0\}$, we have that $N(L^{n_0}) = 0$, and hence $\varphi = L\psi$. Thus $X = L(X)$.

Step 4: we now come to the final step, which is to show that if $L(X) = X$, then $N(L) = 0$, i.e., either case (1) or case (2) of the theorem is true. To show this, we first note that, by the continuity of L, we have that $N(L^n)$ is a closed subspace for $n = 1, 2, \cdots$. An analogous argument to that used in Step 3 shows that there exists an integer n_0 such that $N(L^n) = N(L^{n_0})$ for all $n \geq n_0$. Hence, if $L(X) = X$, then $\varphi \in N(L^{n_0})$ satisfies $\varphi = L^{n_0}\psi$ for some $\psi \in X$, and thus $L^{2n_0}\psi = 0$. Thus $\psi \in N(L^{2n_0}) = N(L^{n_0})$, and hence $\varphi = L^{n_0}\psi = 0$. Since $L\varphi = 0$ implies that $L^{n_0}\varphi = 0$, the proof of Step 4 is now complete.

The fact that $(I - A)^{-1}$ is bounded in case (2) follows from Step 1 since in this case $N(L) = \{0\}$.   □

Let $A : X \to X$ be a compact operator of a normed space into itself. A complex number λ is called an *eigenvalue* of A with *eigenelement* $\varphi \in X$ if there exists $\varphi \in X$, $\varphi \neq 0$, such that $A\varphi = \lambda\varphi$. It is easily seen that eigenelements corresponding to different eigenvalues must be linearly independent. We call the dimension of the null space of $L_\lambda := \lambda I - A$ the multiplicity of λ. If $\lambda \neq 0$ is not an eigenvalue of A, then it follows from Riesz's theorem that the *resolvent operator* $(\lambda I - A)^{-1}$ is a well-defined bounded linear operator mapping X onto itself. On the other hand, if $\lambda = 0$, then A^{-1} cannot be bounded on $A(X)$ unless X is finite dimensional since if it were, then $I = A^{-1}A$ would be compact.

Theorem 1.22. *Let $A : X \to X$ be a compact operator on a normed space X. Then A has at most a countable set of eigenvalues having no limit points, except possibly $\lambda = 0$. Each nonzero eigenvalue has finite multiplicity.*

Proof. Suppose there exists a sequence $\{\lambda_n\}$ of not necessarily distinct nonzero eigenvalues with corresponding linearly independent eigenelements $\{\varphi_n\}_1^\infty$ such that $\lambda_n \to \lambda \neq 0$. Let

$$U_n := \mathrm{span}\{\varphi_1, \cdots, \varphi_n\}.$$

Then, by Riesz's lemma, there exists a sequence $\{\psi_n\}$ such that $\psi_n \in U_n$, $\|\psi_n\| = 1$, and $\|\psi_n - \psi\| \geq \frac{1}{2}$ for every $\psi \in U_{n-1}$, $n = 2, 3, \cdots$. If $n > m$, then we have that

$$\lambda_n^{-1} A\psi_n - \lambda_m^{-1} A\psi_m = \psi_n + \left(-\psi_n + \lambda_n^{-1} A\psi_n - \lambda_m^{-1} A\psi_m\right)$$
$$= \psi_n - \psi,$$

where $\psi \in U_{n-1}$ since if $\psi_n = \sum_1^n \beta_j \varphi_j$, then

$$\psi_n - \lambda_n^{-1} A\psi_n = \sum_1^n \beta_j \left(1 - \lambda_n^{-1}\lambda_j\right) \varphi_j \in U_{n-1}$$

and, similarly, $L_{\lambda_m}\psi_m \in U_{m-1}$. Hence we have that $\left\|\lambda_n^{-1} A\psi_n - \lambda_m^{-1} A\psi_m\right\| \geq \frac{1}{2}$, which, since $\lambda_n \to \lambda \neq 0$, contradicts the compactness of the operator A. Hence our initial assumption is false, and this implies the validity of the theorem.   □

A generalization of Theorems 1.22 and 1.21 is the *analytic Fredholm theorem*. To present this theorem, we first set the following definition.

Definition 1.23. Let D be a domain in the complex plane $\mathbb{C}$ and $f : D \to X$ a function from D into the (complex) Banach space X. f is said to be analytic in D if for every $z_0 \in D$ there exists a power series expansion

$$f(z) = \sum_{m=0}^\infty a_m(z - z_0)^m$$

that converges in the norm on X uniformly for all z in a neighborhood of z_0 and where the coefficients a_m are elements from X.

We can now state the following theorem (for a proof see [54]).

Theorem 1.24. *Let D be a domain in $\mathbb{C}$, and let $A : D \to \mathcal{L}(X)$ be an operator-valued analytic function such that $A(z)$ is compact for each $z \in D$. Then either*

1. *$(I - A(z))^{-1}$ does not exist for any $z \in D$ or*
2. *$(I - A(z))^{-1}$ exists for all $z \in D \setminus S$, where S is a discrete subset of D.*

1.3 Adjoint Operator

We now assume that X is a Hilbert space and first characterize the class of bounded linear functionals on X.

Theorem 1.25 (Riesz Representation Theorem). *Let X be a Hilbert space. Then for each bounded linear functional $F : X \to \mathbb{C}$ there exists a unique $f \in X$ such that*

$$F(\varphi) = (\varphi, f)$$

for every $\varphi \in X$. Furthermore, $\|f\| = \|F\|$.

Proof. We first show the uniqueness of the representation. This is easy since if $(\varphi, f_1) = (\varphi, f_2)$ for every $\varphi \in X$, then $(\varphi, f_1 - f_2) = 0$ for every $\varphi \in X$, and setting $\varphi = f_1 - f_2$ we have that $\|f_1 - f_2\|^2 = 0$. Hence $f_1 = f_2$.

We now turn to the existence of f. If $F = 0$, then we can choose $f = 0$. Hence assume $F \neq 0$ and choose $w \in X$ such that $F(w) \neq 0$. Since F is continuous, $N(F) = \{\varphi \in X : F(\varphi) = 0\}$ is a closed (and hence complete) subspace of X. Hence, by Theorem 1.12, there exists a unique best approximation v to w with respect to $N(F)$, and by Theorem 1.11, we have that $w - v \perp N(F)$. Then for $g := w - v$ we have that

$$(F(g)\varphi - F(\varphi)g, g) = 0$$

for every $\varphi \in X$ since $F(g)\varphi - F(\varphi)g \in N(F)$ for every $\varphi \in X$. Hence

$$F(\varphi) = \left(\varphi, \frac{\overline{F(g)}g}{\|g\|^2}\right)$$

for every $\varphi \in X$, i.e.,

$$f := \frac{\overline{F(g)}g}{\|g\|^2}$$

is the element we are seeking.

Finally, to show that $\|f\| = \|F\|$, we note that by the Cauchy–Schwarz inequality we have that $|F(\varphi)| \leq \|f\| \|\varphi\|$ for every $\varphi \in X$, and hence $\|F\| \leq \|f\|$. On the other hand, $F(f) = (f, f) = \|f\|^2$, and hence $\|f\| \leq \|F\|$. We can now conclude that $\|F\| = \|f\|$. $\qquad\square$

Armed with the Riesz representation theorem we can now define the *adjoint operator* A^* of A.

Theorem 1.26. *Let X and Y be Hilbert spaces, and let $A : X \to Y$ be a bounded linear operator. Then there exists a uniquely determined linear operator $A^* : Y \to X$ such that $(A\varphi, \psi) = (\varphi, A^*\psi)$ for every $\varphi \in X$ and $\psi \in Y$. A^* is called the* adjoint *of A and is a bounded linear operator satisfying $\|A^*\| = \|A\|$.*

Proof. For each $\psi \in Y$ the mapping $\varphi \mapsto (A\varphi, \psi)$ defines a bounded linear functional on X since

$$|(A\varphi, \psi)| \leq \|A\| \, \|\varphi\| \, \|\psi\|.$$

Hence by the Riesz representation theorem we can write $(A\varphi, \psi) = (\varphi, f)$ for some $f \in X$. We now define $A^* : Y \to X$ by $A^*\psi = f$. The operator A^* is unique since if $0 = (\varphi, (A_1^* - A_2^*)\psi)$ for every $\varphi \in X$, then setting $\varphi = (A_1^* - A_2^*)\psi$ we have that $\|(A_1^* - A_2^*)\psi\|^2 = 0$ for every $\psi \in Y$, and hence $A_1^* = A_2^*$. To show that A^* is linear, we observe that

$$
\begin{aligned}
(\varphi, \beta_1 A^*\psi_1 + \beta_2 A^*\psi_2) &= \bar{\beta}_1 \, (\varphi, A^*\psi_1) + \bar{\beta}_2 \, (\varphi, A^*\psi_2) \\
&= \bar{\beta}_1 \, (A\varphi, \psi_1) + \bar{\beta}_2 \, (A\varphi, \psi_2) \\
&= (A\varphi, \beta_1\psi_1 + \beta_2\psi_2) \\
&= (\varphi, A^* (\beta_1\psi_1 + \beta_2\psi_2))
\end{aligned}
$$

for every $\varphi \in X$, $\psi_1, \psi_2 \in Y$, and $\beta_1, \beta_2 \in \mathbb{C}$. Hence $\beta_1 A^*\psi_1 + \beta_2 A^*\psi_2 = A^* (\beta_1\psi_1 + \beta_2\psi_2)$, i.e., A^* is linear. To show that A^* is bounded, we note that by the Cauchy–Schwarz inequality we have that

$$\|A^*\psi\|^2 = (A^*\psi, A^*\psi) = (AA^*\psi, \psi) \leq \|A\| \, \|A^*\psi\| \, \|\psi\|$$

for every $\psi \in Y$. Hence $\|A^*\| \leq \|A\|$. Conversely, since A is the adjoint of A^*, we also have that $\|A\| \leq \|A^*\|$, and hence $\|A^*\| = \|A\|$. $\square$

Theorem 1.27. *Let X and Y be Hilbert spaces, and let $A : X \to Y$ be a compact operator. Then $A^* : Y \to X$ is also a compact operator.*

Proof. Let $\|\psi_n\| \leq C$ for some positive constant C. Then, since A^* is bounded, $AA^* : Y \to Y$ is a compact operator. Hence, by passing to a subsequence if necessary, we may assume that the sequence $\{AA^*\psi_n\}$ converges in Y. But

$$
\begin{aligned}
\|A^* (\psi_n - \psi_m)\|^2 &= (AA^* (\psi_n - \psi_m), \psi_n - \psi_m) \\
&\leq 2C \, \|AA^* (\psi_n - \psi_m)\|,
\end{aligned}
$$

i.e., $\{A^*\psi_n\}$ is a Cauchy sequence and, hence, convergent. We can now conclude that A^* is a compact operator. $\square$

The following theorem will be important to us in the next chapter of the book. We first need a lemma.

Lemma 1.28. *Let U be a closed subspace of a Hilbert space X. Then $U^{\perp\perp}=U$.*

Proof. Since U is a closed subspace, we have that $X = U \oplus U^{\perp}$ and $X = U^{\perp} \oplus U^{\perp\perp}$. Hence for $\varphi \in X$ we have that $\varphi = \varphi_1 + \varphi_2$, where $\varphi_1 \in U$ and $\varphi_2 \in U^{\perp}$ and $\varphi = \psi_1 + \psi_2$, where $\psi_1 \in U^{\perp\perp}$ and $\psi_2 \in U^{\perp}$. In particular, $0 = (\varphi_1 - \psi_1) + (\varphi_2 - \psi_2)$, and since it is easily verified that $U \subseteq U^{\perp\perp}$, we have that $\varphi_1 - \psi_1 = \psi_2 - \varphi_2 \in U^{\perp}$. But $\varphi_1 - \psi_1 \in U^{\perp\perp}$, and hence $\varphi_1 = \psi_1$. We can now conclude that $U^{\perp\perp} = U$. $\qquad\square$

Theorem 1.29. *Let X and Y be Hilbert spaces. Then for a bounded linear operator $A : X \to Y$ we have that if $A(X) := \{y \in Y : y = Ax \text{ for some } x \in X\}$ is the range of A, then*

$$A(X)^{\perp} = N(A^*) \text{ and } N(A^*)^{\perp} = \overline{A(X)}.$$

Proof. We have that $g \in A(X)^{\perp}$ if and only if $(A\varphi, g) = 0$ for every $\varphi \in X$. Since $(A\varphi, g) = (\varphi, A^*g)$, we can now conclude that $A^*g = 0$, i.e., $g \in N(A^*)$. On the other hand, by Lemma 1.28, $\overline{A(X)} = \overline{A(X)}^{\perp\perp} = N(A^*)^{\perp}$ since $A(X)^{\perp} = \overline{A(X)}^{\perp} = N(A^*)$. $\qquad\square$

The next theorem is one of the jewels of functional analysis and will play a central role in the next chapter of the book. We note that a bounded linear operator $A : X \to X$ on a Hilbert space X is said to be *self-adjoint* if $A = A^*$, i.e., $(A\varphi, \psi) = (\varphi, A\psi)$ for all $\varphi, \psi \in X$.

Theorem 1.30 (Hilbert–Schmidt Theorem). *Let $A : X \to X$ be a compact, self-adjoint operator on a Hilbert space X. Then, if $A \neq 0$, A has at least one eigenvalue different from zero, all the eigenvalues of A are real, and X has an orthonormal basis consisting of eigenelements of A.*

Proof. It is a simple consequence of the self-adjointness of A that (1) eigenelements corresponding to different eigenvalues are orthogonal and (2) all eigenvalues are real. Hence the first serious problem we face is to show that $A \neq 0$ has at least one eigenvalue different from zero. To this end, let $\lambda = \|A\| > 0$, and consider the operator $T := \lambda^2 I - A^2$. We will show that $\pm\lambda$ is an eigenvalue of A. To show this, we first note that for all $\varphi \in X$ we have that

$$(T\varphi, \varphi) = ((\lambda^2 I - A^2)\varphi, \varphi) = \lambda^2 \|\varphi\|^2 - (A^2\varphi, \varphi)$$
$$= \lambda^2 \|\varphi\|^2 - \|A\varphi\|^2 \geq 0.$$

Now choose a sequence $\{\varphi_n\} \subset X$ such that $\|\varphi_n\| = 1$ and $\|A\varphi_n\| \to \lambda$ as $n \to \infty$. Then, by the preceding identity, $(T\varphi_n, \varphi_n) \to 0$ as $n \to \infty$. To proceed further, we first define a new inner product $\langle \cdot, \cdot \rangle$ on X by

$$\langle \varphi, \psi \rangle := (T\varphi, \psi).$$

The fact that $\langle \cdot, \cdot \rangle$ defines an inner product follows easily from the fact that A, and hence T, is self-adjoint and the fact that $(T\varphi, \varphi) \geq 0$ for all $\varphi \in X$. We now have from the Cauchy–Schwarz inequality that

$$
\begin{aligned}
\|T\varphi_n\|^2 = (T\varphi_n, T\varphi_n) &= \langle \varphi_n, T\varphi_n \rangle \\
&\leq \langle \varphi_n, \varphi_n \rangle^{\frac{1}{2}} \langle T\varphi_n, T\varphi_n \rangle^{\frac{1}{2}} \\
&= (T\varphi_n, \varphi_n)^{\frac{1}{2}} (T^2\varphi_n, T\varphi)^{\frac{1}{2}} \\
&\leq (T\varphi_n, \varphi_n)^{\frac{1}{2}} \|T^2\varphi_n\|^{\frac{1}{2}} \|T\varphi_n\|^{\frac{1}{2}} \\
&\leq \|T\|^{\frac{3}{2}} (T\varphi_n, \varphi_n)^{\frac{1}{2}}.
\end{aligned}
$$

But $(T\varphi_n, \varphi_n) \to 0$ as $n \to \infty$, and hence, by the foregoing inequality, $T\varphi_n \to 0$ as $n \to \infty$. Since A is compact, by passing to a subsequence if necessary, we may assume that $\{A\varphi_n\}$ converges to a limit φ, which satisfies $\|\varphi\| = \lim_{n\to\infty} \|A\varphi_n\| = \lambda > 0$ and $T\varphi = \lim_{n\to\infty} TA\varphi_n = \lim_{n\to\infty} AT\varphi_n = 0$, i.e., $\varphi \neq 0$ and

$$T\varphi = (\lambda I + A)(\lambda I - A)\varphi = 0.$$

Thus either $A\varphi = \lambda\varphi$ or $\lambda\varphi - A\varphi \neq 0$ and $A\psi = -\lambda\psi$ for $\psi = \lambda\varphi - A\varphi$. Thus either λ or $-\lambda$ is a nonzero eigenvalue of A.

We now complete the theorem by showing that X has an orthonormal basis consisting of eigenvectors of A. We first note that if Y is a subspace of X such that $A(Y) \subset Y$, then, by the self-adjointness of A, we have that $A(Y^\perp) \subset Y^\perp$. In particular, let Y be the closed linear span of all the eigenelements of A. For $\lambda \neq 0$ the restriction of A to the nullspace of $L := \lambda I - A$ is λ times the identity operator on the closed subspace $N(L)$. Since the restriction of A to $N(L)$ is compact from $N(L)$ onto $N(L)$, we can conclude from Theorem 1.20 that $N(L)$ has finite dimension. Now pick an orthonormal basis for each eigenspace of A, including the case $\lambda = 0$, and take their union. Since eigenelements corresponding to different eigenvalues are orthogonal, this union is an orthonormal basis for Y. We now note that $A : Y^\perp \to Y^\perp$ is a compact self-adjoint operator that has no eigenvalues since all the eigenelements of A belong to Y. But this is impossible by the first part of our proof unless either A restricted to $Y^\perp$ is the zero operator or $Y^\perp = \{0\}$. If A restricted to $Y^\perp$ is the zero operator, then $Y^\perp = \{0\}$, since otherwise nonzero elements of $Y^\perp$ would be eigenelements of A corresponding to the eigenvalue zero and, hence, in Y, a contradiction. Thus in either case $Y^\perp = \{0\}$, i.e., $Y = X$, and the proof is complete.   □

1.4 Sobolev Space $H^p[0, 2\pi]$

For the proper study of inverse problems it is necessary to consider function spaces that are larger than the classes of continuous and continuously differentiable functions. In particular, Sobolev spaces are the natural spaces to consider in order to apply the tools of functional analysis presented earlier. Hence, in this and the following section, we will present the rudiments of the theory of Sobolev spaces. Our presentation will closely follow the excellent introductory treatment of such spaces by Kress [111], which avoids the use of Fourier transforms in $L^2(\mathbb{R}^n)$ but instead relies on the elementary theory of Fourier series. This simplification is made possible by restricting our attention to planar domains having C^2 boundaries and has the drawback of not being able to achieve the depth of a more sophisticated treatment such as that presented in [127]. However, the limited results we shall present will be sufficient for the purposes of this book.

We begin with the fact that the orthonormal system $\left\{\frac{1}{\sqrt{2\pi}}e^{imt}\right\}_{-\infty}^{\infty}$ is complete in $L^2[0, 2\pi]$ [11]. Hence, by Theorem 1.13, for $\varphi \in L^2[0, 2\pi]$ we have that, in the sense of mean square convergence,

$$\varphi(t) = \sum_{-\infty}^{\infty} a_m e^{imt},$$

where the Fourier coefficients a_m are given by

$$a_m := \frac{1}{2\pi} \int_0^{2\pi} \varphi(t)e^{-imt}\, dt.$$

If we let $(\cdot, \cdot)$ denote the usual L^2-inner product with associated norm $\|\cdot\|$, then by Parseval's equality we have that

$$\sum_{-\infty}^{\infty} |a_m|^2 = \frac{1}{2\pi} \int_0^{2\pi} |\varphi(t)|^2\, dt$$

$$= \frac{1}{2\pi} \|\varphi\|^2.$$

Now let $0 \le p < \infty$. Then we define $H^p[0, 2\pi]$ as the space of all functions $\varphi \in L^2[0, 2\pi]$ such that

$$\sum_{-\infty}^{\infty} (1 + m^2)^p |a_m|^2 < \infty,$$

where the a_m are the Fourier coefficients of φ. The space $H^p = H^p[0, 2\pi]$ is called a *Sobolev space*. Note that $H^0[0, 2\pi] = L^2[0, 2\pi]$.

Theorem 1.31. *$H^p[0, 2\pi]$ is a Hilbert space with inner product*

$$(\varphi, \psi)_p := \sum_{-\infty}^{\infty} (1 + m^2)^p a_m \bar{b}_m,$$

where the a_m and b_m are the Fourier coefficients of φ and ψ, respectively. The trigonometric polynomials are dense in $H^p[0, 2\pi]$.

Proof. If is easily verified that H^p is a vector space and $(\cdot, \cdot)_p$ is an inner product. Note that the fact that $(\cdot, \cdot)_p$ is well defined follows from the Cauchy–Schwarz inequality

$$\left| \sum_{-\infty}^{\infty} (1 + m^2)^p a_m \bar{b}_m \right|^2 \leq \sum_{-\infty}^{\infty} (1 + m^2)^p |a_m|^2 \sum_{-\infty}^{\infty} (1 + m^2)^p |b_m|^2 .$$

To show that H^p is complete, let $\{\varphi_n\}$ be a Cauchy sequence, i.e.,

$$\sum_{-\infty}^{\infty} (1 + m^2)^p |a_{m,n} - a_{m,k}|^2 < \epsilon^2$$

for all $n, k \geq N = N(\epsilon)$, where $a_{m,n}$ are the Fourier coefficients of φ_n. In particular,

$$\sum_{-M_1}^{M_2} (1 + m^2)^p |a_{m,n} - a_{m,k}|^2 < \epsilon^2 \tag{1.3}$$

for all M_1 and M_2 and $n, k \geq N(\epsilon)$. Since $\mathbb{C}$ is complete, there exists a sequence $\{a_m\}$ in $\mathbb{C}$ such that $a_{m,n} \to a_m$ as $n \to \infty$ for each fixed m. Letting $k \to \infty$ in (1.3) implies that

$$\sum_{-M_1}^{M_2} (1 + m^2)^p |a_{m,n} - a_m|^2 \leq \epsilon^2$$

for all $n \geq N(\epsilon)$ and all M_1 and M_2. Hence

$$\sum_{-\infty}^{\infty} (1 + m^2)^p |a_{m,n} - a_m|^2 \leq \epsilon^2 \tag{1.4}$$

for all $n \geq N(\epsilon)$. Defining

$$f_m(t) := e^{imt}$$

and

$$\varphi := \sum_{-\infty}^{\infty} a_m f_m ,$$

we have by (1.4) and the triangle inequality that

$$\left[\sum_{-\infty}^{\infty} (1 + m^2)^p |a_m|^2 \right]^{\frac{1}{2}} \leq \epsilon + \left[\sum_{-\infty}^{\infty} (1 + m^2)^p |a_{m,n}|^2 \right]^{\frac{1}{2}} < \infty ,$$

i.e., $\varphi \in H^p$. From (1.4) we can conclude that $\|\varphi - \varphi_n\| \to 0$ as $n \to \infty$, and hence H^p is complete.

To prove the last statement of the theorem, let $\varphi \in H^p$ with Fourier coefficients a_m. Then for

$$\varphi_n := \sum_{-n}^{n} a_m f_m$$

we have that

$$\|\varphi - \varphi_n\|_p^2 = \sum_{|m|=n+1}^{\infty} (1+m^2)^p \, |a_m|^2 \to 0$$

as $n \to \infty$ since the full series is convergent. From this we can conclude that the trigonometric polynomials are dense in H^p. $\square$

Theorem 1.32 (Rellich's Theorem). *If $q > p$, then $H^q[0, 2\pi]$ is dense in $H^p[0, 2\pi]$ and the embedding operator $I : H^q \to H^p$ is compact.*

Proof. Since $(1 + m^2)^p \leq (1 + m^2)^q$ for $0 \leq p < q < \infty$, it follows that $H^q \subset H^p$ and $\|\varphi\|_p \leq \|\varphi\|_q$ for every $\varphi \in H^q$. The denseness of H^q in H^p follows from the denseness of trigonometric polynomials in H^p.

To show that $I : H^q \to H^p$ is a compact operator, define $I_n : H^q \to H^p$ by

$$I_n \varphi := \sum_{-n}^{n} a_m f_m$$

for $\varphi \in H^q$ having Fourier coefficients a_m. Then

$$\|(I_n - I)\varphi\|_p^2 = \sum_{|m|=n+1}^{\infty} (1+m^2)^p \, |a_m|^2$$

$$\leq \frac{1}{(1+n^2)^{q-p}} \sum_{|m|=n+1}^{\infty} (1+m^2)^q \, |a_m|^2$$

$$\leq \frac{1}{(1+n^2)^{q-p}} \|\varphi\|_q^2 .$$

Since I_n has finite-dimensional range, I_n is a compact operator, and from the preceding inequality we have that $\|I_n - I\| \leq (1 + n^2)^{\frac{(p-q)}{2}} \to 0$ as $n \to \infty$. Hence I is compact by Theorem 1.17. $\square$

Theorem 1.33 (Sobolev Embedding Theorem). *Let $p > \frac{1}{2}$ and $\varphi \in H^p[0, 2\pi]$. Then φ coincides almost everywhere with a continuous and 2π-periodic function (i.e., the difference between φ and this function is a function η such that $\|\eta\|_p = 0$).*

Proof. For $\varphi \in H^p[0, 2\pi]$ we have that for $p > \frac{1}{2}$

$$\left[\sum_{-\infty}^{\infty} |a_m e^{imt}| \right]^2 \leq \sum_{-\infty}^{\infty} \frac{1}{(1+m^2)^p} \sum_{-\infty}^{\infty} (1+m^2)^p \, |a_m|^2$$

by the Cauchy–Schwarz inequality. Hence the Fourier series for φ is absolutely and uniformly convergent and thus coincides with a continuous 2π-periodic function. Since the Fourier series for φ agrees with φ almost everywhere (as defined in the theorem), the proof is complete.   □

Definition 1.34. For $0 \leq p < \infty$, $H^{-p} = H^{-p}[0, 2\pi]$ is defined as the *dual space* of $H^p[0, 2\pi]$, i.e., the space of bounded linear functionals defined on $H^p[0, 2\pi]$.

Recall that for F, a bounded linear functional defined on $H^p[0, 2\pi]$, the norm of F is defined by

$$\|F\|_p := \sup_{\substack{\varphi \in H^p \\ \|\varphi\|_p = 1}} |F\varphi|.$$

The following theorem gives an explicit expression for $\|F\|$ and a characterization of H^{-p}.

Theorem 1.35. *For $F \in H^{-p}[0, 2\pi]$ the norm is given by*

$$\|F\|_p = \left[\sum_{-\infty}^{\infty} (1 + m^2)^{-p} |c_m|^2 \right]^{\frac{1}{2}},$$

where $c_m = F(f_m)$. Conversely, for each sequence $\{c_m\}$ in $\mathbb{C}$ satisfying

$$\sum_{-\infty}^{\infty} (1 + m^2)^{-p} |c_m|^2 < \infty$$

there exists a bounded linear functional $F \in H^{-p}[0, 2\pi]$ with $F(f_m) = c_m$.

Proof. Assume that $\{c_m\}$ satisfies the inequality of the theorem, and define $F : H^p \to \mathbb{C}$ by

$$F(\varphi) := \sum_{-\infty}^{\infty} a_m c_m$$

for $\varphi \in H^p$ with Fourier coefficients a_m. Then F is well defined since by the Cauchy–Schwarz inequality

$$|F(\varphi)|^2 \leq \sum_{-\infty}^{\infty} (1 + m^2)^{-p} |c_m|^2 \sum_{-\infty}^{\infty} (1 + m^2)^p |a_m|^2,$$

and furthermore

$$\|F\|_p \leq \left[\sum_{-\infty}^{\infty} (1 + m^2)^{-p} |c_m|^2 \right]^{\frac{1}{2}}.$$

On the other hand, let $F \in H^{-p}$ such that $F(f_m) = c_m$, and define φ_n by

$$\varphi_n := \sum_{-n}^{n} (1+m^2)^{-p} \bar{c}_m f_m.$$

Then

$$\|\varphi_n\|_p = \left[\sum_{-n}^{n} (1+m^2)^{-p} |c_m|^2 \right]^{\frac{1}{2}},$$

and hence

$$\|F\|_p \geq \frac{|F(\varphi_n)|}{\|\varphi_n\|_p} = \left[\sum_{-n}^{n} (1+m^2)^{-p} |c_m|^2 \right]^{\frac{1}{2}}.$$

By the calculation in the first part of the theorem we can now conclude that

$$\|F\|_p = \left[\sum_{-\infty}^{\infty} (1+m^2)^{-p} |c_m|^2 \right]^{\frac{1}{2}}.$$

$\square$

It follows from Theorem 1.35 that Rellich's theorem remains valid for $-\infty < p, q < \infty$.

Theorem 1.36. *For $g \in L^2[0, 2\pi]$ the duality pairing*

$$G(\varphi) := \frac{1}{2\pi} \int_0^{2\pi} \varphi(t) g(t) \, dt, \quad \varphi \in H^p,$$

defines a bounded linear functional on $H^p[0, 2\pi]$, i.e., $G \in H^{-p}[0, 2\pi]$. In particular, $L^2[0, 2\pi]$ may be viewed as a subspace of the dual space $H^{-p}[0, 2\pi]$, $0 \leq p < \infty$, and the trigonometric polynomials are dense in $H^{-p}[0, 2\pi]$.

Proof. Let b_m be the Fourier coefficients of g. Then, since $G(f_m) = b_m$, by the second part of Theorem 1.35, we have that $G \in H^{-p}$. Now let $F \in H^{-p}$ with $F(f_m) = c_m$, and define $F_n \in H^{-p}$ by

$$F_n(\varphi) := \frac{1}{2\pi} \int_0^{2\pi} \varphi(t) g_n(t) \, dt,$$

where

$$g_n := \sum_{-n}^{n} c_m \bar{f}_m.$$

Then

$$\|F - F_n\|_p^2 = \sum_{|m|=n+1}^{\infty} (1+m^2)^{-p} |c_m|^2$$

tends to zero as n tends to infinity, which implies that the trigonometric polynomials are dense in $H^{-p}[0, 2\pi]$. $\square$

The preceding duality pairing can be extended to bounded linear functionals in H^{-p}. In particular, for $\varphi \in H^p$ and $g \in H^{-p}$ we define the integral

$$\int_0^{2\pi} \varphi(t)g(t)\, dt$$

to be $g(\varphi)$. We also note that H^{-p} becomes a Hilbert space by extending the inner product previously defined for $p \geq 0$ to $p < 0$.

More generally, if X is a normed space with dual space X^*, then for $g \in X^*$ and $\varphi \in X$ we define the duality pairing $\langle g, \varphi \rangle$ by $\langle g, \varphi \rangle := g(\varphi)$.

1.5 Sobolev Space $H^p(\partial D)$

We now want to define Sobolev spaces on the boundary ∂D of a planar domain D, Sobolev spaces defined on D, and the relationship between these two spaces. To this end, let ∂D be the boundary of a simply connected bounded domain $D \subset \mathbb{R}^2$ such that ∂D is a class C^k, i.e., ∂D has a k-times continuously differentiable 2π-periodic representation $\partial D = \{x(t) : t \in [0, 2\pi), x \in C^k[0, 2\pi]\}$. Then for $0 \leq p \leq k$ we can define the Sobolev space $H^p(\partial D)$ as the space of all functions $\varphi \in L^2(\partial D)$ such that $\varphi(x(t)) \in H^p[0, 2\pi]$. The inner product and norm on $H^p(\partial D)$ are defined via the inner product on $H^p[0, 2\pi]$ by

$$(\varphi, \psi)_{H^p(\partial D)} := (\varphi(x(t)), \psi(x(t)))_{H^p[0,2\pi]}.$$

It can be shown (Theorem 8.14 of [111]) that the foregoing definitions are invariant with respect to parameterization.

The Sobolev space $H^1(D)$ for a bounded domain $D \subset \mathbb{R}^2$ with ∂D of class C^1 is defined as the completion of the space $C^1(\bar{D})$ with respect to the norm

$$\|u\|_{H^1(D)} := \left[\int_D (|u(x)|^2 + |\operatorname{grad} u(x)|^2)\, dx \right]^{\frac{1}{2}}.$$

Note that functions in $H^1(D)$ are in general not differentiable in the classical sense. However, a function $u \in H^1(D)$ will have derivatives $\partial u / \partial x_j \in L^2(D)$, $j = 1, 2$, in the sense that

$$\frac{\partial u}{\partial x_j} = \lim_{n \to \infty} \frac{\partial u}{\partial x_j},$$

where $u_n \in C^1(D)$ is a Cauchy sequence with respect to the norm in $H^1(D)$ and $u_n \to u$ in $L^2(D)$. In particular, if $C_0^1(D)$ is the space of continuously differentiable functions whose support is a compact set of D, then for every $v \in C_0^1(D)$ we have that u satisfies

$$\int_D \frac{\partial u}{\partial x_j} v \, dx = - \int_D u \frac{\partial v}{\partial x_j} \, dx. \tag{1.5}$$

Since any function in $C^1(D)$ clearly satisfies (1.5), we have that functions u in $H^1(D)$ have a *weak derivative*. More generally, we define the Sobolev space $W^{1,2}(D)$ as those functions $u \in L^2(D)$ whose first-order weak derivatives $\partial u / \partial x_j$ are all in $L^2(D)$, i.e., there exists $g \in L^2(D)$ such that for every $v \in C_0^1(D)$ we have that

$$\int_D g v \, dx = - \int_D u \frac{\partial v}{\partial x_j} \, dx.$$

Clearly $W^{1,2}(D)$ is a Hilbert space under the norm used for $H^1(D)$ with the obvious inner product and $H^1(D) \subset W^{1,2}(D)$. It can in fact be shown that $H^1(D) = W^{1,2}(D)$ [70].

It is easily seen that $H^1(D)$ is a subspace of $L^2(D)$. The main purpose of this section is to show that functions in $H^1(D)$ have a meaning when restricted to ∂D, i.e., the *trace* of functions in $H^1(D)$ to the boundary ∂D is well defined. To this end, we will need the following theorem from calculus [11].

Theorem 1.37 (Dini's Theorem). *If $\{\varphi_n\}_1^\infty$ is a sequence of real-valued continuous functions converging pointwise to a continuous limit function φ on a compact set D, and if $\varphi_n(x) \geq \varphi_{n+1}(x)$ for each $x \in D$ and every $n = 1, 2, \cdots$, then $\varphi_n \to \varphi$ uniformly on D.*

Making use of Dini's theorem, we can now prove the following basic result, called the *trace theorem*. In the study of partial differential equations, trace theorems play an important role, and we shall encounter another of these theorems in Chap. 5 of this book.

Theorem 1.38. *Let $D \subset \mathbb{R}^2$ be a simply connected bounded domain with ∂D in class C^2. Then there exists a positive constant C such that*

$$\|u\|_{H^{\frac{1}{2}}(\partial D)} \leq C \|u\|_{H^1(D)}$$

for all $u \in H^1(D)$, i.e., for $u \in H^1(D)$ the operator $u \to u|_{\partial D}$ is well defined and bounded from $H^1(D)$ into $H^{\frac{1}{2}}(\partial D)$.

Proof. We first consider continuously differentiable functions u defined in the strip $\mathbb{R} \times [0,1]$ that are 2π-periodic with respect to the first variable. Let $Q := [0, 2\pi) \times [0,1]$, and for $0 \leq \eta \leq 1$ define

$$a_m(\eta) := \frac{1}{2\pi} \int_0^{2\pi} u(t, \eta) e^{-imt} \, dt.$$

Then by Parseval's equality we have that

$$\sum_{-\infty}^{\infty} |a_m(\eta)|^2 = \frac{1}{2\pi} \int_0^{2\pi} |u(t,\eta)|^2 \, dt, \quad 0 \le \eta \le 1.$$

By Dini's theorem this series is uniformly convergent. Hence we can integrate term by term to obtain

$$\sum_{-\infty}^{\infty} \int_0^1 |a_m(\eta)|^2 \, d\eta = \frac{1}{2\pi} \|u\|_{L^2(Q)}^2.$$

Similarly, from

$$a_m'(\eta) = \frac{1}{2\pi} \int_0^{2\pi} \frac{\partial u}{\partial \eta}(t,\eta) e^{-imt} \, dt$$

and

$$im\, a_m(\eta) = \frac{1}{2\pi} \int_0^{2\pi} \frac{\partial u}{\partial t}(t,\eta) e^{-imt} \, dt$$

we see that

$$\sum_{-\infty}^{\infty} \int_0^1 |a_m'(\eta)|^2 \, d\eta = \frac{1}{2\pi} \left\| \frac{\partial u}{\partial \eta} \right\|_{L^2(Q)}^2$$

and

$$\sum_{-\infty}^{\infty} \int_0^1 m^2 |a_m(\eta)|^2 \, d\eta = \frac{1}{2\pi} \left\| \frac{\partial u}{\partial t} \right\|_{L^2(Q)}^2.$$

We now assume that $u(\cdot,1) = 0$. Then from the Cauchy–Schwarz inequality and the fact that $a_m(1) = 0$ for all m we have that

$$
\begin{aligned}
\|u(\cdot,0)\|_{H^{\frac{1}{2}}[0,2\pi]}^2 &= \sum_{-\infty}^{\infty} (1+m^2)^{\frac{1}{2}} |a_m(0)|^2 \\
&= 2 \sum_{-\infty}^{\infty} (1+m^2)^{\frac{1}{2}} \operatorname{Re} \int_1^0 a_m'(\eta) \overline{a_m(\eta)} \, d\eta \qquad (1.6) \\
&\le 2 \sum_{-\infty}^{\infty} \left[\int_0^1 |a_m'(\eta)|^2 \, d\eta \right]^{\frac{1}{2}} \left[(1+m^2) \int_0^1 |a_m(\eta)|^2 \, d\eta \right]^{\frac{1}{2}} \\
&\le 2 \left[\sum_{-\infty}^{\infty} \int_0^1 |a_m'(\eta)|^2 \, d\eta \right]^{\frac{1}{2}} \left[\sum_{-\infty}^{\infty} (1+m^2) \int_0^1 |a_m(\eta)|^2 \, d\eta \right]^{\frac{1}{2}} \\
&= \frac{1}{\pi} \left\| \frac{\partial u}{\partial \eta} \right\|_{L^2(Q)} \left[\|u\|_{L^2(Q)}^2 + \left\| \frac{\partial u}{\partial t} \right\|_{L^2(Q)}^2 \right]^{\frac{1}{2}} \\
&\le \frac{1}{\pi} \|u\|_{H^1(Q)}^2.
\end{aligned}
$$

We now return to the domain D and choose a parallel strip $D_h := \{x + \eta h \nu(x) : x \in \partial D, \eta \in [0,1]\}$, where ν is the unit inner normal to ∂D, $h > 0$, such that each $y \in D_h$ is uniquely representable through projection onto ∂D in the form $y = x + \eta h \nu(x)$ with $x \in \partial D$, $\eta \in [0,1]$. Let ∂D_h denote the inner boundary of D_h. By parameterizing $\partial D = \{x(t) : 0 \leq t \leq 2\pi\}$ we have a parameterization of D_h in the form

$$x(t, \eta) = x(t) + \eta h \nu(x(t)), \quad 0 \leq t < 2\pi, \quad 0 \leq \eta \leq 1.$$

Inequality (1.6) now shows that for all $u \in C^1(D_h)$ with $u = 0$ on ∂D_h we have that

$$\|u\|_{H^{\frac{1}{2}}(\partial D)} = \|u(x(t))\|_{H^{\frac{1}{2}}[0,2\pi]} \leq \frac{1}{\sqrt{\pi}} \|u(x(t,\eta))\|_{H^1(Q)}$$
$$\leq C \|u\|_{H^1(D_h)},$$

where C is a positive constant depending on the bounds for the first derivatives of the mapping $x(t, \eta)$ and its inverse.

We next extend this estimate to arbitrary $u \in C^1(\bar{D})$. To this end, choose a function $g \in C^1(\bar{D})$ such that $g(y) = 0$ for $y \notin D_h$ and $g(y) = f(\eta)$ for $y = x + \eta h \nu(x) \in D_h$, where

$$f(\eta) := (1 - \eta)^2(1 + 3\eta).$$

Then $f(0) = f'(0) = 1$ and $f(1) = f'(1) = 0$, which implies that

$$\|u\|_{H^{\frac{1}{2}}(\partial D)} = \|gu\|_{H^{\frac{1}{2}}(\partial D)} \leq C \|gu\|_{H^1(D)} \leq C_1 \|u\|_{H^1(D)}$$

for all $u \in C^1(\bar{D})$, where C_1 is a positive constant depending on the bounds for g and its first derivatives.

We have now established the desired inequality for $u \in C^1(\bar{D})$, i.e., $A : u \mapsto u|_{\partial D}$ is a bounded operator from $C^1(\bar{D})$ into $H^{\frac{1}{2}}(\partial D)$. It can be easily shown [115] that if X is a dense subspace of a normed space $\hat{X}$ and Y is a Banach space, then, if $A : X \to Y$ is a bounded linear operator, A can be extended to a bounded linear operator $\hat{A} : \hat{X} \to Y$, where $\|\hat{A}\| = \|A\|$. The desired inequality now follows from this result by extending the operator A from $C^1(\bar{D})$ to $H^1(D)$.   □

We note that in the foregoing proof, ∂D must be in the class C^2 since $\nu = \nu(x)$ must be continuously differentiable.

2

Ill-Posed Problems

For problems in mathematical physics, Hadamard postulated three properties that he deemed to be of central importance:

1. Existence of a solution,
2. Uniqueness of a solution,
3. Continuous dependence of the solution on the data.

A problem satisfying all three of these requirements is called well-posed. To be more precise, we make the following definition: let $A : U \to V$ be an operator from a subset U of a normed space X into a subset V of a normed space Y. The equation $A\varphi = f$ is called *well-posed* if A is bijective and $A^{-1} : V \to U$ is continuous. Otherwise, $A\varphi = f$ is called *ill-posed* or *improperly posed*. Contrary to Hadamard's point of view, in recent years it has become clear that many important problems of mathematical physics are in fact ill-posed! In particular, all of the inverse scattering problems considered in this book are ill-posed, and for this reason we devote a short chapter to the mathematical theory of ill-posed problems. But first we present a simple example of an ill-posed problem.

Example 2.1. Consider the initial-boundary value problem

$$\frac{\partial u}{\partial t} = \frac{\partial^2 u}{\partial x^2} \quad \text{in} \quad [0, \pi] \times [0, T]$$
$$u(0, t) = u(\pi, t) = 0 \quad , \quad 0 \le t \le T$$
$$u(x, 0) = \varphi(x) \quad , \quad 0 \le x \le \pi \,,$$

where $\varphi \in C[0, \pi]$ is a given function. Then, by separation of variables, we obtain the solution

$$u(x, t) = \sum_{1}^{\infty} a_n e^{-n^2 t} \sin nx,$$

$$a_n = \frac{2}{\pi} \int_0^\pi \varphi(y) \sin ny \, dy \,,$$

F. Cakoni and D. Colton, *A Qualitative Approach to Inverse Scattering Theory*,
Applied Mathematical Sciences 188, DOI 10.1007/978-1-4614-8827-9_2,
© Springer Science+Business Media New York 2014

and it is not difficult to show that this solution is unique and depends continuously on the initial data with respect to the maximum norm, i.e.,

$$\max_{[0,\pi]\times[0,T]} |u(x,t)| \le C \max_{[0,\pi]} |\varphi(x)|$$

for some positive constant C [43]. Now consider the *inverse problem* of determining φ from $f := u(\cdot, T)$. In this case,

$$u(x,t) = \sum_1^\infty b_n e^{n^2(T-t)} \sin nx\,,$$

$$b_n = \frac{2}{\pi} \int_0^\pi f(y) \sin ny\, dy\,,$$

and hence

$$\|\varphi\|^2 = \frac{2}{\pi} \sum_1^\infty |b_n|^2\, e^{2n^2 T}\,,$$

which is infinite unless the b_n decay extremely rapidly. Even if this is the case, small perturbations of f (and hence of the b_n) will result in the nonexistence of a solution! Note that the inverse problem can be written as an integral equation of the first kind with smooth kernel:

$$\int_0^\pi K(x,y)\varphi(y)\, dy = f(x)\quad,\quad 0 \le x \le \pi\,,$$

where

$$K(x,y) = \frac{2}{\pi} \sum_1^\infty e^{-n^2 T} \sin nx \sin ny \quad,\quad 0 \le x, y \le \pi.$$

In particular, the preceding integral operator is compact in any reasonable function space, for example, $L^2[0,\pi]$.   □

Theorem 2.2. *Let X and Y be normed spaces, and let $A : X \to Y$ be a compact operator. Then $A\varphi = f$ is ill-posed if X is not of finite dimension.*

Proof. Assume A^{-1} exists and is continuous. Then $I = A^{-1}A : X \to X$ is compact, and hence, by Theorem 1.20 X, is finite dimensional.   □

We will now proceed, again following [111], to present the basic mathematical ideas for treating ill-posed problems. For a more detailed discussion we refer the reader to [71, 98, 111], and, in particular, [68].

2.1 Regularization Methods

Methods for constructing a stable approximate solution to an ill-posed problem are called *regularization* methods. In particular, for A a bounded linear operator, we want to approximate the solution φ of $A\varphi = f$ from a knowledge of a perturbed right-hand side with a known error level

$$\|f - f^\delta\| \leq \delta.$$

When $f \in A(X)$, then, if A is injective, there exists a unique solution φ of $A\varphi = f$. However, in general we cannot expect that $f^\delta \in A(X)$. How do we construct a reasonable approximation φ^δ to φ that depends continuously on f^δ?

Definition 2.3. Let X and Y be normed spaces, and let $A : X \to Y$ be an injective bounded linear operator. Then a family of bounded linear operators $R_\alpha : Y \to X$, $\alpha > 0$, such that

$$\lim_{\alpha \to 0} R_\alpha A\varphi = \varphi$$

for every $\varphi \in X$, is called a *regularization scheme* for A. The parameter α is called the *regularization* parameter .

We clearly have that $R_\alpha f \to A^{-1}f$ as $\alpha \to 0$ for every $f \in A(X)$. The following theorem shows that for compact operators this convergence cannot be uniform.

Theorem 2.4. *Let X and Y be normed spaces, let $A : X \to Y$ be an injective compact operator, and assume X has infinite dimension. Then the operators R_α cannot be uniformly bounded with respect to α as $\alpha \to 0$ and $R_\alpha A$ cannot be norm convergent as $\alpha \to 0$.*

Proof. Assume $\|R_\alpha\| \leq C$ as $\alpha \to 0$. Then, since $R_\alpha f \to A^{-1}f$ as $\alpha \to 0$ for every $f \in A(X)$, we have that $\|A^{-1}f\| \leq C\|f\|$, and hence A^{-1} is bounded on $A(X)$. But this implies $I = A^{-1}A$ is compact on X, which contradicts the fact that X has infinite dimension.

Now assume that $R_\alpha A$ is norm convergent as $\alpha \to 0$, i.e., $\|R_\alpha A - I\| \to 0$ as $\alpha \to 0$. Then there exists $\alpha > 0$ such that $\|R_\alpha A - I\| < \frac{1}{2}$, and hence for every $f \in A(X)$ we have that

$$\begin{aligned}
\|A^{-1}f\| &= \|A^{-1}f - R_\alpha AA^{-1}f + R_\alpha f\| \\
&\leq \|A^{-1}f - R_\alpha AA^{-1}f\| + \|R_\alpha f\| \\
&\leq \|I - R_\alpha A\| \|A^{-1}f\| + \|R_\alpha\| \|f\| \\
&\leq \frac{1}{2} \|A^{-1}f\| + \|R_\alpha\| \|f\| .
\end{aligned}$$

Hence $\|A^{-1}f\| \leq 2\|R_\alpha\| \|f\|$, i.e., $A^{-1} : A(X) \to X$ is bounded and we again have arrived at a contradiction. $\qquad\square$

A regularization scheme approximates the solution φ of $A\varphi = f$ by

$$\varphi_\alpha^\delta := R_\alpha f^\delta.$$

Writing

$$\varphi_\alpha^\delta - \varphi = R_\alpha f^\delta - R_\alpha f + R_\alpha A\varphi - \varphi,$$

we have the estimate

$$\left\|\varphi_\alpha^\delta - \varphi\right\| \le \delta \left\|R_\alpha\right\| + \left\|R_\alpha A\varphi - \varphi\right\|.$$

By Theorem 2.4, the first term on the right-hand side is large for α small, whereas the second term on the right-hand side is large if α is not small! So how do we choose α? A reasonable strategy is to choose $\alpha = \alpha(\delta)$ such that $\varphi_\alpha^\delta \to \varphi$ as $\delta \to 0$.

Definition 2.5. A *strategy* for a regularization scheme R_α, $\alpha > 0$, i.e., a method for choosing the regularization parameter $\alpha = \alpha(\delta)$, is called *regular* if for every $f \in A(X)$ and all $f^\delta \in Y$ such that $\left\|f^\delta - f\right\| \le \delta$ we have that

$$R_{\alpha(\delta)} f^\delta \to A^{-1} f$$

as $\delta \to 0$.

A natural strategy for choosing $\alpha = \alpha(\delta)$ is the *discrepancy principle* of Morozov [130], i.e., the residual $\left\|A\varphi_\alpha^\delta - f^\delta\right\|$ should not be smaller than the accuracy of the measurements of f. In particular, $\alpha = \alpha(\delta)$ should be chosen such that $\left\|AR_\alpha f^\delta - f^\delta\right\| = \gamma\delta$ for some constant $\gamma \ge 1$. Given a regularization scheme, the question, of course, is whether or not such a strategy is regular.

2.2 Singular Value Decomposition

Henceforth X and Y will always be infinite-dimensional Hilbert spaces and $A : X \to Y$, $A \neq 0$, will always be a compact operator. Note that $A^*A : X \to X$ is compact and self-adjoint. Hence, by the Hilbert–Schmidt theorem, there exists at most a countable set of eigenvalues $\{\lambda_n\}_1^\infty$, of A^*A and if $A^*A\varphi_n = \lambda_n \varphi_n$ then $(A^*A\varphi_n, \varphi_n) = \lambda_n \left\|\varphi_n\right\|^2$, i.e., $\left\|A\varphi_n\right\|^2 = \lambda_n \left\|\varphi_n\right\|^2$, which implies that $\lambda_n \ge 0$ for $n = 1, 2, \cdots$. The nonnegative square roots of the eigenvalues of A^*A are called the *singular values* of A.

Theorem 2.6. *Let $\{\mu_n\}_1^\infty$ be the sequence of nonzero singular values of the compact operator $A : X \to Y$ ordered such that*

$$\mu_1 \ge \mu_2 \ge \mu_3 \ge \cdots.$$

Then there exist orthonormal sequences $\{\varphi_n\}_1^\infty$ in X and $\{g_n\}_1^\infty$ in Y such that

$$A\varphi_n = \mu_n g_n \quad, \quad A^* g_n = \mu_n \varphi_n.$$

For every $\varphi \in X$ we have the singular value decomposition

$$\varphi = \sum_{1}^{\infty} (\varphi, \varphi_n)\varphi_n + P\varphi,$$

where $P : X \to N(A)$ is the orthogonal projection operator of X onto $N(A)$ and

$$A\varphi = \sum_{1}^{\infty} \mu_n(\varphi, \varphi_n)g_n.$$

The system (μ_n, φ_n, g_n) is called a singular system *of A.*

Proof. Let $\{\varphi_n\}_1^{\infty}$ be the orthonormal eigenelements of A^*A corresponding to $\{\mu_n\}_1^{\infty}$, i.e.,

$$A^*A\varphi_n = \mu_n^2\varphi_n,$$

and define a second orthonormal sequence by

$$g_n := \frac{1}{\mu_n}A\varphi_n.$$

Then $A\varphi_n = \mu_n g_n$ and $A^*g_n = \mu_n\varphi_n$. The Hilbert–Schmidt theorem implies that

$$\varphi = \sum_{1}^{\infty} (\varphi, \varphi_n)\varphi_n + P\varphi,$$

where $P : X \to N(A^*A)$ is the orthogonal projection operator of X onto $N(A^*A)$. But $\psi \in N(A^*A)$ implies that $(A\psi, A\psi) = (\psi, A^*A\psi) = 0$, and hence $N(A^*A) = N(A)$. Finally, applying A to the preceding expansion (first apply A to the partial sum and then take the limit), we have that

$$A\varphi = \sum_{1}^{\infty} \mu_n(\varphi, \varphi_n)g_n.$$

$\square$

We now come to the main result that will be needed to study compact operator equations of the first kind, i.e., equations of the form $A\varphi = f$, where A is a compact operator.

Theorem 2.7 (Picard's Theorem). *Let $A : X \to Y$ be a compact operator with singular system (μ_n, φ_n, g_n). Then the equation $A\varphi = f$ is solvable if and only if $f \in N(A^*)^{\perp}$ and*

$$\sum_{1}^{\infty} \frac{1}{\mu_n^2}|(f, g_n)|^2 < \infty. \tag{2.1}$$

In this case a solution to $A\varphi = f$ is given by

$$\varphi = \sum_1^\infty \frac{1}{\mu_n}(f,g_n)\varphi_n.$$

Proof. The necessity of $f \in N(A^*)^\perp$ follows from Theorem 1.29. If φ is a solution of $A\varphi = f$, then

$$\mu_n(\varphi,\varphi_n) = (\varphi, A^*g_n) = (A\varphi, g_n) = (f, g_n).$$

But from the singular value decomposition of φ we have that

$$\|\varphi\|^2 = \sum_1^\infty |(\varphi,\varphi_n)|^2 + \|P\varphi\|^2 \, ,$$

and hence

$$\sum_1^\infty \frac{1}{\mu_n^2}|(f,g_n)|^2 = \sum_1^\infty |(\varphi,\varphi_n)|^2 \le \|\varphi\|^2 \, ,$$

which implies the necessity of condition (2.1).

Conversely, assume that $f \in N(A^*)^\perp$ and (2.1) is satisfied. Then from (2.1) we have that

$$\varphi := \sum_1^\infty \frac{1}{\mu_n}(f,g_n)\varphi_n$$

converges in the Hilbert space X. Applying A to this series we have that

$$A\varphi = \sum_1^\infty (f,g_n)g_n.$$

But, since $f \in N(A^*)^\perp$, this is the singular value decomposition of f corresponding to the operator A^*, and hence $A\varphi = f$. □

Note that Picard's theorem illustrates the ill-posed nature of the equation $A\varphi = f$. In particular, setting $f^\delta = f + \delta g_n$ we obtain a solution of $A\varphi^\delta = f^\delta$ given by $\varphi^\delta = \varphi + \delta\varphi_n/\mu_n$. Hence, if $A(X)$ is not finite dimensional, then

$$\frac{\|\varphi^\delta - \varphi\|}{\|f^\delta - f\|} = \frac{1}{\mu_n} \to \infty$$

since, by Theorem 1.14, we have that $\mu_n \to 0$. We say that $A\varphi = f$ is *mildly ill-posed* if the singular values decay slowly to zero and *severely ill-posed* if they decay very rapidly (for example, exponentially). All of the inverse scattering problems considered in this book are severely ill-posed.

Henceforth, to focus on ill-posed problems, we will always assume that $A(X)$ is infinite dimensional, i.e., the set of singular values is an infinite set.

Example 2.8. Consider the case of the backward heat equation discussed in Example 2.1. The problem considered in this example is equivalent to solving the compact operator equation $A\varphi = f$, where

$$(A\varphi)(x) := \int_0^\pi K(x,y)\varphi(y)\,dy \quad , \quad 0 \le x \le \pi,$$

and

$$K(x,y) := \frac{2}{\pi} \sum_1^\infty e^{-n^2 T} \sin nx \sin ny.$$

Then A is easily seen to be self-adjoint with eigenvalues given by $\lambda_n = e^{-n^2 T}$. Hence $\mu_n = \lambda_n$, and the compact operator equation $A\varphi = f$ is severely ill posed.   □

Picard's theorem suggests trying to regularize $A\varphi = f$ by damping or filtering out the influence of the higher-order terms in the solution φ given by

$$\varphi = \sum_1^\infty \frac{1}{\mu_n}(f, g_n)\varphi_n.$$

The following theorem does exactly that. We will subsequently consider two specific regularization schemes by making specific choices of the function q, which appears in the theorem.

Theorem 2.9. *Let $A : X \to Y$ be an injective compact operator with singular system (μ_n, φ_n, g_n), and let $q : (0, \infty) \times (0, \|A\|] \to \mathbb{R}$ be a bounded function such that for every $\alpha > 0$ there exists a positive constant $c(\alpha)$ such that*

$$|q(\alpha, \mu)| \le c(\alpha)\mu \quad , \quad 0 < \mu \le \|A\| ,$$

and

$$\lim_{\alpha \to 0} q(\alpha, \mu) = 1 \quad , \quad 0 < \mu \le \|A\| .$$

Then the bounded linear operators $R_\alpha : Y \to X$, $\alpha > 0$, defined by

$$R_\alpha f := \sum_1^\infty \frac{1}{\mu_n} q(\alpha, \mu_n)(f, g_n)\varphi_n$$

for $f \in Y$, describe a regularization scheme with

$$\|R_\alpha\| \le c(\alpha).$$

Proof. Noting that from the singular value decomposition of f with respect to the operator A^* we have that

$$\|f\|^2 = \sum_1^\infty |(f, g_n)|^2 + \|Pf\|^2,$$

where $P: X \to N(A^*)$ is the orthogonal projection of X onto $N(A^*)$, we see that for every $f \in Y$ we have that

$$\|R_\alpha f\|^2 = \sum_1^\infty \frac{1}{\mu_n^2} |q(\alpha, \mu_n)|^2 |(f, g_n)|^2$$

$$\leq |c(\alpha)|^2 \sum_1^\infty |(f, g_n)|^2$$

$$\leq |c(\alpha)|^2 \|f\|^2,$$

and hence $\|R_\alpha\| \leq c(\alpha)$. From

$$(R_\alpha A\varphi, \varphi_n) = \frac{1}{\mu_n} q(\alpha, \mu_n)(A\varphi, g_n)$$

$$= q(\alpha, \mu_n)(\varphi, \varphi_n)$$

and the singular value decomposition for $R_\alpha A\varphi - \varphi$ we obtain, using the fact that A is injective, that

$$\|R_\alpha A\varphi - \varphi\|^2 = \sum_1^\infty |(R_\alpha A\varphi - \varphi, \varphi_n)|^2$$

$$= \sum_1^\infty |q(\alpha, \mu_n) - 1|^2 |(\varphi, \varphi_n)|^2.$$

Now let $\varphi \in X$, $\varphi \neq 0$, and let M be a bound for q. We first note that for every $\epsilon > 0$ there exists $N = N(\epsilon)$ such that

$$\sum_{N+1}^\infty |(\varphi, \varphi_n)|^2 < \frac{\epsilon}{2(M+1)^2}.$$

Since $\lim_{\alpha \to 0} q(\alpha, \mu) = 1$, there exists $\alpha_0 = \alpha_0(\epsilon)$ such that

$$|q(\alpha, \mu_n) - 1|^2 < \frac{\epsilon}{2\|\varphi\|^2}$$

for $n = 1, 2, \cdots, N$ and all α such that $0 < \alpha \leq \alpha_0$. We now have that, for $0 < \alpha \leq \alpha_0$,

$$\|R_\alpha A\varphi - \varphi\|^2 = \sum_1^N |q(\alpha, \mu_n) - 1|^2 |(\varphi, \varphi_n)|^2$$

$$+ \sum_{N+1}^\infty |q(\alpha, \mu_n) - 1|^2 |(\varphi, \varphi_n)|^2$$

$$\leq \frac{\epsilon}{2\|\varphi\|^2} \sum_1^N |(\varphi, \varphi_n)|^2 + \frac{\epsilon}{2}.$$

But, since A is injective,

$$\|\varphi\|^2 = \sum_1^\infty |(\varphi, \varphi_n)|^2,$$

and hence $\|R_\alpha A\varphi - \varphi\|^2 \le \epsilon$ for $0 < \alpha \le \alpha_0$. We can now conclude that $R_\alpha A\varphi \to \varphi$ as $\alpha \to 0$ for every $\varphi \in X$ and the theorem is proved. $\qquad\square$

A particular choice of q now leads to our first regularization scheme, the *spectral cutoff* method .

Theorem 2.10. *Let* $A : X \to Y$ *be an injective compact operator with singular system* (μ_n, φ_n, g_n). *Then the* spectral cutoff

$$R_m f := \sum_{\mu_n \ge \mu_m} \frac{1}{\mu_n}(f, g_n)\varphi_n$$

describes a regularization scheme with regularization parameter $m \to \infty$ *and* $\|R_m\| = 1/\mu_m$.

Proof. Choose q such that $q(m, \mu) = 1$ for $\mu \ge \mu_m$ and $q(m, \mu) = 0$ for $\mu < \mu_m$. Then, since $\mu_m \to 0$ as $m \to \infty$, the conditions of the previous theorem are clearly satisfied with $c(m) = \frac{1}{\mu_m}$. Hence $\|R_m\| \le \frac{1}{\mu_m}$. Equality follows from the identity $R_m g_m = \varphi_m/\mu_m$. $\qquad\square$

We conclude this section by establishing a discrepancy principle for the spectral cutoff regularization scheme.

Theorem 2.11. *Let* $A : X \to Y$ *be an injective compact operator with dense range in* Y, *and let* $f \in Y$ *and* $\delta > 0$. *Then there exists a smallest integer* m *such that*

$$\|AR_m f - f\| \le \delta.$$

Proof. Since $\overline{A(X)} = Y$, A^* is injective. Hence the singular value decomposition with the singular system (μ_n, g_n, φ_n) for A^* implies that for every $f \in Y$ we have that

$$f = \sum_1^\infty (f, g_n)g_n. \tag{2.2}$$

Hence

$$\|(AR_m - I)f\|^2 = \sum_{\mu_n < \mu_m} |(f, g_n)|^2 \to 0 \tag{2.3}$$

as $m \to \infty$. In particular, there exists a smallest integer $m = m(\delta)$ such that $\|AR_m f - f\| \le \delta$. $\qquad\square$

Note that from (2.2) and (2.3) we have that

$$\|AR_m f - f\|^2 = \|f\|^2 - \sum_{\mu_n \geq \mu_m} |(f, g_n)|^2. \qquad (2.4)$$

In particular, $m(\delta)$ is determined by the condition that $m(\delta)$ is the smallest value of m such that the right-hand side of (2.4) is less than or equal to δ^2. For example, in the case of the backward heat equation (Example 2.1) we have that $g_n(x) = \sqrt{2/\pi} \sin nx$, and hence m is determined by the condition that m is the smallest integer such that

$$\|f\|^2 - \sum_1^m |b_n|^2 \leq \delta^2,$$

where the b_n are the Fourier coefficients of f.

It can be shown that the preceding discrepancy principle for the spectral cutoff method is regular (Theorem 15.26 of [111]).

2.3 Tikhonov Regularization

We now introduce and study the most popular regularization scheme in the field of ill-posed problems.

Theorem 2.12. *Let $A : X \to Y$ be a compact operator. Then for every $\alpha > 0$ the operator $\alpha I + A^* A : X \to X$ is bijective and has a bounded inverse. Furthermore, if A is injective, then*

$$R_\alpha := (\alpha I + A^* A)^{-1} A^*$$

describes a regularization scheme with $\|R_\alpha\| \leq 1/2\sqrt{\alpha}$.

Proof. From

$$\alpha \|\varphi\|^2 \leq (\alpha \varphi + A^* A \varphi, \varphi)$$

for $\varphi \in X$ we can conclude that for $\alpha > 0$ the operator $\alpha I + A^* A$ is injective. Hence, since $A^* A$ is a compact operator, by Riesz's theorem we have that $(\alpha I + A^* A)^{-1}$ exists and is bounded.

Now assume that A is injective, and let (μ_n, φ_n, g_n) be a singular system for A. Then for $f \in Y$ the unique solution φ_α of

$$\alpha \varphi_\alpha + A^* A \varphi_\alpha = A^* f$$

is given by

$$\varphi_\alpha = \sum_1^\infty \frac{\mu_n}{\alpha + \mu_n^2} (f, g_n) \varphi_n,$$

i.e., R_α can be written in the form

$$R_\alpha f = \sum_{1}^{\infty} \frac{1}{\mu_n} q(\alpha, \mu_n)(f, g_n)\varphi_n \,,$$

where

$$q(\alpha, \mu) = \frac{\mu^2}{\alpha + \mu^2}.$$

Since $0 < q(\alpha, \mu) < 1$ and $\sqrt{\alpha}\mu \leq (\alpha + \mu^2)/2$, we have that $|q(\alpha, \mu)| \leq \mu/2\sqrt{\alpha}$, and the theorem follows from Theorem 2.9.   □

The next theorem shows that the function $\varphi_\alpha = R_\alpha f$ can be obtained as the solution of an optimization problem.

Theorem 2.13. *Let $A : X \to Y$ be a compact operator, and let $\alpha > 0$. Then for every $f \in Y$ there exists a unique $\varphi_\alpha \in X$ such that*

$$\|A\varphi_\alpha - f\|^2 + \alpha \|\varphi_\alpha\|^2 = \inf_{\varphi \in X} \left\{ \|A\varphi - f\|^2 + \alpha \|\varphi\|^2 \right\}.$$

The minimizer is the unique solution of $\alpha\varphi_\alpha + A^* A\varphi_\alpha = A^* f$.

Proof. From

$$\|A\varphi - f\|^2 + \alpha \|\varphi\|^2 = \|A\varphi_\alpha - f\|^2 + \alpha \|\varphi_\alpha\|^2$$
$$+ 2\mathrm{Re}(\varphi - \varphi_\alpha, \alpha\varphi_\alpha + A^* A\varphi_\alpha - A^* f)$$
$$+ \|A(\varphi - \varphi_\alpha)\|^2 + \alpha \|\varphi - \varphi_\alpha\|^2 \,,$$

which is valid for every $\varphi, \varphi_\alpha \in X$, we see that if φ_α satisfies $\alpha\varphi_\alpha + A^* A\varphi_\alpha = A^* f$, then φ_α minimizes the *Tikhonov functional*

$$\|A\varphi - f\|^2 + \alpha \|\varphi\|^2 .$$

On the other hand, if φ_α is a minimizer of the Tikhonov functional, then set

$$\psi := \alpha\varphi_\alpha + A^* A\varphi_\alpha - A^* f$$

and assume that $\psi \neq 0$. Then for $\varphi := \varphi_\alpha - t\psi$, with t a real number, we have that

$$\|A\varphi - f\|^2 + \alpha \|\varphi\|^2 = \|A\varphi_\alpha - f\|^2 + \alpha \|\varphi_\alpha\|^2$$
$$- 2t \|\psi\|^2 + t^2(\|A\psi\|^2 + \alpha \|\psi\|^2). \qquad (2.5)$$

The minimum of the right-hand side of (2.5) occurs when

$$t = \frac{\|\psi\|^2}{\|A\psi\|^2 + \alpha \|\psi\|^2} \,,$$

and for this t we have that $\|A\varphi - f\|^2 + \alpha \|\varphi\|^2 < \|A\varphi_\alpha - f\|^2 + \alpha \|\varphi_\alpha\|^2$, which contradicts the definition of φ_α. Hence $\psi = 0$, i.e., $\alpha\varphi_\alpha + A^* A\varphi_\alpha = A^* f$.   □

By the interpretation of Tikhonov regularization as the minimizer of the Tikhonov functional, its solution φ_α keeps the residual $\|A\varphi_\alpha - f\|^2$ small and is stabilized through the penalty term $\alpha \|\varphi_\alpha\|^2$. This suggests the following two constrained optimization problems:

Minimum norm solution: for a given $\delta > 0$ minimize $\|\varphi\|$ such that $\|A\varphi - f\| \leq \delta$.
Quasi-solutions: for a given $\rho > 0$ minimize $\|A\varphi - f\|$ such that $\|\varphi\| \leq \rho$.

We begin with the idea of a minimum norm solution and view this as a discrepancy principle for choosing φ in a Tikhonov regularization.

Theorem 2.14. *Let $A : X \to Y$ be an injective compact operator with dense range in Y, and let $f \in Y$ with $\|f\| > \delta > 0$. Then there exists a unique α such that*

$$\|AR_\alpha f - f\| = \delta.$$

Proof. We must show that

$$F(\alpha) := \|AR_\alpha f - f\|^2 - \delta^2$$

has a unique zero. As in Theorem 2.11, we have that

$$f = \sum_1^\infty (f, g_n) g_n,$$

and for $\varphi_\alpha = R_\alpha f$ we have that

$$\varphi_\alpha = \sum_1^\infty \frac{\mu_n}{\alpha + \mu_n^2} (f, g_n) \varphi_n.$$

Hence

$$F(\alpha) = \sum_1^\infty \frac{\alpha^2}{(\alpha + \mu_n^2)^2} |(f, g_n)|^2 - \delta^2.$$

Since F is a continuous function of α and strictly monotonically increasing with limits $F(\alpha) \to -\delta^2$ as $\alpha \to 0$ and $F(\alpha) \to \|f\|^2 - \delta^2 > 0$ as $\alpha \to \infty$, F has exactly one zero $\alpha = \alpha(\delta)$. $\square$

To prove the regularity of the foregoing discrepancy principle for Tikhonov regularizations, we need to introduce the concept of *weak convergence*.

Definition 2.15. A sequence $\{\varphi_n\}$ in X is said to be *weakly convergent* to $\varphi \in X$ if

$$\lim_{n \to \infty} (\psi, \varphi_n) = (\psi, \varphi)$$

for every $\psi \in X$ and we write $\varphi_n \rightharpoonup \varphi$, $n \to \infty$.

Note that norm convergence $\varphi_n \to \varphi$, $n \to \infty$, always implies weak convergence, but, as the following example shows, the converse is generally false.

Example 2.16. Let ℓ^2 be the space of all sequences $\{a_n\}_1^\infty$, $a_n \in \mathbb{C}$, such that

$$\sum_1^\infty |a_n|^2 < \infty. \tag{2.6}$$

It is easily shown that, with componentwise addition and scalar multiplication, ℓ^2 is a Hilbert space with inner product

$$(a,b) = \sum_1^\infty a_n \bar{b}_n,$$

where $a = \{a_n\}_1^\infty$ and $b = \{b_n\}_1^\infty$. In ℓ^2 we now define the sequence $\{\varphi_n\}$ by $\varphi_n = (0,0,0,\cdots,1,0,\cdots)$, where the one appears in the nth entry. Then $\{\varphi_n\}$ is not norm convergent since $\|\varphi_n - \varphi_m\| = \sqrt{2}$ for $m \neq n$, and hence $\{\varphi_n\}$ is not a Cauchy sequence. On the other hand, for $\psi = \{a_n\} \in \ell^2$ we have that $(\psi, \varphi_n) = a_n \to 0$ as $n \to \infty$ due to the convergence of the series in (2.6). Hence $\{\varphi_n\}$ is weakly convergent to zero in ℓ^2.

Theorem 2.17. *Every bounded sequence in a Hilbert space contains a weakly convergent subsequence.*

Proof. Let $\{\varphi_n\}$ be a bounded sequence, $\|\varphi_n\| \leq C$. Then for each integer m the sequence (φ_m, φ_n) is bounded for all n. Hence by the Bolzano–Weierstrass theorem and a diagonalization process (cf. the proof of Theorem 1.17) we can select a subsequence $\{\varphi_{n(k)}\}$ such that $(\varphi_m, \varphi_{n(k)})$ converges as $k \to \infty$ for every integer m. Thus the linear functional F defined by

$$F(\psi) := \lim_{k \to \infty} (\psi, \varphi_{n(k)})$$

is well defined on $U := \text{span}\{\varphi_m\}$ and, by continuity, on $\bar{U}$. Now let $P : X \to \bar{U}$ be the orthogonal projection operator, and for arbitrary $\psi \in X$ write $\psi = P\psi + (I - P)\psi$. For arbitrary $\psi \in X$ define $F(\psi)$ by

$$F(\psi) := \lim_{k \to \infty} (\psi, \varphi_{n(k)}) = \lim_{k \to \infty} \left[(P\psi, \varphi_{n(k)}) + ((I - P)\psi, \varphi_{n(k)}) \right]$$
$$= \lim_{k \to \infty} (P\psi, \varphi_{n(k)}),$$

where we have used the easily verifiable fact that P is self-adjoint. Thus F is defined on all of X. Furthermore, $\|F\| \leq C$. Hence, by the Riesz representation theorem, there exists a unique $\varphi \in X$ such that $F(\psi) = (\psi, \varphi)$ for every $\psi \in X$. We can now conclude that $\lim_{k \to \infty}(\psi, \varphi_{n(k)}) = (\psi, \varphi)$ for every $\psi \in X$, i.e., $\varphi_{n(k)}$ is weakly convergent to φ as $k \to \infty$. $\qquad \square$

We are now in a position to show that the discrepancy principle of Theorem 2.14 is regular.

Theorem 2.18. *Let $A : X \to Y$ be an injective compact operator with dense range in Y. Let $f \in A(X)$ and $f^\delta \in Y$ satisfy $\left\| f^\delta - f \right\| \leq \delta < \left\| f^\delta \right\|$ with $\delta > 0$. Then there exists a unique $\alpha = \alpha(\delta)$ such that*

$$\left\| A R_{\alpha(\delta)} f^\delta - f^\delta \right\| = \delta$$

and

$$R_{\alpha(\delta)} f^\delta \to A^{-1} f$$

as $\delta \to 0$.

Proof. In view of Theorem 2.14, we only need to establish convergence. Since $\varphi^\delta = R_{\alpha(\delta)} f^\delta$ minimizes the Tikhonov functional, we have that

$$
\begin{aligned}
\delta^2 + \alpha \left\| \varphi^\delta \right\|^2 &= \left\| A\varphi^\delta - f^\delta \right\|^2 + \alpha \left\| \varphi^\delta \right\|^2 \\
&\leq \left\| A A^{-1} f - f^\delta \right\|^2 + \alpha \left\| A^{-1} f \right\|^2 \\
&\leq \delta^2 + \alpha \left\| A^{-1} f \right\|^2 ,
\end{aligned}
$$

and hence $\left\| \varphi^\delta \right\| \leq \left\| A^{-1} f \right\|$. Now let $g \in Y$. Then

$$
\begin{aligned}
\left| (A\varphi^\delta - f, g) \right| &\leq \left(\left\| A\varphi^\delta - f^\delta \right\| + \left\| f^\delta - f \right\| \right) \left\| g \right\| \\
&\leq 2\delta \left\| g \right\| \to 0
\end{aligned}
\tag{2.7}
$$

as $\delta \to 0$. Since A is injective, $A^*(Y)$ is dense in X, and hence for every $\psi \in X$ there exists a sequence $\{ g_n \}$ in Y such that $A^* g_n \to \psi$. Then

$$
(\varphi^\delta - \varphi, \psi) = (\varphi^\delta - \varphi, A^* g_n) + (\varphi^\delta - \varphi, \psi - A^* g_n)
\tag{2.8}
$$

and, for every $\epsilon > 0$,

$$
\left| (\varphi^\delta - \varphi, \psi - A^* g_n) \right| \leq \left\| \varphi^\delta - \varphi \right\| \left\| \psi - A^* g_n \right\| < \frac{\epsilon}{2}
\tag{2.9}
$$

for all $\delta > 0$ and $N > N_0$ since $\left\| \varphi^\delta - \varphi \right\|$ is bounded. Hence for $N > N_0$ and δ sufficiently small we have from (2.7)–(2.9) that

$$
\begin{aligned}
\left| (\varphi^\delta - \varphi, \psi) \right| &\leq \left| (\varphi^\delta - \varphi, A^* g_n) \right| + \left| (\varphi^\delta - \varphi, \psi - A^* g_n) \right| \\
&\leq \left| (A\varphi^\delta - f, g_n) \right| + \frac{\epsilon}{2} \\
&\leq \epsilon ,
\end{aligned}
$$

where we have set $f = A\varphi$. We can now conclude that $\varphi^\delta \rightharpoonup A^{-1} f$ as $\delta \to 0$. Then, again using the fact that $\left\| \varphi^\delta \right\| \leq \left\| A^{-1} f \right\|$, we have that

$$\left\|\varphi^\delta - A^{-1}f\right\|^2 = \left\|\varphi^\delta\right\|^2 - 2\mathrm{Re}\left(\varphi^\delta, A^{-1}f\right) + \left\|A^{-1}f\right\|^2 \qquad (2.10)$$
$$\leq 2\left(\left\|A^{-1}f\right\|^2 - \mathrm{Re}\left(\varphi^\delta, A^{-1}f\right)\right) \to 0$$

as $\delta \to 0$, and the proof is complete. □

Under additional conditions on f, which may be viewed as a regularity condition on f, we can obtain results on the order of convergence.

Theorem 2.19. *Under the assumptions of Theorem 2.18, if $f \in AA^*(Y)$, then*

$$\left\|\varphi^\delta - A^{-1}f\right\| = O\left(\delta^{1/2}\right) \quad , \quad \delta \to 0.s$$

Proof. We have that $A^{-1}f = A^*g$ for some $g \in Y$. Then from (2.10) we have that

$$\left\|\varphi^\delta - A^{-1}f\right\|^2 \leq 2\left(\left\|A^{-1}f\right\|^2 - \mathrm{Re}\left(\varphi^\delta, A^{-1}f\right)\right)$$
$$= 2\mathrm{Re}\left(A^{-1}f - \varphi^\delta, A^{-1}f\right)$$
$$= 2\mathrm{Re}\left(f - A\varphi^\delta, g\right)$$
$$\leq 2\left(\left\|f - f^\delta\right\| + \left\|f^\delta - A\varphi^\delta\right\|\right)\|g\|$$
$$\leq 4\delta\|g\| ,$$

and the theorem follows. □

Tikhonov regularization methods also apply to cases where both the operator and the right-hand side are perturbed, i.e., both the operator and the right-hand side are "noisy." In particular, consider the operator equation $A_h\varphi = f^\delta$, $A_h : X \to Y$, where $\|A_h - A\| \leq h$ and $\|f - f^\delta\| \leq \delta$, respectively. Then the Tikhonov regularization operator is given by

$$R_\alpha := \left(\alpha I + A_h^* A_h\right)^{-1} A_h^*,$$

and the regularization solution $\varphi^\alpha := R_\alpha f^\delta$ is found by minimizing the Tikhonov functional

$$\left\|A_h\varphi - f^\delta\right\| + \alpha \|\varphi\| .$$

The regularization parameter $\alpha = \alpha(\delta, h)$ is determined from the equation

$$\left\|A_h\varphi_\alpha - f^\delta\right\|^2 = \left(\delta + h \|\varphi_\alpha\|^2\right).$$

Then all of the results obtained earlier in the case where A is not noisy can be generalized to the present case where both A and f are noisy. For details we refer the reader to [130].

We now turn our attention to the *method of quasi-solutions*.

Theorem 2.20. *Let $A : X \to Y$ be an injective compact operator and let $\rho > 0$. Then for every $f \in Y$ there exists a unique $\varphi_0 \in X$ with $\|\varphi_0\| \leq \rho$ such that*

$$\|A\varphi_0 - f\| \leq \|A\varphi - f\|$$

for all φ satisfying $\|\varphi\| \leq \rho$. The element φ_0 is called the quasi-solution *of $A\varphi = f$ with constraint ρ.*

Proof. We note that φ_0 is a quasi-solution with constraint ρ if and only if $A\varphi_0$ is a best approximation to f with respect to the set $V := \{A\varphi : \|\varphi\| \leq \rho\}$. Since A is linear, V is clearly convex, i.e., $\lambda\varphi_1 + (1-\lambda)\varphi_2 \in V$ for all $\varphi_1, \varphi_2 \in V$ and $0 \leq \lambda \leq 1$. Suppose there were two best approximations to f, i.e., there exist $v_1, v_2 \in V$ such that

$$\|f - v_1\| = \|f - v_2\| = \inf_{v \in V} \|f - v\|.$$

Then, since V is convex, $\frac{1}{2}(v_1 + v_2) \in V$, and hence

$$\left\| f - \frac{v_1 + v_2}{2} \right\| \geq \|f - v_1\|.$$

By the parallelogram equality we now have that

$$\|v_1 - v_2\|^2 = 2\|f - v_1\|^2 + 2\|f - v_2\|^2$$
$$- 4\left\| f - \frac{v_1 + v_2}{2} \right\|^2$$
$$\leq 0,$$

and hence $v_1 = v_2$. Thus if there were two quasi-solutions φ_1 and φ_2, then $A\varphi_1 = A\varphi_2$. But since A is injective $\varphi_1 = \varphi_2$, i.e., the quasi-solution, if it exists, is unique.

To prove the existence of a quasi-solution, let $\{\varphi_n\}$ be a minimizing sequence, i.e., $\|\varphi_n\| \leq \rho$, and

$$\lim_{n \to \infty} \|A\varphi_n - f\| = \inf_{\|\varphi\| \leq \rho} \|A\varphi - f\|. \tag{2.11}$$

By Theorem 2.17, there exists a weakly convergent subsequence of $\{\varphi_n\}$, and without loss of generality we assume that $\varphi_n \rightharpoonup \varphi_0$ as $n \to \infty$ for some $\varphi_0 \in X$. We will show that $A\varphi_n \to A\varphi_0$ as $n \to \infty$. Since for every $\varphi \in X$ we have that

$$\lim_{n \to \infty}(A\varphi_n, \varphi) = \lim_{n \to \infty}(\varphi_n, A^*\varphi) = (\varphi_0, A^*\varphi) = (A\varphi_0, \varphi),$$

we can conclude that $A\varphi_n \rightharpoonup A\varphi_0$. Now suppose that $A\varphi_n$ does not converge to $A\varphi_0$. Then $\{A\varphi_n\}$ has a subsequence such that $\|A\varphi_{n(k)} - A\varphi_0\| \geq \delta$ for some $\delta > 0$. Since $\|\varphi_n\| \leq \rho$ and A is compact, $\{A\varphi_{n(k)}\}$ has a convergent subsequence that we again call $\{A\varphi_{n(k)}\}$. But since convergent sequences are also weakly convergent and have the same limit, $A\varphi_{n(k)} \rightarrow A\varphi_0$, which is a contradiction. Hence $A\varphi_n \rightarrow A\varphi_0$. From (2.11) we can now conclude that

$$\|A\varphi_0 - f\| = \inf_{\|\varphi\| \leq \rho} \|A\varphi - f\| \,,$$

and since $\|\varphi_0\|^2 = \lim_{n \to \infty} (\varphi_n, \varphi_0) \leq \rho \|\varphi_0\|$, we have that $\|\varphi_0\| \leq \rho$. This completes the proof of the theorem. □

We next show that under appropriate assumptions the method of quasi-solutions is regular.

Theorem 2.21. *Let $A : X \rightarrow Y$ be an injective compact operator with dense range, and let $f \in A(X)$ and $\rho \geq \|A^{-1}f\|$. For $f^\delta \in Y$ with $\|f^\delta - f\| \leq \delta$, let φ^δ be the quasi-solution to $A\varphi = f^\delta$ with constraint ρ. Then $\varphi^\delta \rightharpoonup A^{-1}f$ as $\delta \rightarrow 0$, and if $\rho = \|A^{-1}f\|$, then $\varphi^\delta \rightarrow A^{-1}f$ as $\delta \rightarrow 0$.*

Proof. Let $g \in Y$. Then, since $\|A^{-1}f\| \leq \rho$ and $\|A\varphi^\delta - f^\delta\| \leq \|A\varphi - f^\delta\|$ for $f = A\varphi$, we have that

$$\begin{aligned} \left|(A\varphi^\delta - f, g)\right| &\leq \left(\|A\varphi^\delta - f^\delta\| + \|f^\delta - f\|\right) \|g\| \\ &\leq \left(\|AA^{-1}f - f^\delta\| + \|f^\delta - f\|\right) \|g\| \qquad (2.12) \\ &\leq 2\delta \|g\| \,. \end{aligned}$$

Hence $(A\varphi^\delta - f, g) = (\varphi^\delta - A^{-1}f, A^*g) \rightarrow 0$ as $\delta \rightarrow 0$ for every $g \in Y$. Since A is injective, $A^*(Y)$ is dense in X, and we can conclude that $\varphi^\delta \rightharpoonup A^{-1}f$ as $\delta \rightarrow 0$ (cf. the proof of Theorem 2.18).

When $\rho = \|A^{-1}f\|$, we have (using $\|\varphi^\delta\| \leq \rho = \|A^{-1}f\|$) that

$$\begin{aligned} \|\varphi^\delta - A^{-1}f\|^2 &= \|\varphi^\delta\|^2 - 2\operatorname{Re}(\varphi^\delta, A^{-1}f) + \|A^{-1}f\|^2 \qquad (2.13) \\ &\leq 2\operatorname{Re}(A^{-1}f - \varphi^\delta, A^{-1}f) \rightarrow 0 \end{aligned}$$

as $\delta \rightarrow 0$. □

Note that for regularity we need to know a priori the norm of the solution to the noise-free equation.

Theorem 2.22. *Under the assumptions of Theorem 2.21, if $f \in AA^*(Y)$ and $\rho = \|A^{-1}f\|$, then*

$$\|\varphi^\delta - A^{-1}f\| = O\left(\delta^{1/2}\right) \,, \qquad \delta \rightarrow 0.$$

Proof. We can write $A^{-1}f = A^*g$ for some $g \in Y$. From (2.12) and (2.13) we have that $\|\varphi^\delta - A^{-1}f\|^2 \leq 2\operatorname{Re}(f - A\varphi^\delta, g) \leq 4\delta \|g\|$, and the theorem follows. □

3

Scattering by Imperfect Conductors

In this chapter we consider a very simple scattering problem corresponding to the scattering of a time-harmonic plane wave by an imperfect conductor. Although the problem is simple compared to most problems in scattering theory, its mathematical resolution took many years to accomplish and was the focus of energy of some of the outstanding mathematicians of the twentieth century, in particular Kupradze, Rellich, Vekua, Müller, and Weyl. Indeed, the solution of the full three-dimensional problem was not fully realized until 1981 (cf. Sect. 9.5 of [54]). Here we will content ourselves with the two-dimensional scalar problem and its solution by the method of integral equations. As will be seen, the main difficulty of this approach is the presence of eigenvalues of the interior Dirichlet problem for the Helmholtz equation, and we will overcome this difficulty using the ideas of Jones [96], Ursell [156], and Kleinman and Roach [109].

The plan of this chapter is as follows. We begin by considering Maxwell's equations and then derive the scalar impedance boundary value problem corresponding to the scattering of a time-harmonic plane wave by an imperfectly conducting infinite cylinder. After a brief detour to discuss the relevant properties of Bessel and Hankel functions that will be needed in the sequel, we proceed to show that our scattering problem is well posed by deriving Rellich's lemma and using the method of modified single layer potentials. We will conclude this chapter by giving a brief discussion on weak solutions of the Helmholtz equation. (This theme will be revisited in greater detail in Chap. 5).

3.1 Maxwell's Equations

Consider electromagnetic wave propagation in a homogeneous, isotropic, nonconducting medium in $\mathbb{R}^3$ with electric permittivity ϵ and magnetic permeability μ. A time-harmonic electromagnetic wave with frequency $\omega > 0$ is described by the electric and magnetic fields

F. Cakoni and D. Colton, *A Qualitative Approach to Inverse Scattering Theory*, Applied Mathematical Sciences 188, DOI 10.1007/978-1-4614-8827-9_3, © Springer Science+Business Media New York 2014

$$\mathcal{E}(x,t) = \epsilon^{-1/2} E(x) e^{-i\omega t},$$
$$\mathcal{H}(x,t) = \mu^{-1/2} H(x) e^{-i\omega t}, \qquad (3.1)$$

where $x \in \mathbb{R}^3$ and $\mathcal{E}$ and $\mathcal{H}$ satisfy *Maxwell's equations*

$$\operatorname{curl} \mathcal{E} + \mu \frac{\partial \mathcal{H}}{\partial t} = 0,$$
$$\operatorname{curl} \mathcal{H} - \epsilon \frac{\partial \mathcal{E}}{\partial t} = 0. \qquad (3.2)$$

In particular, from (3.1) and (3.2) we see that E and H must satisfy

$$\operatorname{curl} E - ikH = 0,$$
$$\operatorname{curl} H + ikE = 0, \qquad (3.3)$$

where the *wave number* k is defined by $k = \omega \sqrt{\epsilon \mu}$.

Now assume that a time -monic electromagnetic plane wave (factoring out $e^{-i\omega t}$)

$$E^i(x) = E^i(x; d, p) = \frac{1}{k^2} \operatorname{curl} \operatorname{curl} p e^{ikx \cdot d},$$
$$H^i(x) = H^i(x; d, p) = \frac{1}{ik} \operatorname{curl} p e^{ikx \cdot d}, \qquad (3.4)$$

where d is a constant unit vector and p is the (constant) polarization vector, is an incident field that is scattered by an obstacle D that is an *imperfect conductor*, i.e., the electromagnetic field penetrates D by only a small amount. Let the total fields E and H be given by

$$E = E^i + E^s,$$
$$H = H^i + H^s, \qquad (3.5)$$

where $E^s(x) = E^s(x; d, p)$ and $H^s(x) = H^s(x; d, p)$ are the scattered fields that arise due to the presence of the obstacle D. Then E^s, H^s must be an "outgoing" wave that satisfies the *Silver–Müller radiation condition*

$$\lim_{r \to \infty} (H^s \times x - rE^s) = 0, \qquad (3.6)$$

where $r = |x|$. Since D is an imperfect conductor, on the boundary ∂D the field E must satisfy the boundary condition

$$\nu \times \operatorname{curl} E - i\lambda(\nu \times E) \times \nu = 0, \qquad (3.7)$$

where $\lambda = \lambda(x) > 0$ is the surface impedance defined on ∂D. Then the mathematical problem associated with the scattering of time-harmonic plane waves by an imperfect conductor is to find a solution E, H of Maxwell's equations (3.3) in the exterior of D such that (3.4)–(3.7) are satisfied. In particular, (3.3)–(3.7) define a *scattering problem* for Maxwell's equations.

Now consider the scattering due to an infinite cylinder with cross section D and axis on the x_3-coordinate axis where $x = (x_1, x_2, x_3) \in \mathbb{R}^3$. Assume $E = (0, 0, E_3)$, $p = (0, 0, 1)$, and $d = (d_1, d_2, 0)$, i.e.,

$$E^i(x) = e^{ikx \cdot d} \hat{e}_3,$$

where $\hat{e}_3$ is the unit vector in the positive x_3 direction. Then E and H will be independent of x_3, and from Maxwell's equations we have that $H = (H_1, H_2, 0)$, where E_3, H_1, and H_2 satisfy

$$\frac{\partial E_3}{\partial x_2} = ikH_1,$$

$$\frac{\partial E_3}{\partial x_1} = -ikH_2,$$

$$\frac{\partial H_2}{\partial x_1} - \frac{\partial H_1}{\partial x_2} = -ikE_3.$$

In particular,

$$\Delta E_3 + k^2 E_3 = 0 \quad \text{in } \mathbb{R}^2 \setminus \bar{D}. \tag{3.8}$$

In order for E_3^s to be "outgoing," we require that E_3^s satisfy the *Sommerfeld radiation condition*

$$\lim_{r \to \infty} \sqrt{r} \left(\frac{\partial E_3^s}{\partial r} - ikE_3^s \right) = 0. \tag{3.9}$$

Finally, we need to determine the boundary condition satisfied by

$$E_3(x) = e^{ikx \cdot d} + E_3^s(x), \tag{3.10}$$

where now $x \in \mathbb{R}^2$. To this end, we compute for $E = (0, 0, E_3)$ and $\nu = (\nu_1, \nu_2, 0)$ that $\nu \times \text{curl } E = (0, 0, -\partial E_3/\partial \nu)$ and $(\nu \times E) \times \nu = E$. This then implies that (3.7) becomes

$$\frac{\partial E_3}{\partial \nu} + i\lambda E_3 = 0. \tag{3.11}$$

Equations (3.8)–(3.11) provide the mathematical formulation of the scattering of a time-harmonic electromagnetic plane wave by an imperfectly conducting infinite cylinder, and it is this problem that will concern us for the rest of this chapter.

3.2 Bessel Functions

We begin our study of the scattering problem (3.8)–(3.11) by examining special solutions of the *Helmholtz equation* (3.8). In particular, if we look for solutions to (3.8) in the form

$$E_3(x) = y(kr)e^{in\theta} \quad , n = 0, \pm 1, \pm 2, \cdots ,$$

where (r, θ) are cylindrical coordinates, we find that $y(r)$ is a solution of Bessel's equation

$$y'' + \frac{1}{r}y' + \left(1 - \frac{\nu^2}{r^2}\right)y = 0 \tag{3.12}$$

for $\nu = n$. For arbitrary real ν we see by direct calculation and the ratio test that

$$J_\nu(r) := \sum_{k=0}^{\infty} \frac{(-1)^k}{k!\Gamma(k+\nu+1)} \left(\frac{r}{2}\right)^{\nu+2k} , \tag{3.13}$$

where Γ denotes the gamma function, is a solution of Bessel's equation for $0 \le r < \infty$. J_ν is called a *Bessel function* of order ν. For $\nu = -n, n = 1, 2, \cdots$, the first n terms of (3.13) vanish, and hence

$$J_{-n}(r) = \sum_{k=n}^{\infty} \frac{(-1)^k}{k!(k-n)!} \left(\frac{r}{2}\right)^{-n+2k}$$

$$= \sum_{s=0}^{\infty} \frac{(-1)^{n+s}}{(n+s)!s!} \left(\frac{r}{2}\right)^{n+2s}$$

$$= (-1)^n J_n(r) ,$$

which shows that J_n and J_{-n} are linearly dependent. However, if $\nu \neq n$, then it is easily seen that J_ν and $J_{-\nu}$ are linearly independent solutions of Bessel's equation.

Unfortunately, we are interested precisely in the case where $\nu = n$, and hence we must find a second linearly independent solution of Bessel's equation. This is easily done using Frobenius' method, and for $n = 0, 1, 2, \cdots$ we obtain the desired second solution to be given by

$$Y_n(r) := \frac{2}{\pi} J_n(r) \log \frac{r}{2} - \frac{1}{\pi} \sum_{k=0}^{n-1} \frac{(n-k-1)!}{k!} \left(\frac{r}{2}\right)^{2k-n}$$

$$- \frac{1}{\pi} \sum_{k=0}^{\infty} \frac{(-1)^k \left(\frac{r}{2}\right)^{n+2k}}{k!\,(n+k)!} [\psi(k+1) + \psi(k+n+1)] , \tag{3.14}$$

where $\psi(1) = -\gamma$, $\psi(m+1) = -\gamma + 1 + \frac{1}{2} + \cdots + \frac{1}{m}$ for $m = 1, 2, \cdots$, $\gamma = 0.57721566 \cdots$ is Euler's constant, and the finite sum is set equal to zero if $n = 0$. From (3.13) and (3.14) we see that

$$J_n(r) = \frac{1}{n!} \left(\frac{r}{2}\right)^n [1 + O(r^2)] \quad , \quad r \to 0 , \tag{3.15}$$

and, for $n \geq 1$,

$$Y_n(r) = -\frac{(n-1)!}{\pi} \left(\frac{r}{2}\right)^{-n} \begin{cases} 1 + O(r^2 \log r), & n = 1, \\ 1 + O(r^2), & n > 1, \end{cases} \qquad r \to 0, \qquad (3.16)$$

whereas for $n = 0$ we have that

$$Y_0(r) = \frac{2}{\pi} \log r + O(1) \quad , \quad r \to 0. \tag{3.17}$$

Note that in (3.15) and (3.16) the constant implicit in the order term is independent of n for $n > 1$. Finally, for n a positive integer we define Y_{-n} by

$$Y_{-n}(r) = (-1)^n Y_n(r) \,,$$

which implies that J_n and Y_n are linearly independent for all integers $n = 0, \pm 1, \pm 2, \cdots$. The function Y_n is called the *Neumann function* of order n.

Of considerable importance to us in the sequel are the *Hankel functions* $H_n^{(1)}$ and $H_n^{(2)}$ of the first and second kind of order n, respectively, which are defined by

$$\begin{aligned} H_n^{(1)}(r) &:= J_n(r) + iY_n(r) \,, \\ H_n^{(2)}(r) &:= J_n(r) - iY_n(r) \end{aligned} \tag{3.18}$$

for $n = 0, \pm 1, \pm 2, \cdots, 0 < r < \infty$. $H_n^{(1)}$ and $H_n^{(2)}$ clearly define a second pair of linearly independent solutions to Bessel's equation.

Now let y_1 and y_2 be any two solutions of Bessel's equation

$$(ry_1')' + \left(r - \frac{\nu^2}{r}\right) y_1 = 0 \,, \tag{3.19}$$

$$(ry_2')' + \left(r - \frac{\nu^2}{r}\right) y_2 = 0 \,, \tag{3.20}$$

and define the *Wronskian* by

$$W(y_1, y_2) := \begin{vmatrix} y_1 & y_2 \\ y_1' & y_2' \end{vmatrix} .$$

Then multiplying (3.19) by y_2 and subtracting it from (3.20) multiplied by y_1 we see that

$$\frac{d}{dr}(rW) = 0 \,,$$

and hence

$$W(y_1, y_2) = \frac{C}{r} \,,$$

where C is a constant. The constant C can be computed by

$$C = \lim_{r \to 0} rW(y_1, y_2).$$

In particular, making use of (3.15)–(3.18) we find that

$$W(J_n, H_n^{(1)}) = \frac{2i}{\pi r}, \tag{3.21}$$

$$W(H_n^{(1)}, H_n^{(2)}) = -\frac{4i}{\pi r}. \tag{3.22}$$

We now note that for $0 \le r < \infty$, $0 < |t| < \infty$, we have that

$$e^{rt/2} e^{-r/2t} = \sum_{j=0}^{\infty} \frac{r^j t^j}{2^j j!} \sum_{k=0}^{\infty} \frac{(-1)^k r^k}{2^k t^k k!},$$

and, setting $j - k = n$, we have that

$$e^{r/2(t-1/t)} = \sum_{n=-\infty}^{\infty} \left[\sum_{k=0}^{\infty} \frac{(-1)^k r^{n+2k}}{2^{n+2k}(n+k)!k!} \right] t^n$$

$$= \sum_{-\infty}^{\infty} J_n(r) t^n. \tag{3.23}$$

Setting $t = ie^{i\theta}$ in (3.23) gives the *Jacobi–Anger expansion*

$$e^{ir\cos\theta} = \sum_{-\infty}^{\infty} i^n J_n(r) e^{in\theta}. \tag{3.24}$$

In the remaining chapters of this book we will often be interested in entire solutions of the Helmholtz equation of the form

$$v_g(x) := \int_0^{2\pi} e^{ikr\cos(\theta-\phi)} g(\phi)\, d\phi, \tag{3.25}$$

where $g \in L^2[0, 2\pi]$. The function v_g is called a *Herglotz wave function* with kernel g. These functions were first introduced by Herglotz in a lecture in 1945 in Göttingen and were subsequently studied by Magnus [125], Müller [131], and Hartman and Wilcox [83]. From (3.25) and the Jacobi–Anger expansion, we see that since g has the Fourier expansion

$$g(\phi) = \frac{1}{2\pi} \sum_{-\infty}^{\infty} a_n (-i)^n e^{in\phi},$$

where

$$\sum_{-\infty}^{\infty} |a_n|^2 < \infty, \tag{3.26}$$

v_g is a Herglotz wave function if and only if v_g has an expansion of the form

$$v_g(x) = \sum_{-\infty}^{\infty} a_n J_n(kr) e^{in\theta}$$

such that (3.26) is valid. Note that v_g is identically zero if and only if $g = 0$.

Finally, we note the asymptotic relations [121]

$$
\begin{aligned}
J_n(r) &= \sqrt{\frac{2}{\pi r}} \cos\left(r - \frac{n\pi}{2} - \frac{\pi}{4}\right) + O(r^{-3/2}), \quad r \to \infty, \\
H_n^{(1)}(r) &= \sqrt{\frac{2}{\pi r}} \exp i\left(r - \frac{n\pi}{2} - \frac{\pi}{4}\right) + O(r^{-3/2}), \quad r \to \infty,
\end{aligned}
\tag{3.27}
$$

and the *addition formula* [121]

$$H_0^{(1)}(k\,|x - y|) = \sum_{-\infty}^{\infty} H_n^{(1)}(k\,|x|) J_n(k\,|y|) e^{in\theta}, \tag{3.28}$$

which is uniformly convergent together with its first derivatives on compact subsets of $|x| > |y|$, and θ denotes the angle between x and y.

3.3 Direct Scattering Problem

We will now show that the scattering problem for an imperfect conductor in $\mathbb{R}^2$ is well posed. We will always assume that $D \subset \mathbb{R}^2$ is a bounded domain containing the origin with connected complement such that ∂D is in class C^2. Our aim is to show the existence of a unique solution $u \in C^2(\mathbb{R}^2 \setminus \bar{D}) \cap C(\mathbb{R}^2 \setminus D)$ of the exterior *impedance boundary value problem*

$$\Delta u + k^2 u = 0 \quad \text{in } \mathbb{R}^2 \setminus \bar{D}, \tag{3.29}$$

$$u(x) = e^{ikx \cdot d} + u^s(x), \tag{3.30}$$

$$\lim_{r \to \infty} \sqrt{r}\left(\frac{\partial u^s}{\partial r} - iku^s\right) = 0, \tag{3.31}$$

$$\frac{\partial u}{\partial \nu} + i\lambda u = 0 \quad \text{on } \partial D, \tag{3.32}$$

where (3.32) is assumed in the sense of uniform convergence as $x \to \partial D$, $\lambda \in C(\partial D), \lambda(x) > 0$ for $x \in \partial D$, ν is the unit outward normal to ∂D, and the Sommerfeld radiation condition (3.31) is assumed to hold uniformly in θ, where $k > 0$ is the wave number and (r, θ) are polar coordinates. We also want to show that the solution u of (3.29)–(3.32) depends continuously on the incident field u^i in an appropriate norm.

We define the (radiating) *fundamental solution* to the Helmholtz equation by

$$\Phi(x, y) := \frac{i}{4} H_0^{(1)}(k\,|x - y|) \tag{3.33}$$

and note that $\Phi(x,y)$ satisfies the Sommerfeld radiation condition with respect to both x and y, and as $|x-y| \to 0$ we have that

$$\Phi(x,y) = \frac{1}{2\pi} \log \frac{1}{|x-y|} + O(1). \tag{3.34}$$

Theorem 3.1 (Representation Theorem). *Let $u^s \in C^2(\mathbb{R}^2 \setminus \bar{D}) \cap C(\mathbb{R}^2 \setminus D)$ be a solution of the Helmholtz equation in the exterior of D satisfying the Sommerfeld radiation condition and such that $\partial u/\partial \nu$ exists in the sense of uniform convergence as $x \to \partial D$. Then for $x \in \mathbb{R}^2 \setminus \bar{D}$ we have that*

$$u^s(x) = \int_{\partial D} \left(u^s(y) \frac{\partial}{\partial \nu(y)} \Phi(x,y) - \frac{\partial u^s}{\partial \nu}(y)\Phi(x,y) \right) ds(y).$$

Proof. Let $x \in \mathbb{R}^2 \setminus \bar{D}$, and circumscribe it with a disk

$$\Omega_{x,\epsilon} := \{y : |x-y| < \epsilon\},$$

where $\Omega_{x,\epsilon} \subset \mathbb{R}^2 \setminus \bar{D}$. Let Ω_R be a disk of radius R centered at the origin and containing D and $\Omega_{x,\epsilon}$ in its interior. Then from Green's second identity we have that

$$\int_{\partial D + \partial \Omega_{x,\epsilon} + \partial \Omega_R} \left(u^s(y) \frac{\partial}{\partial \nu(y)} \Phi(x,y) - \frac{\partial u^s}{\partial \nu}(y)\Phi(x,y) \right) ds(y) = 0.$$

From the definition of the Hankel function, we have that

$$\frac{d}{dr} H_0^{(1)}(r) = -H_1^{(1)}(r),$$

and hence on $\partial \Omega_{x,\epsilon}$ we have that

$$\frac{\partial}{\partial \nu(y)} \Phi(x,y) = \frac{1}{2\pi} \frac{1}{|x-y|} + O(|x-y| \log |x-y|). \tag{3.35}$$

Using (3.34) and (3.35) and letting $\epsilon \to 0$ we see that

$$u^s(x) = \int_{\partial D} \left(u^s(y) \frac{\partial}{\partial \nu(y)} \Phi(x,y) - \frac{\partial u^s}{\partial \nu}(y)\Phi(x,y) \right) ds(y)$$
$$- \int_{|y|=R} \left(u^s(y) \frac{\partial}{\partial \nu(y)} \Phi(x,y) - \frac{\partial u^s}{\partial \nu}(y)\Phi(x,y) \right) ds(y), \tag{3.36}$$

where as usual ν is the unit outward normal to the boundary of the (interior) domain. Hence to establish the theorem we must show that the second integral tends to zero as $R \to \infty$.

We first show that

$$\lim_{R\to\infty} \int_{|y|=R} |u^s|^2 \, ds = O(1). \tag{3.37}$$

To this end, from the Sommerfeld radiation condition we have that

$$0 = \lim_{R \to \infty} \int_{|y|=R} \left| \frac{\partial u^s}{\partial r} - iku^s \right|^2 ds$$

$$= \lim_{R \to \infty} \int_{|y|=R} \left(\left| \frac{\partial u^s}{\partial r} \right|^2 + k^2 |u^s|^2 + 2k \operatorname{Im} \left(u^s \frac{\partial \overline{u^s}}{\partial r} \right) \right) ds. \tag{3.38}$$

Green's first identity applied to $D_R = \Omega_R \setminus \bar{D}$ gives

$$\int_{|y|=R} u^s \frac{\partial \overline{u^s}}{\partial r} ds = \int_{\partial D} u^s \frac{\partial \overline{u^s}}{\partial \nu} ds - k^2 \int_{D_R} |u^s|^2 dy + \int_{D_R} |\text{grad } u^s|^2 dy,$$

and hence from (3.38) we have that

$$\lim_{R \to \infty} \int_{|y|=R} \left(\left| \frac{\partial u^s}{\partial r} \right|^2 + k^2 |u^s|^2 \right) ds = -2k \operatorname{Im} \int_{\partial D} u^s \frac{\partial \overline{u^s}}{\partial \nu} ds, \tag{3.39}$$

and from this we can conclude that (3.37) is true.

To complete the proof, we now note the identity

$$\int_{|y|=R} \left(u^s(y) \frac{\partial}{\partial \nu(y)} \Phi(x,y) - \frac{\partial u^s}{\partial \nu}(y) \Phi(x,y) \right) ds(y) =$$

$$= \int_{|y|=R} u^s(y) \left(\frac{\partial}{\partial |y|} \Phi(x,y) - ik\Phi(x,y) \right) ds(y) \tag{3.40}$$

$$- \int_{|y|=R} \Phi(x,y) \left(\frac{\partial u^s}{\partial |y|}(y) - iku^s(y) \right) ds(y).$$

Applying the Cauchy–Schwarz inequality to each of the integrals on the right-hand side of (3.40) and using (3.37), the facts that $\Phi(x,y) = O(1/\sqrt{R})$ and Φ and u^s satisfy the Sommerfeld radiation condition we have that

$$\lim_{R \to \infty} \int_{|y|=R} \left(u^s(y) \frac{\partial}{\partial \nu(y)} \Phi(x,y) - \frac{\partial u^s}{\partial \nu}(y) \Phi(x,y) \right) ds(y) = 0,$$

and the proof is complete.   □

Now let D be a bounded domain with C^2 boundary ∂D and $u \in C^2(D) \cap C^1(\bar{D})$ a solution of the Helmholtz equation in D. Then, using the techniques of the proof of the preceding theorem, it can easily be shown that for $x \in D$ we have the *representation formula*

$$u(x) = \int_{\partial D} \left(\frac{\partial u}{\partial \nu}(y) \Phi(x,y) - u(y) \frac{\partial}{\partial \nu(y)} \Phi(x,y) \right) ds(y). \tag{3.41}$$

Hence, since $\Phi(x,y)$ is a real-analytic function of x_1 and x_2, where $x = (x_1, x_2)$ and $x \neq y$, we have that u is real-analytic in D. This proves the following theorem.

Theorem 3.2. *Solutions of the Helmholtz equation are real-analytic functions of their independent variables.*

The identity theorem for real-analytic functions [95] and Theorem 3.2 imply that solutions of the Helmholtz equation satisfy the *unique continuation principle*, i.e., if u is a solution of the Helmholtz equation in a domain D and $u(x) = 0$ for x in a neighborhood of a point $x_0 \in D$, then $u(x) = 0$ for all x in D.

We are now in a position to show that if a solution to the scattering problem (3.29)–(3.32) exists, then it is unique.

Theorem 3.3. *Let $u^s \in C^2(\mathbb{R}^2 \setminus \bar{D}) \cap C(\mathbb{R}^2 \setminus D)$ be a solution of the Helmholtz equation in $\mathbb{R}^2 \setminus \bar{D}$ satisfying the Sommerfeld radiation condition and the boundary condition $\partial u^s / \partial \nu + i\lambda u^s = 0$ on ∂D (in the sense of uniform convergence as $x \to \partial D$). Then $u^s = 0$.*

Proof. Let Ω be a disk centered at the origin and containing D in its interior. Then from Green's second identity, the fact that R and λ are real, and hence

$$\frac{\partial u^s}{\partial \nu} + i\lambda u^s = \frac{\partial \overline{u^s}}{\partial \nu} - i\lambda \overline{u^s} = 0 \quad \text{on } \partial D,$$

we have that

$$\int_{\partial \Omega} \left(\overline{u^s} \frac{\partial u^s}{\partial r} - u^s \frac{\partial \overline{u^s}}{\partial r} \right) ds = \int_{\partial D} \left(\overline{u^s} \frac{\partial u^s}{\partial \nu} - u^s \frac{\partial \overline{u^s}}{\partial \nu} \right) ds$$

$$= -2i \int_{\partial D} \lambda |u^s|^2 \, ds. \tag{3.42}$$

But since, by Theorem 3.2, $u^s \in C^\infty(\mathbb{R}^2 \setminus \bar{D})$ (in fact real-analytic), we have that, for $x \in \mathbb{R}^2 \setminus \Omega$, u^s can be expanded in a Fourier series

$$u^s(r, \theta) = \sum_{-\infty}^{\infty} a_n(r) e^{in\theta},$$

$$a_n(r) = \frac{1}{2\pi} \int_0^{2\pi} u^s(r, \theta) e^{-in\theta} \, d\theta, \tag{3.43}$$

where the series and its derivatives with respect to r are absolutely and uniformly convergent on compact subsets of $\mathbb{R}^2 \setminus \Omega$. In particular, it can be verified directly that $a_n(r)$ is a solution of Bessel's equation and, since u^s satisfies the Sommerfeld radiation condition,

$$a_n(r) = \alpha_n H_n^{(1)}(kr), \tag{3.44}$$

where the α_n are constants. Substituting (3.43) and (3.44) into (3.42) and integrating termwise, we see from the fact that $H_n^{(1)}(kr) = \overline{H_n^{(2)}(kr)}$ and the Wronskian formula (3.22) that

$$8i \sum_{-\infty}^{\infty} |a_n|^2 = -2i \int_{\partial D} \lambda |u^s|^2 \, ds.$$

Since $\lambda > 0$, we can now conclude that $a_n = 0$ for every integer n, and hence $u^s(x) = 0$ for $x \in \mathbb{R}^2 \setminus \Omega$. By Theorem 3.2 and the identity theorem for real-analytic functions, we can now conclude that $u^s(x) = 0$ for $x \in \mathbb{R}^2 \setminus \bar{D}$.   □

Corollary 3.4. *If the solution of the scattering problem (3.29)–(3.32) exists, then it is unique.*

Proof. If two solutions u_1 and u_2 exist, then their difference $u^s = u_1 - u_2$ satisfies the hypothesis of Theorem 3.3, and hence $u^s = 0$, i.e., $u_1 = u_2$.   □

The next theorem is a classic result in scattering theory that was first proved by Rellich [143] and Vekua [157] in 1943. Due, perhaps, to wartime conditions, Vekua's paper remained unknown in the West, and the result is commonly attributed only to Rellich.

Theorem 3.5 (Rellich's Lemma). *Let $u \in C^2(\mathbb{R}^2 \setminus \bar{D})$ be a solution of the Helmholtz equation satisfying*

$$\lim_{R \to \infty} \int_{|y|=R} |u|^2 \, ds = 0.$$

Then $u = 0$ in $\mathbb{R}^2 \setminus \bar{D}$.

Proof. Let Ω be a disk centered at the origin and containing D in its interior. Then, as in Theorem 3.3, we have that for $x \in \mathbb{R}^2 \setminus \Omega$

$$u(r, \theta) = \sum_{-\infty}^{\infty} a_n(r) e^{in\theta},$$

$$a_n(r) = \frac{1}{2\pi} \int_0^{2\pi} u(r, \theta) e^{-in\theta} \, d\theta,$$

and $a_n(r)$ is a solution of Bessel's equation, i.e.,

$$a_n(r) = \alpha_n H_n^{(1)}(kr) + \beta_n H_n^{(2)}(kr), \tag{3.45}$$

where the α_n and β_n are constants. By Parseval's equality, we have that

$$\int_{|y|=R} |u|^2 \, ds = 2\pi R \sum_{-\infty}^{\infty} |a_n(R)|^2,$$

and hence, from the hypothesis of the theorem,

$$\lim_{R \to \infty} R \, |a_n(R)|^2 = 0. \tag{3.46}$$

From (3.45), the asymptotic expansion of $H_n^{(1)}(kr)$ given by (3.27), and the fact that $\overline{H_n^{(1)}(kr)} = H_n^{(2)}(kr)$, we see from (3.46) that $\alpha_n = \beta_n = 0$ for every n, and hence $u = 0$ in $\mathbb{R}^2 \setminus \Omega$. By Theorem 3.2 and the identity theorem for real-analytic functions, we can now conclude as in Theorem 3.3 that $u(x) = 0$ for $x \in \mathbb{R}^2 \setminus \bar{D}$. $\qquad\square$

Theorem 3.6. *Let $u^s \in C^2(\mathbb{R}^2 \setminus \bar{D}) \cap C(\mathbb{R}^2 \setminus D)$ be a radiating solution of the Helmholtz equation such that $\frac{\partial u}{\partial \nu}(x)$ converges uniformly as $x \to \partial D$ and*

$$\mathrm{Im} \int_{\partial D} u^s \frac{\partial \overline{u^s}}{\partial \nu} \, ds \geq 0.$$

Then $u^s = 0$ in $\mathbb{R}^2 \setminus \bar{D}$.

Proof. This follows from identity (3.39) and Rellich's lemma. $\qquad\square$

We now want to use the method of integral equations to establish the existence of a solution to the scattering problem (3.29)–(3.32). To this end, we note that the *single layer potential*

$$u^s(x) = \int_{\partial D} \varphi(y)\Phi(x,y)\, ds(y), \quad x \in \mathbb{R}^2 \setminus \partial D \qquad (3.47)$$

with continuous density φ satisfies the Sommerfeld radiation condition, is a solution of the Helmholtz equation in $\mathbb{R}^2 \setminus \partial D$, is continuous in $\mathbb{R}^2$, and satisfies the discontinuity property [111, 127]

$$\frac{\partial u^s_\pm}{\partial \nu}(x) = \int_{\partial D} \varphi(y)\frac{\partial}{\partial \nu(x)}\Phi(x,y)\, ds(y) \mp \frac{1}{2}\varphi(x), \quad x \in \partial D,$$

where

$$\frac{\partial u^s_\pm}{\partial \nu}(x) := \lim_{h \to 0} \nu(x) \cdot \nabla u\left(x \pm h\nu(x)\right).$$

(For future reference, we note that these properties of the single layer potential are also valid for $\varphi \in H^{-1/2}(\partial D)$, where the integrals are interpreted in the sense of duality pairing [111,127].) In particular, (3.47) will solve the scattering problem (3.29)–(3.32) provided

$$\varphi(x) - 2\int_{\partial D}\varphi(y)\frac{\partial}{\partial \nu(x)}\Phi(x,y)\, ds(y) - 2i\lambda(x)\int_{\partial D}\varphi(y)\Phi(x,y)\, ds(y)$$
$$= 2\left[\frac{\partial u^i}{\partial \nu}(x) + i\lambda(x)u^i(x)\right], \quad x \in \partial D, \qquad (3.48)$$

where $u^i(x) = e^{ikx \cdot d}$. Hence, to establish the existence of a solution to the scattering problem (3.29)–(3.32), it suffices to show the existence of a solution to (3.48) in the normed space $C(\partial D)$ (Example 1.3).

To this end, we first note that the integral operators in (3.48) are compact. This can easily be shown by approximating each of the kernels $K(x,y)$ in (3.48) by

$$K_n(x,y) := \begin{cases} h(n\,|x-y|)K(x,y), & x \neq y, \\ 0, & x = y, \end{cases}$$

where

$$h(t) := \begin{cases} 0, & 0 \leq t \leq \frac{1}{2}, \\ 2t-1, & \frac{1}{2} \leq t \leq 1, \\ 1, & 1 \leq t < \infty \end{cases}$$

and using Theorem 1.17 and the fact that integral operators with continuous kernels are compact operators on $C(\partial D)$ (cf. Theorem 2.21 of [111]). Hence, by Riesz's theorem, it suffices to show that the homogeneous equation has only a trivial solution. But this is in general not the case! In particular, let k^2 be a Dirichlet eigenvalue, i.e., there exists $u \in C^2(D) \cap C(\bar{D})$, with u not identically zero, such that

$$\Delta u + k^2 u = 0 \quad \text{in } D,$$
$$u = 0 \quad \text{on } \partial D.$$

It can be shown that $u \in C^1(\bar{D})$ [51] and $\partial u/\partial \nu$ is not identically zero since, if it were, then by the representation formula (3.41) u would be identically zero, which it is not by assumption. Hence for $\varphi := \partial u/\partial \nu$ we have from Green's second identity that

$$\int_{\partial D} \varphi(y)\Phi(x,y)\,ds(y) = 0, \quad x \in \mathbb{R}^2 \setminus \bar{D} \tag{3.49}$$

and, by continuity, for $x \in \mathbb{R}^2 \setminus D$. Hence, using the previously stated discontinuity properties for single layer potentials, we have that

$$\varphi(x) - 2\int_{\partial D} \varphi(y)\frac{\partial}{\partial \nu(x)}\Phi(x,y)\,ds(y) = 0, \quad x \in \partial D. \tag{3.50}$$

Equations (3.49) and (3.50) now imply that φ is a nontrivial solution of the homogeneous equation corresponding to (3.48). Thus we cannot use Riesz's theorem to establish the existence of a solution to (3.48).

To obtain an integral equation that is uniquely solvable for all values of the wave number k, we need to modify the kernel of the representation (3.47). We will do this following the ideas of [96, 109, 156]. We begin by defining the function $\chi = \chi(x,y)$ by

$$\chi(x,y) := \frac{i}{4}\sum_{-\infty}^{\infty} a_n H_n^{(1)}(kr)H_n^{(1)}(kr_y)e^{in(\theta-\theta_y)}, \tag{3.51}$$

where x has polar coordinates (r, θ), y has polar coordinates (r_y, θ_y), and the coefficients a_n are chosen such that the series converges for $|x|, |y| > R$, where $\Omega_R := \{x : |x| \leq R\} \subset D$. The fact that this can be done follows from (3.15), (3.16), and (3.18) and the fact that

$$H_{-n}^{(1)}(kr) = (-1)^n H_n^{(1)}(kr)$$

for $n = 0, 1, 2, 3, \cdots$. In particular these equations imply that

$$\left| H_n^{(1)}(kr) \right| = O\left(\frac{2^{|n|} (|n| - 1)!}{(kr)^{|n|}} \right)$$

for $n = \pm 1, \pm 2, \cdots$ and r on compact subsets of $(0, \infty)$. Defining

$$\Gamma(x, y) := \Phi(x, y) + \chi(x, y)$$

we now see that the *modified single layer potential*

$$u^s(x) := \int_{\partial D} \varphi(y) \Gamma(x, y) \, ds(y) \tag{3.52}$$

for continuous density φ and $x \in \mathbb{R}^2 \setminus (\partial D \cup \Omega_R)$ satisfies the Sommerfeld radiation condition, is a solution of the Helmholtz equation in $\mathbb{R}^2 \setminus (\partial D \cup \Omega_R)$, and satisfies the same discontinuity properties as the single layer potential (3.47). Hence (3.52) will solve the scattering problem (3.29)–(3.32) provided φ satisfies (3.48), with Φ replaced by Γ. By Riesz's theorem, a solution of this equation exists if the corresponding homogeneous equation only has a trivial solution.

Let φ be a solution of this homogeneous equation. Then (3.52) will be a solution of (3.29)–(3.32) with $e^{ikx \cdot d}$ set equal to zero and hence, by Corollary 3.4, we have that if u^s is defined by (3.52), then $u^s(x) = 0$ for $x \in \mathbb{R}^2 \setminus \bar{D}$. By the continuity of (3.52) across ∂D, u^s is a solution of the Helmholtz equation in $D \setminus \Omega_R$, $u^s \in C^2(D \setminus \bar{\Omega}_R) \cap C(\bar{D} \setminus \Omega_R)$, and $u^s(x) = 0$ for $x \in \partial D$. From (3.51), (3.52), and the addition formula for Bessel functions, we see that there exist constants α_n such that for $R_1 \leq |x| \leq R_2$, where $R < R_1 < R_2$ and $\{x : |x| < R_2\} \subset D$, we can represent u^s in the form

$$u^s(x) = \sum_{-\infty}^{\infty} \alpha_n \left\{ J_n(kr) + a_n H_n^{(1)}(kr) \right\} e^{in\theta}.$$

Since

$$u_+^s(x) := \lim_{\substack{x \to \partial D \\ x \in D}} u^s(x),$$

$$\frac{\partial u_+^s}{\partial \nu}(x) := \lim_{\substack{x \to \partial D \\ x \in D}} \frac{\partial u^s}{\partial \nu}(x)$$

exist and are continuous, we can apply Green's second identity to u^s and $\bar{u}^s$ over $D \setminus \{x : |x| \leq R_1\}$ and use the Wronskian relations (3.21) and (3.22) to see that

$$0 = \int_{\partial D} \left(u_+^s \frac{\partial \bar{u}_+^s}{\partial \nu} - \bar{u}_+^s \frac{\partial u_+^s}{\partial \nu} \right) ds = \int_{|x|=R_1} \left(u^s \frac{\partial \bar{u}^s}{\partial \nu} - \bar{u}^s \frac{\partial u^s}{\partial \nu} \right) ds$$

$$= 2i \sum_{-\infty}^{\infty} |\alpha_n|^2 \left(1 - |1 + 2a_n|^2 \right).$$

Hence, if either $|1 + 2a_n| < 1$ or $|1 + 2a_n| > 1$ for $n = 0, \pm 1, \pm 2, \cdots$, then $\alpha_n = 0$ for $n = 0, \pm 1, \pm 2, \cdots$, i.e., $u^s(x) = 0$ for $R_1 \leq |x| \leq R_2$. By Theorem 3.2 and the identity theorem for real-analytic functions, we can now conclude that $u^s(x) = 0$ for $x \in D \setminus \Omega_R$. Recalling that $u^s(x) = 0$ for $x \in \mathbb{R}^2 \setminus \bar{D}$, we now see from the discontinuity property of single layer potentials that

$$0 = \frac{\partial u_-^s}{\partial \nu} - \frac{\partial u_+^s}{\partial \nu}(x) = \varphi(x),$$

i.e., the homogeneous equation under consideration only has the trivial solution $\varphi = 0$. Hence, by Riesz's theorem, the corresponding inhomogeneous equation has a unique solution that depends continuously on the right-hand side.

Theorem 3.7. *There exists a unique solution of the scattering problem (3.29)–(3.32) that depends continuously on $u^i(x) = e^{ikx \cdot d}$ in $C^1(\partial D)$.*

It is often important to find a solution of (3.29)–(3.32) in a larger space than $C^2(\mathbb{R}^2 \setminus \bar{D}) \cap C^1(\mathbb{R}^2 \setminus D)$. To this end, let $\Omega_R := \{x : |x| < R\}$, and define the Sobolev spaces

$$H_{loc}^1(\mathbb{R}^2 \setminus \bar{D}) := \{u : u \in H^1 \left((\mathbb{R}^2 \setminus \bar{D}) \cap \Omega_R \right) \text{ for every } R > 0$$
$$\text{such that } (\mathbb{R}^2 \setminus D) \cap \Omega_R \neq \emptyset \},$$
$$H_{com}^1(\mathbb{R}^2 \setminus \bar{D}) := \{u : u \in H^1(\mathbb{R}^2 \setminus \bar{D}), u \text{ is identically}$$
$$\text{zero outside some ball centered at}$$
$$\text{the origin} \}.$$

We recall that $H^{-p}(\partial D)$, $0 \leq p < \infty$, is the dual space of $H^p(\partial D)$ and, for $f \in H^{-p}(\partial D)$ and $v \in H^p(\partial D)$,

$$\int_{\partial D} fv \, ds := f(v)$$

is defined by duality pairing.
Then, for $f \in H^{-1/2}(\partial D)$, a *weak solution* of

$$\Delta u + k^2 u = 0 \quad \text{in } \mathbb{R}^2 \setminus \bar{D}, \tag{3.53}$$

$$\lim_{r \to \infty} \sqrt{r} \left(\frac{\partial u}{\partial r} - iku \right) = 0, \tag{3.54}$$

$$\frac{\partial u}{\partial \nu} + i\lambda u = f \quad \text{on } \partial D \tag{3.55}$$

is defined as a function $u \in H^1_{loc}(\mathbb{R}^2 \setminus \bar{D})$ such that

$$-\int_{\mathbb{R}^2 \setminus \bar{D}} \left(\nabla u \cdot \nabla v - k^2 uv \right) dx + i \int_{\partial D} \lambda uv \, ds = \int_{\partial D} fv \, ds \tag{3.56}$$

for all $v \in H^1_{com}(\mathbb{R}^2 \setminus \bar{D})$ such that u satisfies the Sommerfeld radiation condition (3.54). Note that by the trace theorem we have that $v|_{\partial D} \in H^{1/2}(\partial D)$ is well defined, and hence the integral on the right-hand side of (3.56) is well defined by duality pairing. The radiation condition also makes sense in the weak case since, by regularity results for elliptic equations [127], any weak solution is automatically infinitely differentiable in $\mathbb{R}^2 \setminus \bar{D}$. It is easily verified that if $u \in C^2(\mathbb{R}^2 \setminus \bar{D}) \cap C^1(\mathbb{R}^2 \setminus D)$ is a solution of (3.53)–(3.55), then u is also a weak solution of (3.53)–(3.55), i.e., u satisfies (3.56). The following theorem will be proved in Chap. 8.

Theorem 3.8. *There exists a unique weak solution of the scattering problem (3.53)–(3.55), and the mapping taking the boundary data $f \in H^{-1/2}(\partial D)$ onto the solution $u \in H^1((\mathbb{R}^2 \setminus \bar{D}) \setminus \bar{\Omega}_R)$ is bounded for every R such that $(\mathbb{R}^2 \setminus \bar{D}) \cap \Omega_R \neq \emptyset$.*

In an analogous manner, we can define a weak solution of the Helmholtz equation in a bounded domain D to be any function $u \in H^1(D)$ such that

$$\int_D \left(\nabla u \cdot \nabla v - k^2 uv \right) dx = 0$$

for all $v \in H^1(D)$ such that $v = 0$ on ∂D in the sense of the trace theorem. The following theorems will be useful in the sequel, but we will delay their proofs until Chap. 5, where they will constitute a basic part of the analysis of that chapter.

Theorem 3.9. *Let D be a bounded domain with C^2 boundary ∂D such that k^2 is not a Dirichlet eigenvalue for D. Then for every $f \in H^{1/2}(\partial D)$ there exists a unique weak solution $u \in H^1(D)$ of the Helmholtz equation in D such that $u = f$ on ∂D in the sense of the trace theorem. Furthermore, the mapping taking f onto u is bounded.*

Theorem 3.10. *Let $u \in H^1(D)$ and $\Delta u \in L^2(D)$ in a bounded domain D with C^2 boundary ∂D having unit outward normal ν. Then there exists a positive constant C independent of u such that*

$$\left\|\frac{\partial u}{\partial \nu}\right\|_{H^{-1/2}(\partial D)} \leq C \left\|u\right\|_{H^1(D)}.$$

Finally, we note that Green's identities and the representation formulas for exterior and interior domains remain valid for weak solutions of the Helmholtz equation, and we refer the reader to Chap. 5 for a proof of this fact.

4

Inverse Scattering Problems for Imperfect Conductors

We are now in a position to introduce the inverse scattering problem for an imperfect conductor, in particular given the far-field pattern of the scattered field to determine the support of the scattering object D and the surface impedance λ. Our approach to this problem is based on the *linear sampling method* in inverse scattering theory that was first introduced by Colton and Kirsch [50] and Colton et al. [61]. As will become clear in subsequent chapters, the advantage of this method for solving the inverse scattering problem is that in order to determine the support of the scattering object, it is not necessary to have any a priori information on the physical properties of the scatterer. In particular, the relevant equation that needs to be solved is the same for the case of an imperfect conductor as it is for anisotropic media and partially coated obstacles, which we will consider in the chapters that follow. Of course, for the specific inverse scattering problem we are considering in this chapter, there are alternative approaches to the one we are using, and for one such alternative approach we refer the reader to [113].

The plan of this chapter is as follows. We first introduce the far-field pattern corresponding to the scattering of an incident plane wave by a perfect conductor and prove the reciprocity principle. We then use this principle to show that the far-field operator having a far-field pattern as kernel is injective with dense range. After showing that the solution of the inverse scattering problem is unique, we then use the properties of the far-field operator to establish the linear sampling method for determining the support of the scattering object and conclude by giving a method for determining the surface impedance λ. As we will see in Chap. 8, the methods used in this chapter carry over immediately to the case of partially coated perfect conductors, i.e., the case where the impedance boundary condition is imposed on only a portion of the boundary, with the remaining portion being subject to a Dirichlet boundary condition.

F. Cakoni and D. Colton, *A Qualitative Approach to Inverse Scattering Theory*,   63
Applied Mathematical Sciences 188, DOI 10.1007/978-1-4614-8827-9_4,
© Springer Science+Business Media New York 2014

4.1 Far-Field Patterns

The inverse scattering problems we will be considering in this book all
assume that the given data are the asymptotic behavior of the scattered
field corresponding to an incident plane wave. Hence, our analysis of the
inverse scattering problem must begin with a derivation of precisely what
this asymptotic behavior is. To this end, we first recall the scattering problem
under consideration, i.e., to find $u^s \in C^2(\mathbb{R}^2 \setminus \bar{D}) \cap C(\mathbb{R}^2 \setminus D)$ such that

$$\Delta u + k^2 u = 0 \quad \text{in } \mathbb{R}^2 \setminus \bar{D}, \tag{4.1}$$

$$u(x) = e^{ikx \cdot d} + u^s(x), \tag{4.2}$$

$$\lim_{r \to \infty} \sqrt{r}\left(\frac{\partial u^s}{\partial r} - iku^s\right) = 0, \tag{4.3}$$

$$\frac{\partial u}{\partial \nu} + i\lambda u = 0 \quad \text{on } \partial D, \tag{4.4}$$

where (4.4) is assumed in the sense of uniform convergence as $x \to \partial D$, ν is
the unit outward normal to ∂D, and $\lambda = \lambda(x)$ is a real-valued, positive, and
continuous function defined on ∂D. Then from the asymptotic behavior (3.27)
of the Hankel function, the estimate

$$|x - y| = \left(r^2 - 2rr_y \cos(\theta - \theta_y) + r_y^2\right)^{1/2}$$

$$= r\left(1 - \frac{2r_y}{r}\cos(\theta - \theta_y) + \frac{r_y^2}{r^2}\right)^{1/2}$$

$$= r - r_y \cos(\theta - \theta_y) + O\left(\frac{1}{r}\right),$$

where (r_y, θ_y) are the polar coordinates of y and (r, θ) are the polar coordinates
of x, we see from the Representation Theorem 3.1 that the solution u^s of (4.1)–
(4.4) has the asymptotic behavior

$$u^s(x) = \frac{e^{ikr}}{\sqrt{r}} u_\infty(\theta, \phi) + O(r^{-3/2}), \tag{4.5}$$

where $d = (\cos\phi, \sin\phi)$, k is fixed, and

$$u_\infty(\theta, \phi) = \frac{e^{i\pi/4}}{\sqrt{8\pi k}} \int_{\partial D} \left(u^s \frac{\partial}{\partial \nu_y} e^{-ikr_y \cos(\theta - \theta_y)} - \frac{\partial u^s}{\partial \nu_y} e^{-ikr_y \cos(\theta - \theta_y)}\right) ds(y).$$

$$\tag{4.6}$$

The function u_∞ is called the *far-field pattern* corresponding to the scattering
problem (4.1)–(4.4).

Theorem 4.1. *Suppose the far-field pattern corresponding to (4.1)–(4.4) vanishes identically. Then $u^s(x) = 0$ for $x \in \mathbb{R}^2 \setminus D$.*

Proof. We have that

$$\int_{|y|=R} |u^s|^2 \, ds = \int_{-\pi}^{\pi} |u_\infty(\theta, \phi)|^2 \, d\theta + O\left(\frac{1}{R}\right)$$

as $R \to \infty$. If $u_\infty = 0$, then, by Rellich's lemma, $u^s(x) = 0$ for $x \in \mathbb{R}^2 \setminus \bar{D}$ and by continuity for $x \in \mathbb{R}^2 \setminus D$.   □

We can now consider the inverse scattering problem corresponding to the direct scattering problem (4.1)–(4.4). There are in fact three different inverse scattering problems we could consider:

1. Given u_∞ and λ, determine D.
2. Given u_∞ and D, determine λ.
3. Given u_∞, determine D and λ.

From a practical point of view, the third problem is clearly the most realistic one since in general one cannot expect to know either D or λ a priori. Hence, in what follows, we shall only be concerned with the third problem and will refer to this as the inverse scattering problem. Note that the far-field pattern of

$$u_n(r, \theta) = \frac{1}{n} H_n^{(1)}(kr) e^{in\theta} \quad , n > 0$$

is

$$u_{n,\infty}(\theta) = \frac{1}{n} \sqrt{\frac{2}{k\pi}} e^{-i\pi/4} (-i)^n e^{in\theta} .$$

Hence $u_{n,\infty} \to 0$ as $n \to \infty$ in $L^2[0, 2\pi]$, whereas, since

$$H_n^{(1)}(kr) \sim \frac{-2^n(n-1)!}{\pi(kr)^n} \quad , n \to \infty,$$

u_n will not converge in any reasonable norm. This suggests that the problem of determining u^s from u^∞ is severely ill posed, and in particular we can expect that the inverse scattering problem is also ill posed. Further evidence in this direction is the fact that from (4.6) we see that u_∞ is an infinitely differentiable function of θ, and since in general a measured far-field pattern does not have this property, we have that a solution to the inverse scattering problem does not exist for the case of "noisy" data.

We begin our study of the inverse scattering problem by deriving the following basic property of the far-field pattern.

Theorem 4.2 (Reciprocity Relation). *Let $u_\infty(\theta, \phi)$ be a far-field pattern corresponding to the scattering problem (4.1)–(4.4). Then $u_\infty(\theta, \phi) = u_\infty(\phi + \pi, \theta + \pi)$.*

Proof. For convenience we write $u_\infty(\hat{x}, d) = u_\infty(\theta, \phi)$, where $\hat{x} = x/|x|$, e.g.,

$$e^{-ikr_y \cos(\theta - \theta_y)} = e^{-iky \cdot \hat{x}} = u^i(y, -\hat{x}),$$

where $u^i(x, d) = e^{ikx \cdot d}$ denotes the incident field. Then from Green's second identity we have that

$$\int_{\partial D} \left(u^i(y, d) \frac{\partial}{\partial \nu} u^i(y, -\hat{x}) - u^i(y, -\hat{x}) \frac{\partial}{\partial \nu} u^i(y, d) \right) ds(y) = 0, \qquad (4.7)$$

and, using Green's second identity again, deforming ∂D to $\{x : |x| = r\}$, and letting $r \to \infty$, we have that

$$\int_{\partial D} \left(u^s(y, d) \frac{\partial}{\partial \nu} u^s(y, -\hat{x}) - u^s(y, -\hat{x}) \frac{\partial}{\partial \nu} u^s(y, d) \right) ds(y) = 0. \qquad (4.8)$$

From (4.6) we have that

$$\sqrt{8\pi k}\, e^{-i\pi/4} u_\infty(\hat{x}, d) =$$
$$\int_{\partial D} \left(u^s(y, d) \frac{\partial}{\partial \nu} u^i(y, -\hat{x}) - u^i(y, -\hat{x}) \frac{\partial}{\partial \nu} u^s(y, d) \right) ds(y), \qquad (4.9)$$

and, interchanging the roles of $\hat{x}$ and d,

$$\sqrt{8\pi k}\, e^{-i\pi/4} u_\infty(-d, -\hat{x}) =$$
$$\int_{\partial D} \left(u^s(y, -\hat{x}) \frac{\partial}{\partial \nu} u^i(y, d) - u^i(y, d) \frac{\partial}{\partial \nu} u^s(y, -\hat{x}) \right) ds(y). \qquad (4.10)$$

Now subtract (4.10) from the sum of (4.7)–(4.9) to obtain

$$\sqrt{8\pi k}\, e^{-i\pi/4} \left(u_\infty(\hat{x}, d) - u_\infty(-d, -\hat{x}) \right)$$
$$= \int_{\partial D} \left(u(y, d) \frac{\partial}{\partial \nu} u(y, -\hat{x}) - u(y, -\hat{x}) \frac{\partial}{\partial \nu} u(y, d) \right) ds(y)$$
$$= 0$$

by the boundary condition (4.4). Hence $u_\infty(\hat{x}, d) = u_\infty(-d, -\hat{x})$, and this implies the theorem. $\square$

We now define the *far-field operator* $F : L^2[0, 2\pi] \to L^2[0, 2\pi]$ by

$$(Fg)(\theta) := \int_0^{2\pi} u_\infty(\theta, \phi) g(\phi)\, d\phi. \qquad (4.11)$$

From the representation (4.6) for u_∞ and the fact that u^s depends continuously on u^i in $C^1(\partial D)$ we see that $u_\infty(\theta, \phi)$ is continuous on $[0, 2\pi] \times [0, 2\pi]$.

Theorem 4.3. *The far-field operator corresponding to the scattering problem (4.1)–(4.4) is injective with dense range.*

Proof. Using the reciprocity relation, we see that if F^* denotes the adjoint of F, then

$$
\begin{aligned}
(F^*g)(\theta) &= \int_0^{2\pi} \overline{u_\infty(\phi,\theta)}\, g(\phi)\, d\phi \\
&= \int_0^{2\pi} \overline{u_\infty(\theta+\pi,\phi+\pi)}\, g(\phi)\, d\phi \\
&= \int_0^{2\pi} \overline{u_\infty(\theta+\pi,\phi)}\, g(\phi-\pi)\, d\phi,
\end{aligned}
$$

where we view u_∞ and g as periodic functions of period 2π. We now see that

$$
(F^*g)(\theta) = \overline{(Fh)(\theta+\pi)},
$$

where $h(\phi) = \overline{g(\phi-\pi)}$. Hence F is injective if and only if F^* is injective. By Theorem 1.29, we now see that the theorem will follow if we can show that F is injective.

To this end, suppose $Fg = 0$ for $g \neq 0$. Then, by superposition, there exists a Herglotz wave function v_g with kernel g such that the far-field pattern v_∞ corresponding to this Herglotz wave function as incident field is identically zero. By Rellich's lemma, the scattered field $v^s(x)$ corresponding to v_∞ is identically zero for $x \in \mathbb{R}^2 \setminus \bar{D}$ and the boundary condition (4.4) now implies that

$$
\frac{\partial v_g}{\partial \nu} + i\lambda v_g = 0 \quad \text{on } \partial D.
$$

Since v_g is a solution of the Helmholtz equation in D, we have from Green's second identity applied to v_g and $\bar{v}_g$ that

$$
2i \int_{\partial D} \lambda \left| v_g \right|^2 ds = 0.
$$

Hence $v_g = 0$ on ∂D, and by the boundary condition satisfied by v_g on ∂D, we also have that $\partial v_g / \partial \nu = 0$ on ∂D. The representation formula (3.41) for solutions of the Helmholtz equation in interior domains now shows that $v_g(x) = 0$ for $x \in D$, and hence $g = 0$, a contradiction. Hence $Fg = 0$ implies that $g = 0$, i.e., F is injective, and the theorem follows. $\qquad\square$

4.2 Uniqueness Theorems for Inverse Problem

Our first aim in this section is to show that D is uniquely determined from $u_\infty(\theta, \phi)$ for θ and ϕ in $[0, 2\pi]$ without knowing λ a priori. Our proof is due to Kirsch and Kress [106].

Lemma 4.4. *Assume that k^2 is not a Dirichlet eigenvalue for the bounded domain B with C^2 boundary ∂B and that $\mathbb{R}^2 \setminus \bar{B}$ is connected. Let $u^i(x,d) = e^{ikx \cdot d}$. Then the restriction of $\{u^i(\cdot, d) : |d| = 1\}$ to ∂B is complete in $H^{1/2}(\partial B)$, i.e.,*

$$\overline{\text{span}\,\{u^i(\cdot,d)|_{\partial B} : |d| = 1\}} = H^{1/2}(\partial B).$$

Proof. Let $\varphi \in H^{-1/2}(\partial B)$ satisfy

$$\int_{\partial B} \varphi(y)\, e^{-iky \cdot d}\, ds(y) = 0 \tag{4.12}$$

for all d such that $|d| = 1$. By duality pairing, to prove the lemma, it suffices to show that $\varphi = 0$. To this end, we see that (4.12) implies that the single layer potential

$$u(x) := \int_{\partial B} \varphi(y) \Phi(x, y)\, ds(y) \quad , x \in \mathbb{R}^2 \setminus \partial B$$

has the vanishing far-field pattern $u_\infty = 0$. Hence, by Rellich's lemma, $u(x) = 0$ for $x \in \mathbb{R}^2 \setminus \bar{B}$. It can easily be shown that in this case $\varphi \in C(\partial B)$ (cf. Theorem 4.10 in the next section of this chapter for an analysis in a related case), and since in this case the single layer potential is continuous across ∂B, u satisfies the homogeneous Dirichlet problem in B. Thus, since k^2 is not a Dirichlet eigenvalue for B, $u(x) = 0$ for $x \in B$. From the discontinuity property of the normal derivative of the single layer potential (Sect. 3.3), we can now conclude that

$$0 = \frac{\partial u^-}{\partial \nu} - \frac{\partial u^+}{\partial \nu} = \varphi,$$

and the proof is finished. $\qquad \square$

Theorem 4.5. *Assume that D_1 and D_2 are two scattering obstacles with corresponding surface impedances λ_1 and λ_2 such that for a fixed wave number the far-field patterns for both scatterers coincide for all incident directions d. Then $D_1 = D_2$.*

Proof. By Rellich's lemma, we can conclude that the scattered fields $u^s(\cdot, d)$ corresponding to the incident fields $u^i(x, d) = e^{ikx \cdot d}$ coincide in the unbounded component G of the complement of $\bar{D}_1 \cup \bar{D}_2$. Choose $x_0 \in G$, and consider the two exterior boundary value problems

$$\Delta w_j^s + k^2 w_j^s = 0 \quad \text{in } \mathbb{R}^2 \setminus \bar{D}_j, \tag{4.13}$$

$$\lim_{r \to \infty} \sqrt{r} \left(\frac{\partial w_j^s}{\partial r} - ikw_j^s \right) = 0, \tag{4.14}$$

$$\frac{\partial}{\partial \nu} \left[w_j^s + \Phi(\cdot, x_0) \right] + i\lambda_j \left[w_j^s + \Phi(\cdot, x_0) \right] = 0 \quad \text{on } \partial D_j \tag{4.15}$$

for $j = 1, 2$.

We will first show that $w_1^s(x) = w_2^s(x)$ for $x \in G$. To this end, choose a bounded domain B such that $\mathbb{R}^2 \setminus \bar{B}$ is connected, $\bar{D}_1 \cup \bar{D}_2 \subset B$, $x_0 \notin \bar{B}$, and k^2 is not a Dirichlet eigenvalue for B. Then, by Lemma 4.4, there exists a sequence $\{v_n\}$ in span $\{u^i(\cdot, d) : |d| = 1\}$ such that

$$\|v_n - \Phi(\cdot, x_0)\|_{H^{1/2}(\partial B)} \to 0 \quad , n \to \infty.$$

From Theorem 3.9 one can conclude that $v_n \to \Phi(\cdot, x_0)$ and $\operatorname{grad} v_n \to \operatorname{grad} \Phi(\cdot, x_0)$ as $n \to \infty$ uniformly on $\bar{D}_1 \cup \bar{D}_2$. Since the v_n are linear combinations of plane waves, the corresponding scattered fields $v_{n,1}^s$ and $v_{n,2}^s$ for D_1 and D_2 respectively coincide on G. But from Theorem 3.7 we have that $v_{n,j}^s \to w_j^s$ as $n \to \infty$ uniformly on compact subsets of $\mathbb{R}^2 \setminus \bar{D}_j$ for $j = 1, 2$, and hence $w_1^s(x) = w_2^s(x)$ for $x \in G$.

Now assume that $D_1 \neq D_2$. Then, without loss of generality, there exists $x^* \in \partial G$ such that $x^* \in \partial D_1$ and $x^* \notin \bar{D}_2$ (Fig. 4.1). We can choose $h > 0$ such that

$$x_n := x^* + \frac{h}{n}\nu(x^*) \quad , n = 1, 2, \cdots,$$

is contained in G and consider the solutions $w_{n,j}^s$ to the scattering problem (4.13)–(4.15), with x_0 replaced by x_n. Then $w_{n,1}^s(x) = w_{n,2}^s(x)$ for $x \in G$.

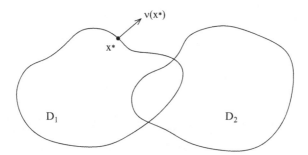

Fig. 4.1. Possible choice of x^*

But, considering $w_n^s = w_{n,2}^s$ as the scattered field corresponding to D_2, we see that

$$\frac{\partial w_n^s}{\partial \nu}(x^*) + i\lambda_1(x^*)w_n^s(x^*) \tag{4.16}$$

remains bounded as $n \to \infty$. On the other hand, considering $w_n^s = w_{n,1}^s$ as the scattered field corresponding to D_1, we have that

$$\frac{\partial w_n^s}{\partial \nu}(x^*) + i\lambda_1(x^*)w_n^s(x^*) = -\left(\frac{\partial \Phi}{\partial \nu}(x^*, x_n) + i\lambda_1(x^*)\Phi(x^*, x_0)\right),$$

and hence (4.16) becomes unbounded as $n \to \infty$. This is a contradiction, and hence $D_1 = D_2$. $\square$

We now want to show that the far-field pattern u_∞ uniquely determines not only D but the surface impedance $\lambda = \lambda(x)$ as well [113]. To this end, we first need the following lemma [98].

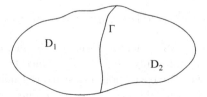

Fig. 4.2. Geometry for Lemma 4.6

Lemma 4.6. *Let $D \subset R^2$ be a domain that is decomposed into two disjoint subdomains D_1 and D_2 with common boundary $\Gamma := \partial D_1 \cap \partial D_2$ (Fig. 4.2). Assume that ∂D is in class C^2. Suppose $u_j \in C^2(D_j) \cap C^1(\bar{D}_j)$ satisfies*

$$\Delta u_j + k^2 u_j = 0 \quad in\ D_j$$

and $u_1 = u_2$ on Γ and $\partial u_1/\partial \nu = \partial u_2/\partial \nu$ on Γ, where ν is the unit outward normal to Γ considered as part of ∂D_1. Then the function

$$u(x) := \begin{cases} u_1(x), & x \in \bar{D}_1, \\ u_2(x), & x \in \bar{D}_2, \end{cases}$$

is a solution to the Helmholtz equation in $D = D_1 \cup D_2 \cup \Gamma$.

Proof. Fix $x_0 \in \Gamma \cap D$, and let $\Omega := \{x : |x - x_0| < \epsilon\} \subset D$. Let $\Omega_j := \Omega \cap D_j$, and let $x \in \Omega_1$. Then by the representation formula (3.41) we have that

$$u_1(x) = \int_{\partial \Omega_1} \left[\frac{\partial u_1}{\partial \nu}(y) \Phi(x,y) - u_1(y) \frac{\partial}{\partial \nu(y)} \Phi(x,y) \right] ds(y)$$

for $x \in \Omega_1$. On the other hand,

$$0 = \int_{\partial \Omega_2} \left[\frac{\partial u_2}{\partial \nu}(y) \Phi(x,y) - u_2(y) \frac{\partial}{\partial \nu(y)} \Phi(x,y) \right] ds(y)$$

for $x \in \Omega_1$. Now add these two equations together, noting that the contributions on $\Gamma \cap \Omega$ cancel, to arrive at

$$u_1(x) = \int_{\partial \Omega} \left[\frac{\partial u}{\partial \nu}(y) \Phi(x,y) - u(y) \frac{\partial}{\partial \nu(y)} \Phi(x,y) \right] ds(y) \qquad (4.17)$$

for $x \in \Omega_1$. Similarly,

$$u_2(x) = \int_{\partial \Omega} \left[\frac{\partial u}{\partial \nu}(y)\Phi(x,y) - u(y)\frac{\partial}{\partial \nu(y)}\Phi(x,y) \right] ds(y) \qquad (4.18)$$

for $x \in \Omega_2$. Now note that the right-hand sides of (4.17) and (4.18) coincide and define a solution of the Helmholtz equation in Ω, and the lemma follows.
□

Theorem 4.7. *Assume that D_1 and D_2 are two scattering obstacles with corresponding surface impedances λ_1 and λ_2 such that for a fixed wave number the far-field patterns coincide for all incident directions d. Then $D_1 = D_2$ and $\lambda_1 = \lambda_2$.*

Proof. By Theorem 4.5, we have that $D_1 = D_2$. Hence it only remains to show that $\lambda_1(x) = \lambda_2(x)$ for $x \in \partial D$, where $D = D_1 = D_2$. Let u_1 and u_2 be the solutions of (4.1)–(4.4) for $\lambda = \lambda_1$ and $\lambda = \lambda_2$, respectively. Then, by Rellich's lemma, $u_1(x) = u_2(x)$ for $x \in \mathbb{R}^2 \setminus \bar{D}$ and hence $u_1 = u_2$ and $\partial u_1/\partial \nu = \partial u_2/\partial \nu$ on ∂D. From the boundary conditions

$$\frac{\partial u_j}{\partial \nu} + i\lambda_j u_j = 0 \quad \text{on } \partial D \qquad (4.19)$$

for $j = 1, 2$ we have that

$$(\lambda_1 - \lambda_2)u_1 = 0 \quad \text{on } \partial D. \qquad (4.20)$$

Now suppose that $u_1 = 0$ on an arc $\Gamma \subset \partial D$. Then from (4.19) we have that $\partial u_1/\partial \nu = 0$ on Γ, and by Lemma 4.6 we have that

$$u(x) = \begin{cases} u_1(x), & x \in \mathbb{R}^2 \setminus D, \\ 0, & x \in D, \end{cases}$$

defines a solution of the Helmholtz equation in $(\mathbb{R}^2 \setminus D) \cup \Gamma \cup D$. By the fact that solutions of the Helmholtz equation are real analytic, we can now conclude that $u_1(x) = 0$ for $x \in \mathbb{R}^2 \setminus D$. But

$$u_1(x) = e^{ikx \cdot d} + u_1^s(x),$$

and u_1^s satisfies the Sommerfeld radiation condition, but $e^{ikx \cdot d}$ does not. This is a contradiction, and hence u_1 cannot vanish on any arc $\Gamma \subset \partial D$. Thus, if $x \in \partial D$, then there exists a sequence $\{x_n\} \subset \partial D$ such that $x_n \to x$ as $n \to \infty$ and $u_1(x_n) \neq 0$ for every n. From (4.20) we have that $\lambda_1(x_n) = \lambda_2(x_n)$ for every n and, since λ_1 and λ_2 are continuous functions, we have that $\lambda_1(x) = \lambda_2(x)$. Since $x \in \partial D$ was an arbitrary point, the theorem is proved.
□

4.3 Linear Sampling Method

We shall now give an algorithm for determining the scattering obstacle D from
a knowledge of the far-field pattern corresponding to the scattering problem

$$\Delta u + k^2 u = 0 \quad \text{in } \mathbb{R}^2 \setminus \bar{D}, \tag{4.21}$$

$$u(x) = e^{ikx \cdot d} + u^s(x), \tag{4.22}$$

$$\lim_{r \to \infty} \sqrt{r} \left(\frac{\partial u^s}{\partial r} - iku^s \right) = 0, \tag{4.23}$$

$$\frac{\partial u}{\partial \nu} + i\lambda u = 0 \quad \text{on } \partial D, \tag{4.24}$$

where $\lambda \in C(\partial D)$, $\lambda(x) > 0$, for $x \in \partial D$, and it is not assumed that λ is
known a priori. The algorithm we have in mind is the *linear sampling method*
and was first introduced by Colton and Kirsch [50] and Colton et al. [61]. For
survey papers discussing this method we refer the reader to [44] and [49].

We begin our discussion of the linear sampling method by considering the
general scattering problem

$$\Delta w + k^2 w = 0 \quad \text{in } \mathbb{R}^2 \setminus \bar{D}, \tag{4.25}$$

$$\lim_{r \to \infty} \sqrt{r} \left(\frac{\partial w}{\partial r} - ikw \right) = 0, \tag{4.26}$$

$$\frac{\partial w}{\partial \nu} + i\lambda w = f \quad \text{on } \partial D, \tag{4.27}$$

where $f \in H^{-1/2}(\partial D)$, i.e., we are considering weak solutions of (4.25)–
(4.27). The *boundary operator* $B : H^{-1/2}(\partial D) \to L^2[0, 2\pi]$ is now defined as
the linear operator mapping f onto the far-field pattern w_∞ corresponding
to (4.25)–(4.27).

Theorem 4.8. *The boundary operator B is compact and injective and has
dense range in $L^2[0, 2\pi]$.*

Proof. By representing w in the form of a modified single layer potential

$$w(x) = \int_{\partial D} \varphi(y) \Gamma(x, y) \, ds(y) \tag{4.28}$$

as discussed in Sect. 3.3 and generalizing the analysis given there for $\varphi \in
C(\partial D)$ to the present case $\varphi \in H^{-1/2}(\partial D)$, it can be shown by Riesz's the-
orem that there exists a unique density $\varphi \in H^{-1/2}(\partial D)$ such that w, as
defined by (4.28), satisfies (4.25)–(4.27) and the mapping $f \to \varphi$ is bounded
in $H^{-1/2}(\partial D)$. From (4.28) we have that the far-field pattern w_∞ is given by

$$w_\infty(\hat{x}) = \int_{\partial D} \varphi(y) \Gamma_\infty(\hat{x}, y) \, ds(y), \tag{4.29}$$

where $\hat{x} = x/|x|$ and Γ_∞ is the far-field pattern of Γ. Viewing $\Gamma_\infty(\hat{x}, \cdot)$ as a function in $H^1(\partial D)$, we see that for $\varphi \in H^{-1}(\partial D)$ we have

$$\left| \int_{\partial D} \varphi(y) \left[\Gamma_\infty(\hat{x}_1, y) - \Gamma_\infty(\hat{x}_2, y) \right] ds(y) \right|$$

$$\leq \|\varphi\|_{H^{-1}(\partial D)} \|\Gamma_\infty(\hat{x}_1, \cdot) - \Gamma_\infty(\hat{x}_2, \cdot)\|_{H^1(\partial D)},$$

and hence (4.29) defines a bounded operator from $H^{-1}(\partial D)$ to $C[0, 2\pi]$. Parameterizing ∂D and using Rellich's theorem (which is also valid for $p, q < 0$), we see that the embedding operator from $H^{-1/2}(\partial D)$ to $H^{-1}(\partial D)$ is compact and (4.29) defines a compact operator from $H^{-1/2}(\partial D)$ to $C[0, 2\pi]$. This implies that (4.29) is also compact from $H^{-1/2}(\partial D)$ to $L^2[0, 2\pi]$. Since $f \to \varphi$ is bounded in $H^{-1/2}(\partial D)$, we can now conclude that $B : H^{-1/2}(\partial D) \to L^2[0, 2\pi]$ is compact.

Now suppose that the far-field pattern w_∞ corresponding to (4.25)–(4.27) vanishes. Then by Rellich's lemma we have that $w(x) = 0$ for $x \in \mathbb{R}^2 \setminus \bar{D}$, and from the weak formulation (3.56) we see that

$$\int_{\partial D} f v \, ds = 0$$

for all $v \in H^1_{com}(\mathbb{R}^2 \setminus \bar{D})$, i.e., from the trace theorem, for every $v \in H^{1/2}(\partial D)$. Hence, by duality pairing, $f = 0$, and this implies that B is injective.

To show that B has dense range, let

$$u_{n,\infty}(\theta) = \sum_{-n}^{n} a_l e^{il\theta}.$$

Then $u_{n,\infty}$ is the far-field pattern of

$$u_n(r, \theta) = \sum_{-n}^{n} a_l \gamma_l^{-1} H_l^{(1)}(kr) e^{il\theta},$$

where

$$\gamma_l = \sqrt{\frac{2}{k\pi}} \exp \left[-i \left(\frac{l\pi}{2} + \frac{\pi}{4} \right) \right]$$

and u_n satisfies (4.25)–(4.27) for

$$f = \left(\frac{\partial u_n}{\partial \nu} + i\lambda u_n \right) \bigg|_{\partial D}.$$

Since f is continuous, and hence in $H^{-1/2}(\partial D)$, we can conclude by the completeness of the trigonometric polynomials in $L^2[0, 2\pi]$ that B has dense range. □

The following theorem will provide the key ingredient of the linear sampling method for determining D from the far-field pattern u_∞.

Theorem 4.9. *If $\Phi_\infty(\hat{x}, z)$ is the far-field pattern of the fundamental solution $\Phi(x, z)$, then $\Phi_\infty(\hat{x}, z)$ is in the range of B if and only if $z \in D$.*

Proof. If $z \in D$, then $\Phi(\cdot, z)$ is the solution of (4.25)–(4.27), with

$$f = \left(\frac{\partial \Phi}{\partial \nu} + i\lambda \Phi \right)\bigg|_{\partial D} \qquad (4.30)$$

and $Bf = \Phi_\infty$. If $z \in \mathbb{R}^2 \setminus D$ and Φ_∞ is in the range of B, then , by Rellich's lemma, $\Phi(\cdot, z)$ is a weak solution of (4.25)–(4.27), with f again given by (4.30). But Φ is not in $H^1_{\text{loc}}(\mathbb{R}^2 \setminus \bar{D})$, and hence this is not possible. Thus, if $z \in \mathbb{R}^2 \setminus D$, then Φ_∞ is not in the range of B. $\square$

Now let v_g be a Herglotz wave function with kernel $g \in L^2[0, 2\pi]$, and define the operator $H : L^2[0, 2\pi] \to H^{-1/2}(\partial D)$ by

$$Hg := \left(\frac{\partial v_g}{\partial \nu} + i\lambda v_g \right)\bigg|_{\partial D} .$$

The importance of the operator H follows from the fact that the far-field operator F is easily seen to have the factorization

$$F = -BH.$$

The following theorem was first proved in [63] (see also [55]).

Theorem 4.10. *The operator H is bounded and injective and has dense range in $H^{-1/2}(\partial D)$.*

Proof. From the definition of H and v_g, H is clearly bounded and injectivity follows from the uniqueness of the solution to the interior impedance problem (see the end of Sect. 4.1). To show that the range is dense, it suffices to show that if

$$u_n(x) := J_n(kr)e^{in\theta},$$

then the set

$$\left\{ \left(\frac{\partial u_n}{\partial \nu} + i\lambda u_n \right)\bigg|_{\partial D} : n = 0, \pm 1, \pm 2, \cdots \right\}$$

is complete in $H^{-1/2}(\partial D)$. By duality pairing, this requires us to show that if $g \in H^{1/2}(\partial D)$ and

$$\int_{\partial D} g(y) \left(\frac{\partial}{\partial \nu} + i\lambda \right) u_n(y)\, ds(y) = 0 \qquad (4.31)$$

for $n = 0, \pm 1, \pm 2, \cdots$, then $g = 0$.

Suppose that (4.31) is valid for some $g \in H^{1/2}(\partial D)$, and let Ω_R be a disk centered at the origin of radius R and containing D in its interior. Then from (4.31) and the addition formula for Bessel functions, we can conclude that

$$u(x) := \int_{\partial D} g(y) \left(\frac{\partial}{\partial \nu(y)} + i\lambda \right) \Phi(x,y)\, ds(y) \qquad (4.32)$$

is identically zero for $x \in \mathbb{R}^2 \setminus \bar{\Omega}_R$. By Theorem 3.2, we can conclude that $u(x) = 0$ for $x \in \mathbb{R}^2 \setminus \bar{D}$. We now make use of the fact that the double layer potential

$$v(x) := \int_{\partial D} \varphi(y) \frac{\partial}{\partial \nu(y)} \Phi(x,y)\, ds(y) \quad , x \in \mathbb{R}^2 \setminus \partial D$$

with continuous density φ satisfies the discontinuity property

$$v_{\pm}(x) = \int_{\partial D} \varphi(y) \frac{\partial}{\partial \nu(y)} \Phi(x,y)\, ds(y) \pm \frac{1}{2}\varphi(x) \quad , x \in \partial D,$$

where $\pm$ denotes the limits as $x \to \partial D$ from outside and inside D, respectively, and that

$$\frac{\partial v_+}{\partial \nu}(x) = \frac{\partial v_-}{\partial \nu}(x) \quad , x \in \partial D\,.$$

Furthermore, these properties remain valid for $\varphi \in H^{1/2}(\partial D)$, where the integrals are interpreted in the sense of duality pairing [111,127]. Hence, since $u(x) = 0$ for $x \in \mathbb{R}^2 \setminus \bar{D}$, we have that

$$0 = g(x) + 2 \int_{\partial D} g(y) \left(\frac{\partial}{\partial \nu(y)} + i\lambda \right) \Phi(x,y)\, ds(y) \quad , x \in \partial D\,.$$

Since $\partial/\partial\nu(y)\Phi(x,y)$ is continuous and $\Phi(x,y) = O(\log|x-y|)$, we can now easily verify that g is continuous.

We now return to (4.32) and use the discontinuity properties of double and single layer potentials with continuous densities to conclude that

$$u_+ - u_- = g$$
$$\frac{\partial u_+}{\partial \nu} - \frac{\partial u_-}{\partial \nu} = -i\lambda g,$$

and, since $u_+ = \partial u_+ / \partial \nu = 0$, we have that

$$\frac{\partial u_-}{\partial \nu} + i\lambda u_- = 0 \quad \text{on } \partial D\,.$$

We can now conclude, as we did at the end of Sect. 4.1, that $u(x) = 0$ for $x \in D$, and since $u(x) = 0$ for $x \in \mathbb{R}^2 \setminus \bar{D}$, we now have that $0 = u_+ - u_- = g$, and the theorem follows. $\square$

To derive an algorithm for determining D, we now introduce the *far-field equation*

$$\int_0^{2\pi} u_\infty(\theta, \phi) g(\phi) \, d\phi = \gamma \exp\left(-ikr_z \cos(\theta - \theta_z)\right), \qquad (4.33)$$

where (r_z, θ_z) are the polar coordinates of a point $z \in \mathbb{R}^2$ and

$$\gamma = \frac{e^{i\pi/4}}{\sqrt{8\pi k}},$$

or, in simpler notation,

$$(Fg)(\hat{x}) = \Phi_\infty(\hat{x}, z),$$

where $\hat{x} = (\cos\theta, \sin\theta)$ and

$$\Phi_\infty(\hat{x}, z) = \gamma e^{-ik\hat{x}\cdot z}$$

is the far-field pattern of the fundamental solution $\Phi(x, z)$. The following theorem provides the mathematical basis of the linear sampling method [14, 54]. To state the theorem, we first notice that for $h \in H^{-1/2}(\partial D)$ there exists a unique weak solution $u \in H^1(D)$ of the interior impedance problem (Chap. 8)

$$\Delta u + k^2 u = 0 \quad \text{in } D, \qquad (4.34)$$

$$\frac{\partial u}{\partial \nu} + i\lambda u = h \quad \text{on } \partial D, \qquad (4.35)$$

where, as before, $\lambda = \lambda(x) \in C(\partial D)$, $\lambda(x) > 0$, for $x \in \partial D$. This solution depends continuously in the $H^1(d)$-norm on the boundary data $h \in H^{1/2}(\partial D)$.

Theorem 4.11. *Let u_∞ be the far-field pattern corresponding to the scattering problem (4.21)–(4.24) with associated far-field operator F, and $\lambda \neq 0$. Then the following holds:*

1. *For $z \in D$ and a given $\epsilon > 0$ there exists a function $g_z^\epsilon \in L^2[0, 2\pi]$ such that*

$$\|Fg_z^\epsilon - \Phi_\infty(\cdot, z)\|_{L^2[0,2\pi]} < \epsilon,$$

and the Herglotz wave function $v_{g_z^\epsilon}$ with kernel g_z^ϵ converges in $H^1(D)$ to the unique solution $u := u_z$ of (4.34)–(4.35) with

$$h := -\left(\frac{\partial}{\partial \nu}\Phi(\cdot, z) + i\lambda\Phi(\cdot, z)\right)$$

as $\epsilon \to 0$.

2. For $z \notin D$ and a given $\epsilon > 0$ every function $g_z^\epsilon \in L^2[0, 2\pi]$ that satisfies

$$\|Fg_z^\epsilon - \Phi_\infty(\cdot, z)\|_{L^2[0,2\pi]} < \epsilon$$

is such that

$$\lim_{\epsilon \to 0} \|v_{g_z^\epsilon}\|_{H^1(D)} = \infty.$$

Proof. Assume $z \in D$. Then, by Theorem 4.9, there exists $f_z \in H^{-1/2}(\partial D)$ such that $Bf_z = -\Phi_\infty(\cdot, z)$. By Theorem 4.10, we see that for a given arbitrary $\epsilon > 0$ there exists a Herglotz wave function with kernel $g_z^\epsilon \in L^2[0, 2\pi]$ such that

$$\|Hg_z^\epsilon - f_z\|_{H^{-1/2}(\partial D)} < \frac{\epsilon}{\|B\|}, \tag{4.36}$$

and consequently

$$\|BHg_z^\epsilon - Bf_z\|_{L^2[0,2\pi]} < \epsilon.$$

Hence, since $F = -BH$, we have that

$$\|Fg_z^\epsilon - \Phi_\infty\|_{L^2[0,2\pi]} < \epsilon.$$

Next, notice that the Herglotz wave function $v_{g_z^\epsilon}$ with kernel g_z^ϵ satisfies (4.34)–(4.35) with $h := -Hg_z^\epsilon$. From (4.36) and the continuity of the solution of (4.34)–(4.35) in terms of boundary data we see that $v_{g_z^\epsilon}$ converges to the solution u_z of (4.34)–(4.35), with $h := -\left(\frac{\partial}{\partial \nu}\Phi(\cdot, z) + i\lambda \Phi(\cdot, z)\right)$ as $\epsilon \to 0$.

Now let $z \notin D$, and assume to the contrary that there exists a sequence $\{\epsilon_n\} \to 0$ and corresponding g_n satisfying $\|Fg_n - \Phi_\infty(\cdot, z)\|_{L^2[0,2\pi]} < \epsilon_n$ such that $\|v_n\|_{H^1(D)}$ remain bounded, where $v_n := v_{g_n}$ is the Herglotz wave function with kernel g_n. From the trace theorem $\|Hg_n\|_{H^{-1/2}(\partial D)}$ also remain bounded. Then, without loss of generality, we may assume weak convergence $Hg_n \rightharpoonup h \in H^{-1/2}(\partial D)$ as $n \to \infty$. Since $B : H^{-1/2}(\partial D) \to L^2[0, 2\pi]$ is a bounded linear operator, we also have that $BHg_n \rightharpoonup Bh$ in $L^2[0, 2\pi]$. But by construction, $BHg_n \to -\Phi_\infty(\cdot, z)$, which means that $Bh = -\Phi_\infty(\cdot, z)$. This contradicts Theorem 4.9, and the second statement of the theorem follows. $\qquad \square$

The *linear sampling method* is based on numerically determining the function g_z in the preceding theorem and, hence, the scattering object D. However, at this point, the numerical scheme that is used is rather ad hoc since in general the far-field equation has no solution even in the case of "noise-free" data u_∞. Nevertheless, the procedure that has been used to determine g_z has been proven to be numerically quite successful and is as follows:

1. Select a grid of "sampling points" in a region known to contain D.
2. Use Tikhonov regularization and the Morozov discrepancy principle to compute an approximate solution g_z to the far-field equation for each z in the foregoing grid. In the case where $\lambda = 0$, a justification for using such a procedure to construct g_z was given in [6], but the general case remains open. It is, of course, possible to use other regularization schemes to reconstruct g_z, and investigations in this direction have reported in [155].
3. Choose a cutoff value C, and assert that $z \in D$ if and only if $\|g_z\| \leq C$. The choice of C is heuristic but becomes empirically easier to choose when the frequency becomes higher [42].

We note that the arguments used to establish Theorem 4.11 do not depend in an essential way on the fact that the obstacle satisfies the impedance boundary condition. In particular, the conclusion of the theorem remains valid for the Neumann boundary condition (which is a particular case of our problem with $\lambda = 0$) or Dirichlet boundary condition (which is a particular case of our problem with $\lambda = \infty$), provided that k^2 is not a Neumann eigenvalue or Dirichlet eigenvalue of $-\Delta$ in D. The main property used in the justification of the linear sampling method is that the interior boundary value problem corresponding to the direct scattering problem is well posed. This discussion reveals that the linear sampling method for solving the inverse scattering problem does not depend on knowing the boundary condition a priori. An additional important feature of this method is that the number of components of the scatterer does not have to be known in advance. In Chap. 8, we revisit the linear sampling method for obstacles with mixed boundary conditions where we also provide numerical examples using the preceding numerical strategy to reconstruct the boundary of the scattering object.

A problem with the linear sampling method as described earlier is that, in general, there does not exist a solution of

$$Fg = \Phi_\infty(\cdot, z)$$

for noise-free data, and hence it is not clear what solution is obtained using Tikhonov regularization. In particular, it is not clear whether Tikhonov regularization indeed leads to the approximations predicted by Theorem 4.11. This question has been addressed and clarified by Arens and Lechleiter [6,7] for the case of the scattering problem with a Dirichlet boundary condition, and it will be discussed in Sect. 7.3. Their approach uses Kirsch's factorization method, which is the subject of our discussion in Chap. 7.

4.4 Determination of Surface Impedance

Having determined the scattering object D (without needing to know λ a priori!) we now want to determine λ. We shall do this following the ideas of [17] and note that the method we will present is also valid when the impedance boundary condition is only imposed on part of the boundary and on the other part the total field u is required to satisfy a Dirichlet boundary condition corresponding to that portion of the boundary that is a perfect conductor (Chap. 8).

We begin by defining

$$w_z := u_z + \Phi(\cdot, z),$$

where $u_z \in H^1(D)$ is the unique weak solution of the interior impedance problem (assuming that $\lambda \neq 0$)

$$\Delta u_z + k^2 u_z = 0 \quad \text{in } D, \tag{4.37}$$

$$\frac{\partial u_z}{\partial \nu} + i\lambda u_z = -\left(\frac{\partial}{\partial \nu}\Phi(\cdot, z) + i\lambda\Phi(\cdot, z)\right) \quad \text{on } \partial D, \tag{4.38}$$

where $z \in D$ and, as before, $\lambda = \lambda(x) \in C(\partial D)$, $\lambda(x) > 0$, for $x \in \partial D$.

We recall that, from the first part of Theorem 4.11, we have that, for a given $\epsilon > 0$ and $z \in D$, there exists a function $g_z^\epsilon \in L^2[0, 2\pi]$ such that

$$\|Fg_z^\epsilon - \Phi_\infty(\cdot, z)\|_{L^2[0,2\pi]} < \epsilon,$$

and the Herglotz wave function $v_{g_z^\epsilon}$ with kernel g_z^ϵ converges in $H^1(D)$ to u_z.

Lemma 4.12. *For every $z_1, z_2 \in D$ we have that*

$$2\int_{\partial D} w_{z_1} \lambda \bar{w}_{z_2} \, ds = -4\pi k \, |\gamma|^2 \, J_0(k\,|z_1 - z_2|)$$

$$- i\left(\overline{u_{z_2}(z_1)} - u_{z_1}(z_2)\right),$$

where $\gamma = e^{i\pi/4}/\sqrt{8\pi k}$ and J_0 is a Bessel function of order zero.

Proof. We previously noted that Green's second identity remains valid for weak solutions of the Helmholtz equation. In particular,

$$2i\int_{\partial D} w_{z_1} \lambda \bar{w}_{z_2} \, ds = \int_{\partial D} \left(w_{z_1}\frac{\partial \bar{w}_{z_2}}{\partial \nu} - \bar{w}_{z_2}\frac{\partial w_{z_1}}{\partial \nu}\right) ds$$

$$= \int_{\partial D} \left(\Phi(\cdot, z_1)\frac{\partial}{\partial \nu}\overline{\Phi(\cdot, z_2)} - \overline{\Phi(\cdot, z_2)}\frac{\partial}{\partial \nu}\Phi(\cdot, z_1)\right) ds$$

$$+ \int_{\partial D} \left(u_{z_1}\frac{\partial}{\partial \nu}\overline{\Phi(\cdot, z_2)} - \overline{\Phi(\cdot, z_2)}\frac{\partial u_{z_1}}{\partial \nu}\right) ds$$

$$+ \int_{\partial D} \left(\Phi(\cdot, z_1)\frac{\partial \bar{u}_{z_2}}{\partial \nu} - \bar{u}_{z_2}\frac{\partial}{\partial \nu}\Phi(\cdot, z_1)\right) ds.$$

But

$$\int_{\partial D} \left(\Phi(\cdot, z_1) \frac{\partial}{\partial \nu} \overline{\Phi(\cdot, z_2)} - \overline{\Phi(\cdot, z_2)} \frac{\partial}{\partial \nu} \Phi(\cdot, z_1) \right) ds =$$

$$= -2ik \int_{|\hat{x}|=1} \Phi_\infty(\hat{x}, z_1) \overline{\Phi_\infty(\hat{x}, z_2)} \, ds(\hat{x})$$

$$= -2ik \, |\gamma|^2 \int_{|\hat{x}|=1} e^{-ik\hat{x}\cdot z_1} e^{ik\hat{x}\cdot z_2} \, ds(\hat{x})$$

$$= -4ik\pi \, |\gamma|^2 \, J_0(k \, |z_1 - z_2|)$$

from the Jacobi–Anger expansion (3.24). From the representation formula (3.41) we now obtain

$$2i \int_{\partial D} w_{z_1} \lambda \bar{w}_{z_2} \, ds = -4ik\pi \, |\gamma|^2 \, J_0(k \, |z_1 - z_2|) + \overline{u_{z_2}(z_1)} - u_{z_1}(z_2),$$

and the lemma follows by dividing both sides by i. □

Setting $z = z_1 = z_2$ in Lemma 4.12 we arrive at the following integral equation for the determination of λ:

$$\int_{\partial D} \lambda(x) |u_z(x) + \Phi(x, z)|^2 = -\frac{1}{4} - \mathrm{Im}(u_z(z)), \qquad z \in D, \qquad (4.39)$$

where u_z is defined by (4.37)–(4.38).

Now assume D is connected, let $\Omega_r \subset D$ be a disk of radius r contained in D, and define the set W by

$$W := \{ f \in L^2(\partial D) : f = w_z|_{\partial D} \, , z \in \Omega_r \} \, .$$

We want to show that W is complete in $L^2(\partial D)$. To this end, let $\varphi \in L^2(\partial D)$ be such that for every $z \in \Omega_r$ and $w_z :=: u_z + \Phi(\cdot, z)|_{\partial D}$ we have that

$$\int_{\partial D} w_z \varphi \, ds = 0.$$

To show that W is complete, we need to show that $\varphi = 0$. To this end, let v be the (weak) solution of the interior impedance problem

$$\Delta v + k^2 v = 0 \quad \text{in } D$$

$$\frac{\partial v}{\partial \nu} + i\lambda v = \varphi \quad \text{on } \partial D \, .$$

Then for every $z \in \Omega_r$ we have that

$$0 = \int_{\partial D} w_z \varphi \, ds = \int_{\partial D} w_z \left(\frac{\partial v}{\partial \nu} + i\lambda v \right) ds$$

$$= \int_{\partial D} \left(u_z \frac{\partial v}{\partial \nu} + i\lambda u_z v + \Phi(\cdot, z) \frac{\partial v}{\partial \nu} + i\lambda \Phi(\cdot, z) v \right) ds$$

$$= \int_{\partial D} \left(u_z \frac{\partial v}{\partial \nu} + v \left(-\frac{\partial u_z}{\partial \nu} - \frac{\partial}{\partial \nu} \Phi(\cdot, z) - i\lambda \Phi(\cdot, z) \right) \right) ds$$

$$+ \int_{\partial D} \left(\Phi(\cdot, z) \frac{\partial v}{\partial \nu} + i\lambda v \Phi(\cdot, z) \right) ds$$

$$= \int_{\partial D} \left(\Phi(\cdot, z) \frac{\partial v}{\partial \nu} - v \frac{\partial}{\partial \nu} \Phi(\cdot, z) \right) ds$$

$$= v(z) \, .$$

Since v is a solution of the Helmholtz equation in D, and hence real-analytic by Theorem 3.2, we now have that $v(z) = 0$ for every $z \in D$, and hence, by the trace theorem and Theorem 3.10, we have that $\varphi = 0$.

Remark 4.13. If D is not connected, then the foregoing completeness result remains true if we replace Ω_r by a union of disks where each component contains one disk from the union.

We now return to the integral equation (4.39). By the completeness of W, we see that the left-hand side of this equation is an injective compact integral operator with positive kernel defined on $L^2(\partial D)$. Given that D is known (e.g., by the linear sampling method) we can now approximate u_z by the Herglotz wave function v_{g_z}, with kernel g_z being the approximate solution of the far-field equation given by the first part of Theorem 4.11. Using Tikhonov regularization techniques (cf. [68]) it is now possible to determine λ by finding a regularized solution of (4.39) in $L^2(\partial D)$ with noisy kernel and noisy right hand; for numerical examples of the reconstruction of λ using this approach we refer the reader to [25] and [26].

In the special case where $\lambda(x)$ is a positive constant $\lambda > 0$, from (4.39) we arrive at

$$\lambda = \frac{-2\pi k |\gamma|^2 - \mathrm{Im}(u_{z_0}(z_0))}{\|u_{z_0} + \Phi(\cdot, z)\|^2_{L^2(\partial D)}} \, .$$

Numerical examples using this formula will be provided in Chap. 8 when we consider mixed boundary value problems in scattering theory for which the same formula is valid.

4.5 Limited Aperture Data

In many cases of practical interest, the far-field data $u_\infty(\theta, \phi) = u_\infty(\hat{x}, d)$, where $\hat{x} = (\cos \theta, \sin \theta)$ and $d = (\cos \phi, \sin \phi)$, are only known for $\hat{x}$ and d on subsets of the unit circle, i.e., we are concerned with limited aperture scattering data. To handle this case, we note that from the proof of

Theorem 4.11 the function $g_z \in L^2[0, 2\pi]$ of this theorem is the kernel of a Herglotz wave function that approximates a solution to the Helmholtz equation in D with respect to the $H^1(D)$ norm. Therefore, to treat the case of limited aperture far-field data, it suffices to show that if Ω_R is a disk of radius R centered at the origin, then a Herglotz wave function can be approximated in $H^1(\Omega_R)$ by a Herglotz wave function with kernel supported in a subset Γ_0 of $L^2[0, 2\pi]$. This new Herglotz wave function and its kernel can now be used in place of g_z and v_{g_z} in Theorem 4.11, where $\|Fg_z - \Phi_\infty(\cdot, x)\|_{L^2[0,2\pi]}$ is replaced by $\|Fg_z - \Phi_\infty(\cdot, z)\|_{\Gamma_1}$, where Γ_1 is a subset of $L^2[0, 2\pi]$ and $\|g_z\|_{L^2[0,2\pi]}$ is replaced by $\|g_z\|_{\Gamma_0}$. In particular, the far-field equation (4.33) now becomes

$$\int_{\Gamma_0} u_\infty(\theta, \phi) g(\phi) \, d\phi = \gamma \exp\left(-ikr_z \cos(\theta - \theta_z)\right) \ , \ \theta \in \Gamma_1 \ .$$

We now proceed to prove the foregoing approximation property [16]. Assuming that k^2 is not a Dirichlet eigenvalue for the disk Ω_R (this is not a restriction since we can always find a disk containing D that has this property), by the trace theorem, it suffices to show that the set of functions

$$v_g(x) := \int_{|d|=1} g(d) e^{ikx \cdot d} \, ds(d),$$

where g is a square integrable function on the unit circle with support in some subinterval of the unit circle, is complete in $H^{1/2}(\partial \Omega_R)$. With a slight abuse of notation we call this subinterval Γ_0. Hence, using duality pairing we must show that if $\varphi \in H^{-1/2}(\partial \Omega_R)$ satisfies

$$\int_{\partial \Omega_R} \varphi(x) \left[\int_{\Gamma_0} g(d) e^{ikx \cdot d} \, ds(d) \right] ds(x) = 0$$

for every $g \in L^2(\Gamma_0)$, then $\varphi = 0$. To this end, we interchange the order of integration [which is valid for $\varphi \in H^{-1/2}(\partial \Omega_R)$ and $g \in L^2(\Gamma_0)$ since φ is a bounded linear functional on $H^{1/2}(\partial \Omega_R)$] to arrive at

$$\int_{\Gamma_0} g(d) \left[\int_{\partial \Omega_R} \varphi(x) e^{ikx \cdot d} \, ds(x) \right] ds(d) = 0$$

for every $g \in L^2(\Gamma_0)$. This in turn implies (taking conjugates) that the far-field pattern $(S\bar{\varphi})_\infty$ of the single layer potential

$$(S\bar{\varphi})(y) := \int_{\partial \Omega_R} \overline{\varphi(x)} \Phi(x, y) \, ds(x) \ , \ y \in \mathbb{R}^2 \setminus \bar{\Omega}_R$$

satisfies

$$(S\bar{\varphi})_\infty(d) := \gamma \int_{\partial \Omega_R} \overline{\varphi(x)} e^{-ikx \cdot d} \, ds(x) = 0$$

for $d \in \Gamma_0$, where

$$\gamma = \frac{e^{i\pi/4}}{\sqrt{8\pi k}} \; .$$

By analyticity, we can conclude that $(S\bar{\varphi})_\infty = 0$ for all vectors d on the unit circle. Arguing now as in the proof of Lemma 4.4, we can conclude that $\bar{\varphi} = 0$ and, hence, $\varphi = 0$. $\qquad\qquad\qquad\qquad\qquad\qquad\qquad\qquad\qquad\qquad\qquad\qquad\square$

In conclusion, we mention that it is also possible to consider inverse scattering problems for D in a piecewise homogeneous background medium instead of only a homogeneous background [44, 49, 67]. To do this requires a knowledge of Green's function for a piecewise homogeneous background medium. In some circumstances, however, the need to know Green's function can be avoided, and for partial progress in this direction we refer the reader to [30] and [47].

4.6 Near-Field Data

Throughout this chapter we have always assumed that the incident field is a plane wave and the measured data are far-field data. The support of the scattering object is then determined using the linear sampling method as applied to the far-field equation

$$\int_0^{2\pi} u_\infty(\theta, \phi)g(\phi)\,d\phi = \Phi_\infty(\theta, z), \qquad \theta \in [0, 2\pi] \text{ and } g \in L^2[0, 2\pi].$$

However, in many applications, the scattering object is typically interrogated using point sources as incident fields and the scattered field is measured near the scatterer. Such inverse scattering problems can again be handled by the linear sampling method, which is now applied to the *near-field equation*

$$\int_{C_0} u^s(x, y)g(y)\,ds(y) = \Phi(x, z), \qquad x \in C_1 \text{ and } g \in L^2(C_0),$$

where Φ is the fundamental solution to the Helmholtz equation defined by (3.33), C_0 and C_1 are simple closed curves containing the scattering object in their interiors, and the scattered field $u^s(x, y)$ corresponding to an incident point source at $y \in C_0$ is measured at a point $x \in C_1$. (Note that C_0 and C_1 can be the same curve.) The entire preceding analysis in this chapter (as well as the analysis in the following chapters) proceeds in a straightforward fashion, including in the case where C_0 and C_1 are segments of simple closed curves C_0 and C_1 such that C_1 is analytic, i.e., the case of limited aperture data. For numerical examples with near-field data we refer the reader to [58].

5

Scattering by Orthotropic Media

Until now the reader has been introduced only to the scattering of time
-harmonic electromagnetic waves by an imperfect conductor. We will now
consider the scattering of electromagnetic waves by a penetrable orthotropic
inhomogeneity embedded in a homogeneous background. As in the previous
chapter, we will confine ourselves to the scalar case that corresponds to the
scattering of electromagnetic waves by an orthotropic infinite cylinder. The direct scattering problem is now modeled by a transmission problem for the
Helmholtz equation outside the scatterer and an equation with nonconstant
coefficients inside the scatterer. This chapter is devoted to the analysis of the
solution to the direct problem.

After a brief discussion of the derivation of the equations that govern
the scattering of electromagnetic waves by an orthotropic infinite cylinder, we
proceed to the solution to the corresponding transmission problem. The integral equation method used by Piana [136] and Potthast [137] to solve the forward problem in this case is only valid under restrictive assumptions. Hence,
following [81], we propose here a variational method and find a solution to the
problem in a larger space than the space of twice continuously differentiable
functions. To build the analytical frame work for this variational method, we
first extend the discussion of Sobolev spaces and weak solutions initiated in
Sects. 1.5 and 3.3. This is followed by a proof of the celebrated Lax–Milgram
lemma and an investigation of the Dirichlet-to-Neumann map. Included are
several simple examples of the use of variational methods for solving boundary value problems. We conclude our chapter with a solvability result for the
direct problem.

5.1 Maxwell Equations for an Orthotropic Medium

We begin by considering electromagnetic waves propagating in an inhomoge-
neous anisotropic medium in $\mathbb{R}^3$ with electric permittivity $\epsilon = \epsilon(x)$, magnetic
permeability $\mu = \mu(x)$, and electric conductivity $\sigma = \sigma(x)$. As the reader

F. Cakoni and D. Colton, *A Qualitative Approach to Inverse Scattering Theory*,
Applied Mathematical Sciences 188, DOI 10.1007/978-1-4614-8827-9_5,
© Springer Science+Business Media New York 2014

knows from Chap. 3, the electromagnetic wave is described by the electric field $\mathcal{E}$ and the magnetic field $\mathcal{H}$ satisfying the *Maxwell equations*

$$\operatorname{curl} \mathcal{E} + \mu \frac{\partial \mathcal{H}}{\partial t} = 0, \qquad \operatorname{curl} \mathcal{H} - \epsilon \frac{\partial \mathcal{E}}{\partial t} = \sigma \mathcal{E}.$$

For time-harmonic electromagnetic waves of the form

$$\mathcal{E}(x,t) = \tilde{E}(x)e^{-i\omega t}, \qquad \mathcal{H}(x,t) = \tilde{H}(x)e^{-i\omega t}$$

with frequency $\omega > 0$, we deduce that the complex-valued space-dependent parts $\tilde{E}$ and $\tilde{H}$ satisfy

$$\operatorname{curl} \tilde{E} - i\omega\mu(x)\tilde{H} = 0,$$
$$\operatorname{curl} \tilde{H} + (i\omega\epsilon(x) - \sigma(x))\tilde{E} = 0.$$

Now let us suppose that the inhomogeneity occupies an infinitely long conducting cylinder. Let D be the cross section of this cylinder having a C^2 boundary ∂D, with ν being the unit outward normal to ∂D. We assume that the axis of the cylinder coincides with the z-axis. We further assume that the conductor is imbedded in a nonconducting homogeneous background, i.e., the electric permittivity $\epsilon_0 > 0$, and the magnetic permeability $\mu_0 > 0$ of the background medium is a positive constants, while the conductivity $\sigma_0 = 0$. Next we define

$$\tilde{E}^{int,ext} = \frac{1}{\sqrt{\epsilon_0}} E^{int,ext}, \qquad \tilde{H}^{int,ext} = \frac{1}{\sqrt{\mu_0}} H^{int,ext}, \qquad k^2 = \epsilon_0\mu_0\omega^2,$$

$$\mathcal{A}(x) = \frac{1}{\epsilon_0}\left(\epsilon(x) + i\frac{\sigma(x)}{\omega}\right), \qquad \mathcal{N}(x) = \frac{1}{\mu_0}\mu(x),$$

where $\tilde{E}^{ext}, \tilde{H}^{ext}$ and $\tilde{E}^{int}, \tilde{H}^{int}$ denote the electric and magnetic fields in the exterior medium and inside the conductor, respectively. For an orthotropic medium we have that the matrices $\mathcal{A}$ and $\mathcal{N}$ are independent of the z-coordinate and are of the form

$$\mathcal{A} = \begin{pmatrix} a_{11} & a_{12} & 0 \\ a_{21} & a_{22} & 0 \\ 0 & 0 & a \end{pmatrix}, \qquad \mathcal{N} = \begin{pmatrix} n_{11} & n_{12} & 0 \\ n_{21} & n_{22} & 0 \\ 0 & 0 & n \end{pmatrix}.$$

In particular, the field E^{int}, H^{int} inside the conductor satisfies

$$\operatorname{curl} E^{int} - ik\mathcal{N}H^{int} = 0, \qquad \operatorname{curl} H^{int} + ik\mathcal{A}E^{int} = 0, \qquad (5.1)$$

and the field E^{ext}, H^{ext} outside the conductor satisfies

$$\operatorname{curl} E^{ext} - ikH^{ext} = 0, \qquad \operatorname{curl} H^{ext} + ikE^{ext} = 0. \qquad (5.2)$$

Across the boundary of the conductor we have the continuity of the tangential component of both the electric and magnetic fields. Assuming that $\mathcal{A}$ is invertible, and using $ikE^{int} = \mathcal{A}^{-1}\mathrm{curl}\, H^{int}$ and $ikE^{ext} = \mathrm{curl}\, H^{ext}$, the Maxwell equations become

$$\mathrm{curl}\,\mathcal{A}^{-1}\mathrm{curl}\, H^{int} - k^2 \mathcal{N} H^{int} = 0 \qquad (5.3)$$

for the magnetic field inside the conductor and

$$\mathrm{curl}\,\mathrm{curl}\, H^{ext} - k^2 H^{ext} = 0 \qquad (5.4)$$

for the magnetic field outside the conductor. If the scattering is due to a given time-harmonic incident field E^i, H^i, then we have that

$$E^{ext} = E^s + E^i, \qquad H^{ext} = H^s + H^i,$$

where E^s, H^s denotes the scattered field. In general the incident field E^i, H^i is an entire solution to (5.2). In particular, in the case of incident plane waves, E^i, H^i is given by (3.4). The scattered field E^s, H^s satisfies the Silver–Müller radiation condition

$$\lim_{r \to \infty} (H^s \times x - rE^s) = 0$$

uniformly in $\hat{x} = x/|x|$ and $r = |x|$.

Now let us assume that the incident wave propagates perpendicular to the axis of the cylinder and is polarized perpendicular to the axis of the cylinder such that

$$H^i(x) = (0,\, 0,\, u^i), \qquad H^s(x) = (0,\, 0,\, u^s), \qquad H^{int}(x) = (0,\, 0,\, v).$$

By elementary vector analysis, it can be seen that (5.3) is equivalent to

$$\nabla \cdot A\nabla v + k^2 n v = 0 \qquad \text{in } D, \qquad (5.5)$$

where

$$A := \frac{1}{a_{11}a_{22} - a_{12}a_{21}} \begin{pmatrix} a_{11} & a_{21} \\ a_{12} & a_{22} \end{pmatrix}.$$

Analogously, (5.4) is equivalent to the Helmholtz equation

$$\Delta u^s + k^2 u^s = 0 \qquad \text{in } \mathbb{R}^2 \setminus \overline{D}. \qquad (5.6)$$

The transmission conditions $\nu \times (H^s + H^i) = \nu \times H^{int}$ and $\nu \times \mathrm{curl}\,(H^s + H^i) = \nu \times \mathcal{A}^{-1}\mathrm{curl}\, H^{int}$ on the boundary of the conductor become

$$v - u^s = u^i \quad \text{and} \quad \nu \cdot A\nabla v - \nu \cdot \nabla u^s = \nu \cdot \nabla u^i \qquad \text{on } \partial D. \qquad (5.7)$$

Finally, the $\mathbb{R}^2$ analog of the Silver–Müller radiation condition is the Sommerfeld radiation condition

$$\lim_{r \to \infty} \sqrt{r}\left(\frac{\partial u^s}{\partial r} - iku^s \right) = 0,$$

which holds uniformly in $\hat{x} = x/|x|$.

Summarizing the foregoing discussion we have that the scattering of incident time-harmonic electromagnetic waves by an orthotropic cylindrical conductor is modeled by the following transmission problem in $\mathbb{R}^2$. Let $D \subset \mathbb{R}^2$ be a nonempty, open, and bounded set having C^2 boundary ∂D such that the exterior domain $\mathbb{R}^2 \setminus \bar{D}$ is connected. The unit normal vector to ∂D, which is directed into the exterior of D, is denoted by ν. On $\bar{D}$ we have a matrix-valued function $A : \bar{D} \to \mathbb{C}^{2 \times 2}$, $A = (a_{jk})_{j,k=1,2}$, with continuously differentiable functions $a_{jk} \in C^1(\bar{D})$. By $\mathrm{Re}(A)$ we mean the matrix-valued function having as entries the real parts $\mathrm{Re}(a_{jk})$, and we define $\mathrm{Im}(A)$ similarly. We suppose that $\mathrm{Re}(A(x))$ and $\mathrm{Im}(A(x))$, $x \in \bar{D}$, are symmetric matrices that satisfy $\bar{\xi} \cdot \mathrm{Im}(A)\, \xi \leq 0$ and $\bar{\xi} \cdot \mathrm{Re}(A)\, \xi \geq \gamma |\xi|^2$ for all $\xi \in \mathbb{C}^3$ and $x \in \bar{D}$, where γ is a positive constant. Note that due to the symmetry of A, $\mathrm{Im}\left(\bar{\xi} \cdot A\, \xi \right) = \bar{\xi} \cdot \mathrm{Im}(A)\, \xi$ and $\mathrm{Re}\left(\bar{\xi} \cdot A\, \xi \right) = \bar{\xi} \cdot \mathrm{Re}(A)\, \xi$. We further assume that $n \in C(\bar{D})$, with $\mathrm{Im}(n) \geq 0$.

For functions $u \in C^1(\mathbb{R}^2 \setminus D)$ and $v \in C^1(\bar{D})$ we define the normal and conormal derivative by

$$\frac{\partial u}{\partial \nu}(x) = \lim_{h \to +0} \nu(x) \cdot \nabla u(x + h\nu(x)), \qquad x \in \partial D$$

and

$$\frac{\partial v}{\partial \nu_A}(x) = \lim_{h \to +0} \nu(x) \cdot A(x) \nabla v(x - h\nu(x)), \qquad x \in \partial D,$$

respectively. Then the scattering of a time-harmonic incident field u^i by an orthotropic inhomogeneity in $\mathbb{R}^2$ can be mathematically formulated as the problem of finding v, u such that

$$\nabla \cdot A\nabla v + k^2 n\, v = 0 \qquad \text{in} \quad D, \tag{5.8}$$

$$\Delta u^s + k^2\, u^s = 0 \qquad \text{in} \quad \mathbb{R}^2 \setminus \bar{D}, \tag{5.9}$$

$$v - u^s = u^i \qquad \text{on} \quad \partial D, \tag{5.10}$$

$$\frac{\partial v}{\partial \nu_A} - \frac{\partial u^s}{\partial \nu} = \frac{\partial u^i}{\partial \nu} \qquad \text{on} \quad \partial D, \tag{5.11}$$

$$\lim_{r \to \infty} \sqrt{r} \left(\frac{\partial u^s}{\partial r} - iku^s \right) = 0. \tag{5.12}$$

The aim of this chapter is to establish the existence of a unique solution to the scattering problem (5.8)–(5.12). In most applications the material properties of the inhomogeneity do not change continuously to those of the background medium, and hence the integral equation methods used in [136] and [137] are not applicable. Therefore, we will introduce a variational method to solve our problem. Since variational methods are well suited to Hilbert spaces, in the next section we reformulate our scattering problem in appropriate Sobolev spaces. To this end, we need to extend the discussion on Sobolev spaces given in Sect. 1.5.

5.2 Mathematical Formulation of Direct Scattering Problem

In the context of variational methods, one naturally seeks a solution to a linear second-order elliptic boundary value problem in the space of functions that are square integrable and have square integrable first partial derivatives. Let D be an open, nonempty, bounded, simply connected subset of $\mathbb{R}^2$ with smooth boundary ∂D. In Sect. 1.5 we introduced the Sobolev spaces $H^1(D)$, $H^{\frac{1}{2}}(\partial D)$, and $H^{-\frac{1}{2}}(\partial D)$. The reader has already encountered the connection between $H^{\frac{1}{2}}(\partial D)$ and $H^1(D)$, that is, $H^{\frac{1}{2}}(\partial D)$ is the trace space of $H^1(D)$. More specifically, for functions defined in $\bar{D}$ the values on the boundary are defined and the restriction of the function to the boundary ∂D is called the *trace*. The operator mapping a function onto its trace is called the *trace operator*. Theorem 1.38 states that the trace operator can be extended as a continuous mapping $\gamma_0 : H^1(D) \to H^{\frac{1}{2}}(\partial D)$, and this extension has a continuous right inverse (see also Theorem 3.37 in [127]). The latter means that for any $f \in H^{\frac{1}{2}}(\partial D)$ there exists a $u \in H^1(D)$ such that $\gamma_0 u = f$ and $\|u\|_{H^1(D)} \leq C \|f\|_{H^{\frac{1}{2}}(\partial D)}$, where C is a positive constant independent of f. (Map D in a one-to-one manner onto the unit disk, and use separation of variables to determine u as a solution to the Dirichlet problem for Laplace's equation. Then map back to D.)

For any integer $r \geq 0$ we let

$$C^r(D) := \{u : \partial^\alpha u \text{ exists and is continuous on } D \text{ for } |\alpha| \leq r\},$$
$$C^r(\bar{D}) := \{u|_{\bar{D}} : u \in C^r(\mathbb{R}^2)\}$$

and put

$$C^\infty(D) = \bigcap_{r \geq 0} C^r(D) \qquad C^\infty(\bar{D}) = \bigcap_{r \geq 0} C^r(\bar{D}).$$

In Sect. 1.5, $H^1(D)$ is naturally defined as the completion of $C^1(\bar{D})$ with respect to the norm

$$\|u\|^2_{H^1(D)} := \|u\|^2_{L^2(D)} + \|\nabla u\|^2_{L^2(D)}.$$

Note that $H^1(D)$ is a Hilbert space with the inner product

$$(u, v)_{H^1(D)} := (u, v)_{L^2(D)} + (\nabla u, \nabla v)_{L^2(D)}.$$

It can be shown that $C^\infty(\bar{D})$ is dense in $H^1(D)$. The proof of this result can be found in [127].

Since $H^1(D)$ is a subspace of $L^2(D)$, we can consider the *embedding* map $\mathcal{I} : H^1(D) \to L^2(D)$ defined by $\mathcal{I}(u) = u \in L^2(D)$ for $u \in H^1(D)$. Obviously, $\mathcal{I}$ is a bounded linear operator. The following two lemmas are particular cases of the well-known *Rellich compactness theorem*.

Lemma 5.1. *The embedding* $\mathcal{I} : H^1(D) \to L^2(D)$ *is compact.*

In the sequel, we also need to consider the Sobolev space $H^2(D)$, which is the space of functions $u \in H^1(D)$ such that u_x and u_y are also in $H^1(D)$. Similarly, $H^2(D)$ can be defined as the completion of $C^2(\bar{D})$ [or $C^\infty(\bar{D})$] with respect to the norm

$$\|u\|^2_{H^2(D)} = \|u\|^2_{L^2(D)} + \|\nabla u\|^2_{L^2(D)} + \|u_{xx}\|^2_{L^2(D)} + \|u_{xy}\|^2_{L^2(D)} + \|u_{yy}\|^2_{L^2(D)}.$$

Lemma 5.2. *The embedding* $\mathcal{I} : H^2(D) \to H^1(D)$ *is a compact operator.*

The proof of the Rellich compactness theorem can be found, for instance, in [72] or [127]. For the special case of $H^p[0, 2\pi]$ this result is proved in Theorem 1.32.

We now define

$$C_0^\infty(D) := \{u : u \in C_K^\infty(D) \text{ for some compact subset } K \text{ of } D\},$$

where

$$C_K^\infty(D) := \{u \in C^\infty(D) : \operatorname{supp} u \subseteq K\}$$

and the support of u, denoted by supp u, is the closure in D of the set $\{x \in D : u(x) \neq 0\}$. The completion of $C_0^\infty(D)$ in $H^1(D)$ is denoted by $H_0^1(D)$ and can be characterized by

$$H_0^1(D) := \{u \in H^1(D) : u|_{\partial D} = 0\},$$

where $u|_{\partial D}$ is understood in the sense of the trace operator $\gamma_0 u$. This space equipped with the inner product of $H^1(D)$ is also a Hilbert space. The following inequality, known as *Poincaré's inequality*, holds for functions in $H_0^1(D)$.

Theorem 5.3 (Poincaré's Inequality). *There exists a positive constant M such that for every $u \in H_0^1(D)$ we have*

$$\int_D |u|^2 \, dx \leq M \int_D \|\nabla u\|^2 \, dx,$$

where M is independent of u but depends on D.

Proof. We first assume that $u \in C_0^1(D)$. Since D is bounded, it can be enclosed in a square $\Gamma := \{|x_i| \leq a, i = 1, 2\}$, and u will continue to be identically zero outside D. Then for any $x = (x_1, x_2) \in \Gamma$ we have, using the Cauchy–Schwarz inequality, that

$$|u(x)|^2 = \left| \int_{-a}^{x_1} u_{x_1}(\xi_1, x_2) \, d\xi_1 \right|^2$$

$$\leq (x_1 + a) \int_{-a}^{x_1} |u_{x_1}|^2 \, d\xi_1$$

$$\leq 2a \int_{-a}^{x_1} |u_{x_1}|^2 \, d\xi_1,$$

and hence

$$\int_{-a}^{a} |u(x)|^2 \, dx_1 \leq 4a^2 \int_{-a}^{a} |u_{x_1}|^2 \, d\xi_1.$$

Now integrate with respect to x_2 from $-a$ to a to obtain

$$\int_{\Gamma} |u(x)|^2 \, dx \leq 4a^2 \int_{\Gamma} |u_{x_1}|^2 \, dx$$

$$\leq 4a^2 \int_{\Gamma} |\nabla u|^2 \, dx.$$

The theorem now follows from the fact that $C_0^1(D)$ is dense in $H_0^1(D)$. $\square$

Remark 5.4. It can be shown that the optimal constant M in the preceding Poincaré's inequality is equal to $1/\lambda_0(D)$, where $\lambda_0(D)$ is the first Dirichlet eigenvalue for $-\Delta$ in D (cf. [95]).

Remark 5.5. Our presentation of Sobolev spaces is by no means complete. A systematic treatment of Sobolev spaces requires the use of the Fourier transform and distribution theory, and we refer the reader to Chap. 3 in [127] for this material.

For later use we recall the following classical result from real analysis.

Lemma 5.6. *Let G be a closed subset of $\mathbb{R}^2$. For each $\epsilon > 0$ there exists a $\chi_\epsilon \in C^\infty(\mathbb{R}^2)$ satisfying*

$$
\begin{aligned}
\chi_\epsilon(x) &= 1 && \text{if } \quad x \in G, \\
0 \leq \chi_\epsilon(x) &\leq 1 && \text{if } \quad 0 < dist(x, G) < \epsilon, \\
\chi_\epsilon(x) &= 0 && \text{if } \quad dist(x, G) > \epsilon,
\end{aligned}
$$

where $dist(x, G)$ denotes the distance of x from G.

The function $\chi_\epsilon(x)$ defined in the preceding lemma is called a *cutoff function* for G. It is used to smooth out the characteristic function of a set.

Keeping in mind the solution to the scattering problem in Sect. 5.1, we now extend the definition of the conormal derivative $\partial u/\partial \nu_A$ to functions $u \in H^1(D, \Delta_A)$, where

$$H^1(D, \Delta_A) := \{u \in H^1(D) : \nabla \cdot A\nabla u \in L^2(D)\},$$

equipped with the graph norm

$$\|u\|_{H^1(D,\Delta_A)}^2 := \|u\|_{H^1(D)}^2 + \|\nabla \cdot A\nabla u\|_{L^2(D)}^2.$$

In particular, we have the following *trace theorem.*

Theorem 5.7. *The mapping* $\gamma_1 : u \to \partial u/\partial \nu_A := \nu \cdot A\nabla u$ *defined in* $C^\infty(\bar{D})$
can be extended by continuity to a linear and continuous mapping, still denoted
by γ_1, *from* $H^1(D, \Delta_A)$ *to* $H^{-\frac{1}{2}}(\partial D)$.

Proof. Let $\phi \in C^\infty(\bar{D})$ and $u \in C^\infty(\bar{D})$. The divergence theorem then
becomes

$$\int_{\partial D} \phi\nu \cdot A\nabla u \, ds = \int_D \nabla\phi \cdot A\nabla u \, dx + \int_D \phi \nabla \cdot A\nabla u \, dx.$$

Because $C^\infty(\bar{D})$ is dense in $H^1(D)$, this equality is still valid for $\phi \in H^1(D)$
and $u \in C^\infty(\bar{D})$. Therefore,

$$\left| \int_{\partial D} \phi\nu \cdot A\nabla u \, ds \right| \leq C\|u\|_{H^1(D,\Delta_A)} \|\phi\|_{H^1(D)} \quad \forall \phi \in H^1(D), \quad \forall u \in C^\infty(\bar{D}),$$

where C is a positive constant independent of ϕ and u but dependent on A
and D. Now let f be an element of $H^{\frac{1}{2}}(\partial D)$. There exists a $\phi \in H^1(D)$ such
that $\gamma_0 \phi = f$, where γ_0 is the trace operator on ∂D. Then the preceding
inequality implies that

$$\left| \int_{\partial D} f\nu \cdot A\nabla u \, ds \right| \leq C\|u\|_{H^1(D,\Delta_A)} \|f\|_{H^{\frac{1}{2}}(\partial D)} \quad \forall f \in H^{\frac{1}{2}}(\partial D), \quad \forall u \in C^\infty(\bar{D}).$$

Therefore, the mapping

$$f \to \int_{\partial D} f\nu \cdot A\nabla u \, ds \qquad f \in H^{\frac{1}{2}}(\partial D)$$

defines a continuous linear functional and

$$\|\nu \cdot A\nabla u\|_{H^{-\frac{1}{2}}(\partial D)} \leq C\|u\|_{H^1(D,\Delta_A)}.$$

Thus, the linear mapping $\gamma_1 : u \to \nu \cdot A\nabla u$ defined on $C^\infty(\bar{D})$ is continuous
with respect to the norm of $H^1(D, \Delta_A)$. Since $C^\infty(\bar{D})$ is dense in $H^1(D, \Delta_A)$,
γ_1 can be extended by continuity to a bounded linear mapping (still called
γ_1) from $H^1(D, \Delta_A)$ to $H^{-\frac{1}{2}}(\partial D)$. $\square$

As a consequence of the preceding theorem we can now extend the divergence
theorem to a wider space of functions.

Corollary 5.8. *Let* $u \in H^1(D)$ *such that* $\nabla \cdot A\nabla u \in L^2(D)$ *and* $v \in H^1(D)$.
Then

$$\int_D \nabla v \cdot A\nabla u \, dx + \int_D v \nabla \cdot A\nabla u \, dx = \int_{\partial D} v\nu \cdot A\nabla u \, ds.$$

Remark 5.9. With the help of a cutoff function for a neighborhood of ∂D we can, in a way similar to that in Theorem 5.7, define $\partial u / \partial \nu_A$ for $u \in H^1_{loc}(\mathbb{R}^2 \setminus \bar{D})$ such that $\nabla \cdot A \nabla v \in L^2_{loc}(\mathbb{R}^2 \setminus \bar{D})$ (see Sect. 3.3 for the definition of H^1_{loc}-spaces).

Remark 5.10. Setting $A = I$ in Theorem 5.7 and Corollary 5.8 we have that $\partial u / \partial \nu$ is well defined in $H^{-\frac{1}{2}}(\partial D)$ for functions $u \in H^1(D, \Delta) := \{u \in H^1(D) : \Delta u \in L^2(D)\}$. Furthermore, the following Green's identity holds:

$$\int_D \nabla v \cdot \nabla u \, dx + \int_D v \, \Delta u \, dx = \int_{\partial D} v \frac{\partial u}{\partial \nu} \, ds \qquad u \in H^1(D, \Delta),\ v \in H^1(D).$$

In particular, Theorem 3.1 and Eq. (3.41) are valid for H^1-solutions to the Helmholtz equation.

We are now ready to formulate the direct scattering problem for an orthotropic medium in $\mathbb{R}^2$ in suitable Sobolev spaces. Assume that A, n, and D satisfy the assumptions of Sect. 5.1. Given $f \in H^{\frac{1}{2}}(\partial D)$ and $h \in H^{-\frac{1}{2}}(\partial D)$, find $u \in H^1_{loc}(\mathbb{R}^2 \setminus \bar{D})$ and $v \in H^1(D)$ such that

$$\nabla \cdot A \nabla v + k^2 n\, v = 0 \qquad \text{in} \quad D, \tag{5.13}$$

$$\Delta u + k^2\, u = 0 \qquad \text{in} \quad \mathbb{R}^2 \setminus \bar{D}, \tag{5.14}$$

$$v - u = f \qquad \text{on} \quad \partial D, \tag{5.15}$$

$$\frac{\partial v}{\partial \nu_A} - \frac{\partial u}{\partial \nu} = h \qquad \text{on} \quad \partial D, \tag{5.16}$$

$$\lim_{r \to \infty} \sqrt{r} \left(\frac{\partial u}{\partial r} - iku \right) = 0. \tag{5.17}$$

The scattering problem (5.8)–(5.12) is a special case of (5.13)–(5.17). In particular, the scattered field u^s and the interior field v satisfy (5.13)–(5.17) with $u = u^s$, $f = u^i|_{\partial D}$, and $h := \left. \dfrac{\partial u^i}{\partial \nu} \right|_{\partial D}$, where the incident wave u^i is such that

$$\Delta u^i + k^2 u^i = 0 \qquad \text{in } \mathbb{R}^2.$$

Note that the boundary conditions (5.15) and (5.16) are assumed in the sense of the trace operator, as discussed previously, and u and v satisfy (5.13) and (5.14), respectively, in the weak sense. The reader already encountered in Sect. 3.3 the concept of a weak solution in the context of the impedance boundary value problem for the Helmholtz equation. In the next section we provide a more systematic discussion of weak solutions and variational methods for finding weak solutions of boundary value problems.

5.3 Variational Methods

We will start this section with an important result from functional analysis, namely, the *Lax–Milgram lemma*. Let X be a Hilbert space with norm $\| \cdot \|$ and inner product $(\cdot, \cdot)$.

Definition 5.11. A mapping $a(\cdot, \cdot) : X \times X \to \mathbb{C}$ is called a *sesquilinear form* if

$$a(\lambda_1 u_1 + \lambda_2 u_2, v) = \lambda_1 a(u_1, v) + \lambda_2 a(u_2, v)$$
$$\text{for all } \lambda_1, \lambda_2 \in \mathbb{C}, \ u_1, u_2, v \in X,$$
$$a(u, \mu_1 v_1 + \mu_2 v_2) = \bar{\mu}_1 a(u, v_1) + \bar{\mu}_2 a(u, v_2)$$
$$\text{for all } \mu_1, \mu_2 \in \mathbb{C}, \ u, v_1, v_2 \in X,$$

with the bar denoting the complex conjugation.

Definition 5.12. A mapping $F : X \to \mathbb{C}$ is called a *conjugate linear functional* if

$$F(\mu_1 v_1 + \mu_2 v_2) = \bar{\mu}_1 F(v_1) + \bar{\mu}_2 F(v_2) \text{ for all } \mu_1, \mu_2 \in \mathbb{C}, \ v_1, v_2 \in X.$$

As will be seen later, we will be interested in solving the following problem: *given a conjugate linear functional* $F : X \to \mathbb{C}$ *and a sesquilinear form* $a(\cdot, \cdot)$ *on* $X \times X$, *find* $u \in X$ *such that*

$$a(u, v) = F(v) \qquad \text{for all} \ \ v \in X. \tag{5.18}$$

The solution to this problem is provided by the following lemma.

Theorem 5.13 (Lax–Milgram Lemma). *Assume that* $a : X \times X \to \mathbb{C}$ *is a sesquilinear form (not necessarily symmetric) for which there exist constants* $\alpha, \beta > 0$ *such that*

$$|a(u, v)| \leq \alpha \|u\| \, \|v\| \qquad \text{for all} \ \ u \in X, \ v \in X \tag{5.19}$$

and

$$|a(u, u)| \geq \beta \|u\|^2 \qquad \text{for all} \ \ u \in X. \tag{5.20}$$

Then for every bounded conjugate linear functional $F : X \to \mathbb{C}$ *there exists a unique element* $u \in X$ *such that*

$$a(u, v) = F(v) \qquad \text{for all} \ \ v \in X. \tag{5.21}$$

Furthermore, $\|u\| \leq C\|F\|$, *where* $C > 0$ *is a constant independent of* F.

Proof. For each fixed element $u \in X$ the mapping $v \to a(u, v)$ is a bounded conjugate linear functional on X, and hence the Riesz representation theorem asserts the existence of a unique element $w \in X$ satisfying

$$a(u, v) = (w, v) \qquad \text{for all} \ \ v \in X.$$

Thus we can define an operator $A : X \to X$ mapping u to w such that

$$a(u, v) = (Au, v) \qquad \text{for all} \ \ u, v \in X.$$

1. We first claim that $A : X \to X$ is a bounded linear operator. Indeed, if $\lambda_1, \lambda_2 \in \mathbb{C}$ and $u_1, u_2 \in X$, then we see, using the properties of the inner product in a Hilbert space, that for each $v \in X$ we have

$$
\begin{aligned}
(A(\lambda_1 u_1 + \lambda_2 u_2), v) &= a((\lambda_1 u_1 + \lambda_2 u_2), v) \\
&= \lambda_1 a(u_1, v) + \lambda_2 a(u_2, v) \\
&= \lambda_1 (A u_1, v) + \lambda_2 (A u_2, v) \\
&= (\lambda_1 A u_1 + \lambda_2 A u_2, v) \, .
\end{aligned}
$$

Since this holds for arbitrary $u_1, u_2, v \in X$, and $\lambda_1, \lambda_2 \in \mathbb{C}$, we have established linearity. Furthermore,

$$
\|Au\|^2 = (Au, Au) = a(u, Au) \le \alpha \|u\| \, \|Au\|.
$$

Consequently, $\|Au\| \le \alpha \|u\|$ for all $u \in X$, and so A is bounded.

2. Next we show that A is one-to-one and the range of A is equal to X. To prove this, we compute

$$
\beta \|u\|^2 \le |a(u, u)| = |(Au, u)| \le \|Au\| \, \|u\|.
$$

Hence, $\beta \|u\| \le \|Au\|$. This inequality implies that A is one-to-one and the range of A is closed in X. Now let $w \in A(X)^\perp$, and observe that $\beta \|w\|^2 \le a(w, w) = (Aw, w) = 0$, which implies that $w = 0$. Since $A(X)$ is closed, we can now conclude that $A(X) = X$.

3. Next, once more from the Riesz representation theorem, there exists a unique $\tilde{w} \in X$ such that

$$
F(v) = (\tilde{w}, v) \qquad \text{for all} \quad v \in X
$$

and $\|\tilde{w}\| = \|F\|$. We then use part 2 of this proof to find a $u \in X$ satisfying $Au = \tilde{w}$. Then

$$
a(u, v) = (Au, v) = (\tilde{w}, v) = F(v) \qquad \text{for all} \quad v \in X,
$$

which proves the solvability of (5.21). Furthermore, we have that

$$
\|u\| \le \frac{1}{\beta} \|Au\| = \frac{1}{\beta} \|\tilde{w}\| = \frac{1}{\beta} \|F\|.
$$

4. Finally, we show that there is at most one element $u \in X$ satisfying (5.21). If there exist $u \in X$ and $\tilde{u} \in X$ such that

$$
a(u, v) = F(v) \quad \text{and} \quad a(\tilde{u}, v) = F(v) \qquad \text{for all} \quad v \in X,
$$

then

$$
a(u - \tilde{u}, v) = 0 \qquad \text{for all} \quad v \in X.
$$

Hence, setting $v = u - \tilde{u}$ we obtain

$$
\beta \|u - \tilde{u}\|^2 \le a(u - \tilde{u}, u - \tilde{u}) = 0,
$$

whence $u = \tilde{u}$.

□

Remark 5.14. If a sesquilinear form $a(\cdot, \cdot)$ satisfies (5.19), then it is said that $a(\cdot, \cdot)$ is *continuous*. A sesquilinear form $a(\cdot, \cdot)$ satisfying (5.20) is called *strictly coercive*.

Example 5.15. As an example of an application of the Lax–Milgram lemma we consider the existence of a unique weak solution to the Dirichlet problem for the Poisson equation: given $f \in H^{\frac{1}{2}}(\partial D)$ and $\rho \in L^2(D)$, find $u \in H^1(D)$ such that

$$
\begin{cases}
\Delta u = -\rho & \text{in } D, \\
u = f & \text{on } \partial D.
\end{cases}
\tag{5.22}
$$

To motivate the definition of a $H^1(D)$ weak solution to the preceding Dirichlet problem, let us consider first $u \in C^2(D) \cap C^1(\bar{D})$ satisfying $\Delta u = -\rho$. Multiplying $\Delta u = -\rho$ by $\bar{v} \in C_0^\infty(D)$ and using Green's first identity we obtain

$$
\int_D \nabla u \cdot \nabla \bar{v} \, dx = \int_D \rho \bar{v} \, dx,
\tag{5.23}
$$

which makes sense for $u \in H^1(D)$ and $v \in H_0^1(D)$ as well. Note that the boundary terms disappear when we apply Green's identity due to the fact that $v = 0$ on ∂D. Now we will use (5.23) to define a weak solution. To this end, we set $X = H_0(D)$ and define

$$
a(w, v) = (\nabla w, \nabla v)_{L^2(D)}, \qquad w, v \in X.
$$

In particular, it is clear that

$$
|a(w, v)| \le \|\nabla w\|_{L^2(D)} \|\nabla v\|_{L^2(D)} \le \|w\|_{H^1(D)} \|v\|_{H^1(D)}.
$$

Furthermore, from Poincaré's inequality there exists a constant $C > 0$ depending only on D such that

$$
a(w, w) = \|\nabla w\|_{L^2(D)}^2 \ge C \|w\|_{H^1(D)}^2,
$$

whence $a(\cdot, \cdot)$ satisfies the assumptions of the Lax–Milgram lemma.

Now let $u_0 \in H^1(D)$ be such that $u_0 = f$ on ∂D and $\|u_0\|_{H^1(D)} \le C\|f\|_{H^{\frac{1}{2}}(\partial D)}$. If $u = f$ on ∂D, then $u - u_0 \in H_0^1(D)$. Next we examine the following problem.

Find $u \in H^1(D)$ such that

$$
\begin{cases}
u - u_0 \in H_0^1(D), \\
a(u - u_0, v) = -a(u_0, v) + (\rho, v)_{L^2(D)} & \text{for all } v \in H_0^1(D).
\end{cases}
\tag{5.24}
$$

A solution to (5.24) is called a *weak solution* of the Dirichlet problem (5.22), and (5.24) is called the *variational form* of (5.22).

Since $a(\cdot, \cdot)$ is continuous, the mapping $F : v \to -a(u_0, v) + (\rho, v)_{L^2(D)}$ is a bounded conjugate linear functional on $H_0^1(D)$. Therefore, from the Lax–Milgram lemma, (5.24) has a unique solution $u \in H^1(D)$ that satisfies

$$\|u\|_{H^1(D)} \leq C(\|u_0\|_{H^1(D)} + \|\rho\|_{L^2(D)}) \leq \tilde{C}(\|f\|_{H^{\frac{1}{2}}(\partial D)} + \|\rho\|_{L^2(D)}),$$

where the constant $\tilde{C} > 0$ is independent of f and ρ.

Obviously, any $C^2(D) \cap C^1(\bar{D})$ solution to the Dirichlet problem is a weak solution. Conversely, if the weak solution u is smooth enough (which depends on the smoothness of ∂D, f, and ρ – see [127]), then the weak solution satisfies (5.22) pointwise. Indeed, taking a function $v \in C_0^\infty(D)$ in (5.24) we see that

$$\int_D (\Delta u + \rho)\, v\, dx = 0 \qquad \text{for all } v \in C_0^\infty(D),$$

and hence $\Delta u = -\rho$ almost everywhere in D. Furthermore, $u - u_0 \in H_0^1(D)$ if and only if $u = u_0$ on ∂D, whence $u = f$ on ∂D.

We now return to the abstract variational problem (5.18) and consider it in the following form: find $u \in X$ such that

$$a(u, v) + b(u, v) = F(v) \qquad \text{for all} \quad v \in X, \tag{5.25}$$

where X is a Hilbert space, $a, b : X \times X \to \mathbb{C}$ are two continuous sesquilinear forms, and F is a bounded conjugate linear functional on X. In addition:

1. Assume that the continuous sesquilinear form $a(\cdot, \cdot)$ is strictly coercive, i.e., $a_1(u, u) \geq \alpha \|u\|^2$ for some positive constant α. From the Lax–Milgram lemma we then have that there exists a bijective bounded linear operator $A : X \to X$ with bounded inverse satisfying

$$a(u, v) = (Au, v) \qquad \text{for all} \quad v \in X.$$

2. Let us denote by B the bounded linear operator from X to X defined by

$$b(u, v) = (Bu, v) \qquad \text{for all} \quad v \in X.$$

The existence and the continuity of B are guaranteed by the Riesz representation theorem (see also the first part of the proof of the Lax–Milgram lemma). We further assume that the operator B is compact.

3. Finally, let $w \in X$ be such that

$$F(v) = (w, v) \qquad \text{for all} \quad v \in X,$$

which is uniquely provided by the Riesz representation theorem.

Under assumptions 1–3, (5.25) equivalently reads as follows:

$$\text{Find } u \in X \text{ such that } Au + Bu = w. \tag{5.26}$$

Theorem 5.16. *Let X and Y be two Hilbert spaces, and let $A : X \to Y$ be a bijective bounded linear operator with bounded inverse $A^{-1} : Y \to X$, and $B : X \to Y$ a compact linear operator. Then $A + B$ is injective if and only if it is surjective. If $A + B$ is injective (and hence bijective), then the inverse $(A + B)^{-1} : Y \to X$ is bounded.*

Proof. Since A^{-1} exists, we have that $A + B = A(I - (-A^{-1})B)$. Furthermore, since A is a bijection, $(I - (-A^{-1})B)$ is injective and surjective if and only if $A + B$ is injective and surjective. Next we observe that $(-A^{-1})B$ is a compact operator since it is the product of a compact operator and a bounded operator. The result of the theorem now follows from Theorem 1.21 and the fact that $(A + B)^{-1} = (I - (-A^{-1})B)^{-1}A^{-1}$. $\qquad\square$

Example 5.17. Consider now the Dirichlet problem for the Helmholtz equation in a bounded domain D: Given $f \in H^{\frac{1}{2}}(\partial D)$, find $u \in H^1(D)$ such that

$$\begin{cases} \Delta u + k^2 u = 0 & \text{in } D, \\ u = f & \text{on } \partial D, \end{cases} \tag{5.27}$$

where k is real. Following Example 5.15, we can write this problem in the following variational form: *find $u \in H^1(D)$ such that*

$$\begin{cases} u - u_0 \in H_0^1(D), \\ a(u - u_0, v) = -a(u_0, v) & \text{for all } v \in H_0^1(D), \end{cases} \tag{5.28}$$

where u_0 is a function in $H^1(D)$ such that $u_0 = f$ on ∂D and $\|u_0\|_{H^1(D)} \leq C\|f\|_{H^{\frac{1}{2}}(\partial D)}$, and the sesquilinear form $a(\cdot, \cdot)$ is defined by

$$a(w, v) := \int_D \left(\nabla w \cdot \nabla \bar{v} - k^2 w \bar{v} \right) dx, \qquad w, v \in H_0^1(D).$$

Obviously, $a(\cdot, \cdot)$ is continuous but not strictly coercive. Defining

$$a_1(w, v) := \int_D \nabla w \cdot \nabla \bar{v} \, dx, \qquad w, v \in H_0^1(D)$$

and

$$a_2(w, v) := -k^2 \int_D w \bar{v} \, dx, \qquad w, v \in H_0^1(D)$$

we have that
$$a(w, v) = a_1(w, v) + a_2(w, v),$$

where now $a_1(\cdot, \cdot)$ is strictly coercive in $H_0^1(D) \times H_0^1(D)$ (Example 5.15). Let $A : H_0^1(D) \to H_0^1(D)$ and $B : H_0^1(D) \to H_0^1(D)$ be bounded linear operators defined by $(Au, v) = a_1(u, v)$ and

$$(Bu, v) = \int_D u\bar{v}\, dx \qquad \text{for all} \quad v \in H_0^1(D),$$

respectively. In particular, A is bounded and has a bounded inverse. We claim that $B : H_0^1(D) \to H_0^1(D)$ is compact. To see this, we first note that

$$\|Bu\|_{H^1(D)}^2 = (Bu, Bu) = \int_D u\overline{Bu}\, dx \leq \|u\|_{L^2(D)} \|Bu\|_{L^2(D)}$$
$$\leq \|u\|_{L^2(D)} \|Bu\|_{H^1(D)},$$

and hence $\|Bu\|_{H^1(D)} \leq \|u\|_{L^2(D)}$. Now let $\{u_j\} \subset H_0^1(D)$ be such that $\|u_j\|_{H^1(D)} \leq C$ for some positive constant C independent of j. Then, since by Rellich's theorem $H^1(D)$, and hence $H_0^1(D)$, is compactly embedded in $L^2(D)$, we have that there exists a subsequence, still denoted by $\{u_j\}$, such that $\{u_j\}$ is strongly convergent in $L^2(D)$, i.e., $\{u_j\}$ is a Cauchy sequence in $L^2(D)$. Since $\|Bu\|_{H^1(D)}$ is bounded by $\|u\|_{L^2(D)}$, we have that $\{Bu_j\}$ is a Cauchy sequence in $H_0^1(D)$, and hence $\{Bu_j\}$ is strongly convergent. This now implies that B is compact, as claimed.

We can now apply Theorem 5.16 to (5.28). In particular, the injectivity of $A - k^2 B$ implies the existence of a unique solution to (5.28). The injectivity of $A - k^2 B$ is equivalent to the fact that the only function $u \in H_0^1(D)$ that satisfies
$$a(u, v) = 0 \qquad \text{for all} \quad v \in H_0^1(D)$$

is $u \equiv 0$. This is the uniqueness question for a weak solution to the Dirichlet boundary value problem for the Helmholtz equation. The values of k^2 for which there exists a nonzero function $u \in H_0^1(D)$ satisfying

$$\Delta u + k^2 u = 0 \qquad \text{in} \ \ D$$

(in the weak sense) are called the *Dirichlet eigenvalues* of $-\Delta$ and the corresponding nonzero solutions are called the *eigensolutions* for $-\Delta$. Note that the zero boundary condition is incorporated in the space $H_0^1(D)$.

Summarizing the preceding analysis, we have shown that if k^2 is not a Dirichlet eigenvalue for $-\Delta$, then (5.27) has a unique solution in $H^1(D)$.

Theorem 5.18. *There exists an orthonormal basis u_j for $H_0^1(D)$ consisting of eigensolutions for $-\Delta$. The corresponding eigenvalues k^2 are all positive and accumulate only at $+\infty$.*

Proof. In Example 5.17 we showed that $u \in H_0^1(D)$ satisfies

$$\Delta u + k^2 u = 0 \qquad \text{in } D$$

if and only if u is a solution to the operator equation $Au - k^2 Bu = 0$, where $A : H_0^1(D) \to H_0^1(D)$ and $B : H_0^1(D) \to H_0^1(D)$ are the bijective operator and compact operator, respectively, constructed in Example 5.17. Since A is a positive definite operator, the equation $Au - k^2 Bu = 0$ can be written as (see [115] for the existence of the operator $A^{\frac{1}{2}}$)

$$\left(\frac{1}{k^2} I - A^{-\frac{1}{2}} B A^{-\frac{1}{2}} \right) u = 0 \qquad u \in H_0^1(D).$$

It is easily verified that A (and hence $A^{-\frac{1}{2}}$) is self-adjoint. Since B is self-adjoint, we can conclude that $A^{-\frac{1}{2}} B A^{-\frac{1}{2}}$ is self-adjoint. Now noting that $A^{-\frac{1}{2}} B A^{-\frac{1}{2}} : H_0^1(D) \to H_0^1(D)$ is compact since it is a product of a compact operator and bounded operators, the result follows from the Hilbert–Schmidt theorem.   □

Remark 5.19. The results of Examples 5.15 and 5.17 are valid as well if D is not simply connected, i.e., $\mathbb{R}^2 \setminus \bar{D}$ is not connected.

The boundary value problems arising in scattering theory are formulated in unbounded domains. To solve such problems using variational techniques developed in this section, we need to write them as equivalent problems in a bounded domain. In particular, introducing a large open disk Ω_R centered at the origin that contains $\bar{D}$, where D is the support of the scatterer, we first solve the problem in $\Omega_R \setminus \bar{D}$ (or in Ω_R in the case of transmission problems) using variational methods. Having solved this problem, we then want to extend the solution outside Ω_R to a solution to the original problem. The main question here is what boundary condition should we impose on the artificial boundary $\partial \Omega_R$ to enable such an extension. To find the appropriate boundary conditions on $\partial \Omega_R$, we introduce the *Dirichlet-to-Neumann map*. We first formalize the definition of a *radiating solution* to the Helmholtz equation.

Definition 5.20. A solution u to the Helmholtz equation whose domain of definition contains the exterior of some disk is called radiating if it satisfies the Sommerfeld radiation condition

$$\lim_{r \to \infty} \sqrt{r} \left(\frac{\partial u}{\partial r} - iku \right) = 0,$$

where $r = |x|$ and the limit is assumed to hold uniformly in all directions $x/|x|$.

Definition 5.21. The Dirichlet-to-Neumann map T is defined by

$$T : w \to \frac{\partial w}{\partial \nu} \qquad \text{on } \partial \Omega_R,$$

where w is a radiating solution to the Helmholtz equation $\Delta w + k^2 w = 0$, $\partial \Omega_R$ is the boundary of some disk of radius R, and ν is the outward unit normal to $\partial \Omega_R$.

Taking advantage of the fact that Ω_R is a disk, by separating variables as in Sect. 3.2 we can find a solution to the exterior Dirichlet problem outside Ω_R in the form of a series expansion involving Hankel functions. Making use of this expansion we can establish the following important properties of the Dirichlet-to-Neumann map.

Theorem 5.22. *The Dirichlet-to-Neumann map T is a bounded linear operator from $H^{\frac{1}{2}}(\partial \Omega_R)$ to $H^{-\frac{1}{2}}(\partial \Omega_R)$. Furthermore, there exists a bounded operator $T_0 : H^{\frac{1}{2}}(\partial \Omega_R) \to H^{-\frac{1}{2}}(\partial \Omega_R)$ satisfying*

$$- \int_{\partial \Omega_R} T_0 w \, \overline{w} \, ds \geq C \|w\|^2_{H^{\frac{1}{2}}(\partial \Omega_R)} \tag{5.29}$$

for some constant $C > 0$ such that $T - T_0 : H^{\frac{1}{2}}(\partial \Omega_R) \to H^{-\frac{1}{2}}(\partial \Omega_R)$ is compact.

Proof. Let w be a radiating solution to the Helmholtz equation outside Ω_R, and let (r, θ) denote polar coordinates in $\mathbb{R}^2$. Then from Sect. 3.2 we have that

$$w(r, \theta) = \sum_{-\infty}^{\infty} \alpha_n H_n^{(1)}(kr) e^{in\theta}, \quad r \geq R \text{ and } 0 \leq \theta \leq 2\pi,$$

where $H_n^{(1)}(kr)$ are the Hankel functions of the first kind of order n. Hence T maps the Dirichlet data of $w|_{\partial \Omega_R}$ given by

$$w|_{\partial \Omega_R} = \sum_{-\infty}^{\infty} a_n e^{in\theta}$$

with coefficients $a_n := \alpha_n H_n^{(1)}(kR)$ onto the corresponding Neumann data given by

$$Tw = \sum_{-\infty}^{\infty} a_n \gamma_n e^{in\theta},$$

where

$$\gamma_n := \frac{k H_n^{(1)'}(kR)}{H_n^{(1)}(kR)}, \quad n = 0, \pm 1, \dots.$$

The Hankel functions and their derivatives do not have real zeros since otherwise the Wronskian (3.22) would vanish. From this we observe that T is bijective. In view of the asymptotic formulas for the Hankel functions developed in Sect. 3.2 we see that

$$c_1|n| \leq |\gamma_n| \leq c_2|n|, \qquad n = \pm 1, \pm 2, \ldots$$

and some constants $0 < c_1 < c_2$. From this the boundness of $T : H^{\frac{1}{2}}(\partial\Omega_R) \to H^{-\frac{1}{2}}(\partial\Omega_R)$ is obvious since from Theorem 1.33 for $p \in \mathbb{R}$ the norm on $H^p(\partial\Omega_R)$ can be described in terms of the Fourier coefficients

$$\|w\|_{H^p(\partial\Omega_R)}^2 = \sum_{-\infty}^{\infty}(1+n^2)^p|a_n|^2.$$

For the limiting operator $T_0 : H^{\frac{1}{2}}(\partial\Omega_R) \to H^{-\frac{1}{2}}(\partial\Omega_R)$ given by

$$T_0 w = -\sum_{-\infty}^{\infty} \frac{|n|}{R} a_n e^{in\theta}$$

we clearly have

$$-\int_{\Omega_R} T_0 w\, \overline{w}\, ds = \sum_{-\infty}^{\infty} 2\pi|n||a_n|^2,$$

with the integral to be understood as the duality pairing between $H^{\frac{1}{2}}(\partial\Omega_R)$ and $H^{-\frac{1}{2}}(\partial\Omega_R)$. Hence

$$-\int_{\partial\Omega_R} T_0 w\, \overline{w}\, ds \geq C\|w\|_{H^{\frac{1}{2}}(\partial\Omega_R)}^2$$

for some constant $C > 0$. Finally, from the series expansions for the Bessel and Neumann functions (Sect. 3.2) for fixed k we derive

$$\gamma_n = -\frac{|n|}{R}\left\{1 + O\left(\frac{1}{|n|}\right)\right\}, \qquad n \to \pm\infty.$$

This implies that $T - T_0$ is compact from $H^{\frac{1}{2}}(\partial\Omega_R)$ into $H^{-\frac{1}{2}}(\partial\Omega_R)$ since it is bounded from $H^{\frac{1}{2}}(\partial\Omega_R)$ into $H^{\frac{1}{2}}(\partial\Omega_R)$ and the embedding from $H^{\frac{1}{2}}(\partial\Omega_R)$ into $H^{-\frac{1}{2}}(\partial\Omega_R)$ is compact by Rellich's Theorem 1.32. This proves the theorem. $\square$

Example 5.23. We consider the problem of finding a weak solution to the exterior Dirichlet problem for the Helmholtz equation: given $f \in H^{\frac{1}{2}}(\partial D)$, find $u \in H^1_{loc}(\mathbb{R}^2 \setminus \bar{D})$ such that

$$\begin{cases} \Delta u + k^2 u = 0 & \text{in } \mathbb{R}^2 \setminus \bar{D}, \\ u = f & \text{on } \partial D, \\ \lim_{r \to \infty} \sqrt{r}\left(\dfrac{\partial u}{\partial r} - iku\right) = 0. \end{cases} \tag{5.30}$$

Instead of (5.30) we solve an equivalent problem in the bounded domain $\Omega_R \setminus \bar{D}$, that is, we find $u \in H^1(\Omega_R \setminus \bar{D})$ such that

$$\begin{cases} \Delta u + k^2 u = 0 & \text{in } \Omega_R \setminus \bar{D}, \\ u = f & \text{on } \partial D, \\ \dfrac{\partial u}{\partial \nu} = Tu & \text{on } \partial \Omega_R, \end{cases} \tag{5.31}$$

where $f \in H^{\frac{1}{2}}(\partial D)$ is the given boundary data, T is the Dirichlet-to-Neumann map, and Ω_R is a large disk containing $\bar{D}$.

Lemma 5.24. *Problems (5.30) and (5.31) are equivalent.*

Proof. First let $u \in H^1_{loc}(\mathbb{R}^2 \setminus \bar{D})$ be a solution to (5.30). Then the restriction of u to $\Omega_R \setminus \bar{D}$ is in $H^1(\Omega_R \setminus \bar{D})$ and is a solution to (5.31). Conversely, let $u \in H^1(\Omega_R \setminus \bar{D})$ be a solution to (5.31). To define u in all of $\mathbb{R}^2 \setminus \bar{D}$, we construct the radiating solution $\tilde{u}$ of the Helmholtz equation outside Ω_R such that $\tilde{u} = u$ on $\partial \Omega_R$. This solution can be constructed in the form of a series expansion in terms of Hankel functions in the same way as in the proof of Theorem 5.22. Hence we have that $Tu = \dfrac{\partial \tilde{u}}{\partial \nu}$. Using Green's second identity for the radiating solution $\tilde{u}$ and the fundamental solution $\Phi(x, y)$ (which is also a radiating solution) we obtain that

$$\int_{\partial \Omega_R} \left[(Tu)(y)\Phi(x, y) - u(y)\frac{\partial \Phi(x, y)}{\partial \nu} \right] ds_y = 0, \qquad x \in \Omega_R.$$

Consequently, the representation formula (3.41) (Remark 6.29) and the fact that $\dfrac{\partial u}{\partial \nu} = Tu$ imply

$$u(x) = \int_{\partial D} \left[u(y)\frac{\partial \Phi(x, y)}{\partial \nu} - \frac{\partial u}{\partial \nu}\Phi(x, y) \right] ds_y$$

$$- \int_{\partial \Omega_R} \left[u(y)\frac{\partial \Phi(x, y)}{\partial \nu} - \frac{\partial u}{\partial \nu}\Phi(x, y) \right] ds_y$$

$$= \int_{\partial D} \left[u(y)\frac{\partial \Phi(x, y)}{\partial \nu} - \frac{\partial u}{\partial \nu}\Phi(x, y) \right] ds_y.$$

Therefore, u coincides with the radiating solution to the Helmholtz equation in the exterior of $\bar{D}$. Hence a solution of (5.30) can be derived from a solution to (5.31). $\qquad\square$

Next we formulate (5.31) as a variational problem. To this end, we define the Hilbert space

$$X := \{u \in H^1(\Omega_R \setminus \bar{D}) : u = 0 \ \text{ on } \partial D\}$$

and the sesquilinear from $a(\cdot, \cdot)$ by

$$a(u, v) = \int_{\Omega_R \setminus \bar{D}} \left(\nabla u \cdot \nabla \bar{v} - k^2 u \bar{v} \right) dx - \int_{\partial \Omega_R} T u \bar{v} \, ds,$$

which is obtained by multiplying the Helmholtz equation in (5.31) by a test function $v \in X$, integrating by parts, and using the boundary condition $\partial u / \partial \nu = T u$ on $\partial \Omega_R$ and the zero boundary condition on ∂D. Now let $u_0 \in H^1(\Omega_R \setminus \bar{D})$ be such that $u_0 = f$ on ∂D. Then the variational formulation of (5.31) reads: *find* $u \in H^1(\Omega_R \setminus \bar{D})$ *such that*

$$\begin{cases} u - u_0 \in X, \\ \\ a(u - u_0, v) = -a(u_0, v) \qquad \text{for all } v \in X. \end{cases} \tag{5.32}$$

To analyze (5.32) we define

$$a_1(w, v) = \int_{\Omega_R \setminus \bar{D}} \left(\nabla w \cdot \nabla \bar{v} + w \bar{v} \right) dx - \int_{\partial \Omega_R} T_0 w \bar{v} \, ds$$

and

$$a_2(w, v) = -(k^2 + 1) \int_{\Omega_R \setminus \bar{D}} w \bar{v} \, dx - \int_{\partial \Omega_R} (T - T_0) w \bar{v} \, ds,$$

where T_0 is the operator defined in Theorem 5.22, and write the equation in (5.32) as

$$a_1(u - u_0, v) + a_2(u - u_0, v) = F(v), \qquad \text{for all } v \in X,$$

with $F(v) := a(u_0, v)$. Since T is a bounded operator from $H^{\frac{1}{2}}(\partial \Omega_R)$ to $H^{-\frac{1}{2}}(\partial \Omega_R)$, F is a bounded conjugate linear functional on X and both $a_1(\cdot, \cdot)$ and $a_2(\cdot, \cdot)$ are continuous on $X \times X$. In addition, using (5.29), we see that

$$a_1(w, w) \geq C \|w\|^2_{H^1(\Omega_R \setminus \bar{D})}.$$

Note that including a L^2-inner product term in $a_1(\cdot, \cdot)$ is important since the Poincaré inequality no longer holds in X. Furthermore, due to the compact embedding of $H^1(\Omega_R \setminus \bar{D})$ into $L^2(\Omega_R \setminus \bar{D})$ and the fact that $T - T_0 : H^{\frac{1}{2}}(\partial \Omega_R) \to H^{-\frac{1}{2}}(\partial \Omega_R)$ is compact, $a_2(\cdot, \cdot)$ gives rise to a compact operator $B : X \to X$ (Example 5.17). Hence from Theorem 5.16 we conclude that the uniqueness of a solution to (5.31) implies the existence of a solution to (5.31) and, consequently, from Lemma 5.24 the existence of a weak solution to (5.30). To prove the uniqueness of a solution to (5.31) we first observe that according to Lemma 5.24 a solution to the homogeneous problem (5.31) ($f = 0$) can be extended to a solution to the homogeneous problem (5.30). Now let u be a solution to the homogeneous problem (5.30). Then Green's first identity and the boundary condition imply

$$\int_{\partial\Omega_R} \frac{\partial u}{\partial \nu}\overline{u}\,ds = \int_{\partial D} \frac{\partial u}{\partial \nu}\overline{u}\,ds + \int_{\Omega_R\setminus\bar{D}} \left(|\nabla u|^2 - k^2|u|^2\right)\,dx \qquad (5.33)$$

$$= \int_{\Omega_R\setminus\bar{D}} \left(|\nabla u|^2 - k^2|u|^2\right)\,dx, \qquad (5.34)$$

whence

$$\mathrm{Im}\left(\int_{\partial\Omega_R} \frac{\partial u}{\partial \nu}\overline{u}\,ds\right) = 0.$$

From Theorem 3.6 we conclude that $u = 0$ in $\mathbb{R}^2 \setminus \bar{D}$, which proves the uniqueness and, therefore, the existence of a unique weak solution to the exterior Dirichlet problem for the Helmholtz equation. Note that in the preceding proof of uniqueness we have used the fact that off the boundary an $H^1_{loc}(\mathbb{R}^2 \setminus \bar{D})$ solution to the Helmholtz equation is real-analytic. This can be seen from the Green representation formula as in Theorem 3.2, which is also valid for radiating solutions to the Helmholtz equation in $H^1_{loc}(\mathbb{R}^2 \setminus \bar{D})$ (Remark 6.29).

In this section we have developed variational techniques for finding weak solutions to boundary value problems for partial differential equations. As the reader has already seen, in scattering problems the boundary conditions are typically the traces of real-analytic solutions, for example, plane waves. Hence, provided that the boundary of the scattering object is smooth, one would expect that the scattered field would not, in fact, be smooth. It can be shown that if the boundary, the boundary conditions, and the coefficients of the equations are smooth enough, then a weak solution is in fact C^2 inside the domain and C^1 up to the boundary. This general statement falls in the class of so-called regularity results for the solutions of boundary value problems for elliptic partial differential equations. Precise formulation of such results can be found in any classic book of partial differential equations (cf. [72] and [127]).

5.4 Solution of Direct Scattering Problem

We now turn our attention to the main goal of this chapter, the solution to the scattering problem (5.13)–(5.17). Following Hähner [81], we shall use the variational techniques developed in Sect. 5.3 to find a solution to this problem. To arrive at a variational formulation of (5.13)–(5.17), we introduce a large open disk Ω_R centered at the origin containing $\bar{D}$ and consider the following problem: given $f \in H^{\frac{1}{2}}(\partial D)$ and $h \in H^{-\frac{1}{2}}(\partial D)$, find $u \in H^1(\Omega_R \setminus \bar{D})$ and $v \in H^1(D)$ such that

$$\nabla \cdot A\nabla v + k^2 n\, v = 0 \quad \text{in} \quad D, \tag{5.35}$$

$$\Delta u + k^2\, u = 0 \quad \text{in} \quad \Omega_R \setminus \bar{D}, \tag{5.36}$$

$$v - u = f \quad \text{on} \quad \partial D, \tag{5.37}$$

$$\frac{\partial v}{\partial \nu_A} - \frac{\partial u}{\partial \nu} = h \quad \text{on} \quad \partial D, \tag{5.38}$$

$$\frac{\partial u}{\partial \nu} = Tu \quad \text{on} \quad \partial \Omega_R, \tag{5.39}$$

where T is the Dirichlet-to-Neumann operator defined in Definition 5.21.

We note that exactly in the same way as in the proof of Lemma 5.24 one can show that a solution u, v to (5.35)–(5.39) can be extended to a solution to the scattering problem (5.13)–(5.17) and, conversely, a solution u, v to the scattering problem (5.13)–(5.17) is such that v and u restricted to $\Omega_R \setminus \bar{D}$ solve (5.35)–(5.39).

Next let $u_f \in H^1(\Omega_R \setminus \bar{D})$ be the unique solution to the following Dirichlet boundary value problem:

$$\Delta u_f + k^2 u_f = 0 \quad \text{in } \Omega_R \setminus \bar{D}, \qquad u_f = f \quad \text{on } \partial D, \qquad u_f = 0 \quad \text{on } \partial \Omega_R.$$

The existence of a unique solution to this problem is shown in Example 5.17 (see also Remark 5.19). Note that we can always choose Ω_R such that k^2 is not a Dirichlet eigenvalue for $-\Delta$ in $\Omega_R \setminus \bar{D}$. An equivalent variational formulation of (5.35)–(5.39) is as follows: find $w \in H^1(\Omega_R)$ such that

$$\int_D \left(\nabla \bar{\phi} \cdot A\nabla w - k^2 n\bar{\phi} w \right) dx + \int_{\Omega_R \setminus \bar{D}} \left(\nabla \bar{\phi} \cdot \nabla w - k^2 \bar{\phi} w \right) dx \tag{5.40}$$

$$- \int_{\partial \Omega_R} \bar{\phi} Tw\, ds = \int_{\partial D} \bar{\phi} h\, ds - \int_{\partial \Omega_R} \bar{\phi} Tu_f\, ds + \int_{\Omega_R \setminus \bar{D}} \left(\nabla \bar{\phi} \cdot \nabla u_f - k^2 \bar{\phi} u_f \right) dx$$

for all $\phi \in H^1(\Omega_R)$. With the help of Green's first identity (Corollary 5.8 and Remark 6.29) it is easy to see that $v := w|_D$ and $u := w|_{\Omega_R \setminus \bar{D}} - u_f$ satisfy (5.35)–(5.39). Conversely, multiplying the equations in (5.35)–(5.39) by a test function and using the transmission conditions one can show that $w = v$ in D and $w = u + u_f$ in $\Omega_R \setminus \bar{D}$ is such that $w \in H^1(\Omega_R)$ and satisfies (6.68), where v, u solve (5.35)–(5.39).

Next we define the following continuous sesquilinear forms on $H^1(\Omega_R) \times H^1(\Omega_R)$:

$$a_1(\psi, \phi) := \int_D \left(\nabla \bar{\phi} \cdot A\nabla \psi + \bar{\phi}\, \psi \right) dx + \int_{\Omega_R \setminus \bar{D}} \left(\nabla \bar{\phi} \cdot \nabla \psi + \bar{\phi}\, \psi \right) dx$$

$$- \int_{\partial \Omega_R} \bar{\phi} T_0 \psi\, ds \qquad \phi, \psi \in H^1(\Omega_R)$$

and

$$a_2(\psi, \phi) := -\int_D (nk^2 + 1)\overline{\phi}\psi \, dx - \int_{\Omega_R \setminus \bar{D}} (k^2 + 1)\overline{\phi}\,\psi \, dx$$

$$-\int_{\partial\Omega_R} \overline{\phi}\,(T - T_0)\psi \, ds \qquad \phi, \psi \in H^1(\Omega_R),$$

where the operator T_0 is the operator defined in Theorem 5.22. Furthermore, we define the bounded conjugate linear functional F on $H^1(\Omega_R)$ by

$$F(\phi) := \int_{\partial D} \overline{\phi}h \, ds - \int_{\partial\Omega_R} \overline{\phi}\,Tu_f \, ds + \int_{\Omega_R \setminus \bar{D}} \left(\nabla\overline{\phi} \cdot \nabla u_f - k^2\overline{\phi}u_f\right) dx.$$

Then (6.68) can be written as the problem of finding $w \in H^1(\Omega_R)$ such that

$$a_1(w, \phi) + a_2(w, \phi) = F(\phi) \qquad \text{for all } \phi \in H^1(\Omega_R).$$

From the assumption $\bar{\xi}\cdot\text{Re}(A)\,\xi \geq \gamma|\xi|^2$ for all $\xi \in \mathbb{C}^3$ and $x \in \bar{D}$ and (5.29) we can conclude that the sesquilinear form $a_1(\cdot, \cdot)$ is strictly coercive. Hence, as a consequence of the Lax–Milgram lemma, the operator $A : H^1(\Omega_R) \to H^1(\Omega_R)$ defined by $a_1(w, \phi) = (Aw, \phi)_{H^1(\Omega_R)}$ is invertible with bounded inverse. Furthermore, due to the compact embedding of $H^1(\Omega_R)$ into $L^2(\Omega_R)$ and the fact that $T - T_0 : H^{\frac{1}{2}}(\partial\Omega_R) \to H^{-\frac{1}{2}}(\partial\Omega_R)$ is compact (Theorem 5.22), we can show exactly in the same way as in Example 5.17 that the operator $B : H^1(\Omega_R) \to H^1(\Omega_R)$ defined by $a_2(w, \phi) = (Bw, \phi)_{H^1(\Omega_R)}$ is compact. Finally, by Theorem 5.16, the uniqueness of a solution to (5.35)–(5.39) implies that a solution exists.

Lemma 5.25. *The problems (5.35)–(5.39) and (5.13)–(5.17) have at most one solution.*

Proof. According to our previous remarks, a solution to the homogeneous problem (5.35)–(5.39) ($f = h = 0$) can be extended to a solution $v \in H^1(D)$ and $u \in H^1_{loc}(\mathbb{R}^2 \setminus \bar{D})$ to the homogeneous problem (5.13)–(5.17). Therefore, it suffices to prove uniqueness for (5.13)–(5.17). Green's first identity and the transmission conditions imply that

$$\int_{\partial\Omega_R} \bar{u}\frac{\partial u}{\partial\nu} \, ds = \int_{\partial D} \bar{u}\frac{\partial u}{\partial\nu} \, ds + \int_{\Omega_R \setminus \bar{D}} \left(|\nabla u|^2 - k^2|u|^2\right)^2 dx$$

$$= \int_D \left(\nabla\bar{v} \cdot A\nabla v - k^2 n|v|^2\right)^2 dx + \int_{\Omega_R \setminus \bar{D}} \left(|\nabla u|^2 - k^2|u|^2\right)^2 dx.$$

Now since $\bar{\xi} \cdot \text{Im}(A)\,\xi \leq 0$ for all $\xi \in \mathbb{C}^2$ and $\text{Im}(n) > 0$ for $x \in \bar{D}$, we conclude that

$$\text{Im} \left(\int_{\partial \Omega_R} \overline{u} \frac{\partial u}{\partial \nu} \, ds \right) \leq 0,$$

which from Theorem 3.6 implies that $u = 0$ in $\mathbb{R}^2 \setminus \bar{D}$. From the transmission conditions we can now conclude that $v = 0$ and $\partial v / \partial \nu_A = 0$ on ∂D.

To conclude that $v = 0$ in D, we employ a unique continuation principle. To this end, we extend $\text{Re}(A)$ to a real, symmetric, positive definite, and continuously differentiable matrix-valued function in $\overline{\Omega}_R$ and $\text{Im}(A)$ to a real, symmetric, continuously differentiable, matrix-valued function that is compactly supported in Ω_R. We also choose a continuously differentiable extension of n into $\overline{\Omega}_R$ and define $v = 0$ in $\Omega_R \setminus \bar{D}$. Since $v = 0$ and $\partial v / \partial \nu_A = 0$ on ∂D, then $v \in H^1(\Omega_R)$ and satisfies $\nabla \cdot A\nabla v + k^2 n v = 0$ in Ω_R. Then, by the regularity result in the interior of Ω_R (Theorem 5.27), v is smooth enough to apply the unique continuation principle (Theorem 17.2.6 in [89]). In particular, since $v = 0$ in $\Omega_R \setminus \bar{D}$, then $v = 0$ in Ω_R. This proves the uniqueness. $\square$

Summarizing the preceding analysis, we have proved the following theorem on the existence, uniqueness, and continuous dependence on the data of a solution to the direct scattering problem for an orthotropic medium in $\mathbb{R}^2$.

Theorem 5.26. *Assume that D, A, and n satisfy the assumptions in Sect. 5.1, and let $f \in H^{\frac{1}{2}}(\partial D)$ and $h \in H^{-\frac{1}{2}}(\partial D)$ be given. Then the transmission problem (5.13)–(5.17) has a unique solution $v \in H^1(D)$ and $u \in H^1_{loc}(\mathbb{R}^2 \setminus \bar{D})$, which satisfy*

$$\|v\|_{H^1(D)} + \|u\|_{H^1(\Omega_R \setminus \bar{D})} \leq C \left(\|f\|_{H^{\frac{1}{2}}(\partial D)} + \|h\|_{H^{-\frac{1}{2}}(\partial D)} \right), \qquad (5.41)$$

with $C > 0$ a positive constant independent of f and h.

Note that the a priori estimate (5.41) is obtained using the fact that by a duality argument $\|F\|$ is bounded by $\|h\|_{H^{-\frac{1}{2}}(\partial D)}$ and $\|u_f\|_{H^1(\Omega_R \setminus \bar{D})}$, which in turn is bounded by $\|f\|_{H^{\frac{1}{2}}(\partial D)}$ (Example 5.17).

We end this section by stating two regularity results from the general theory of partial differential equations formulated for our transmission problem. The proofs of these results are rather technical and beyond the scope of this book.

Let D_1 and D_2 be bounded, open subsets of $\mathbb{R}^2$ such that $\bar{D}_1 \subset D_2$, and assume that A is a matrix-valued function with continuously differentiable entries $a_{jk} \in C^1(\bar{D}_2)$ and $n \in C^1(\bar{D}_2)$. Furthermore, suppose that A is symmetric and satisfies $\bar{\xi} \cdot \text{Re}(A)\xi \geq \gamma |\xi|^2$ for all $\xi \in \mathbb{C}^3$ and $x \in \bar{D}_2$ for some constant $\gamma > 0$.

Theorem 5.27. *If $u \in H^1(D_2)$ and $q \in L^2(D_2)$ satisfy*

$$\nabla \cdot A\nabla u + k^2 n u = q,$$

then $u \in H^2(D_1)$ and

$$\|u\|_{H^2(D_1)} \leq C \left(\|u\|_{H^1(D_2)} + \|q\|_{L^2(D_2)}\right),$$

where $C > 0$ depends only on γ, D_1 and D_2.

For a proof of this theorem in a more general formulation see Theorem 4.16 in [127] or Theorem 15.1 in [70]. Note also that a more general interior regularity theorem shows that if the entries of A and n are smoother than C^1 and q is smoother than L^2, then one can improve the regularity of u, and this eventually leads to a C^2 solution in the interior of D_2.

For later use, in the next theorem we state a local boundary regularity result for the solution to the transmission problem (5.13)–(5.17). By $\Omega_\epsilon(z)$ we denote an open ball centered at $z \in \mathbb{R}^2$ of radius ϵ.

Theorem 5.28. *Assume $z \in \partial D$, and let $u^i \in H^1(D)$ such that $\Delta u^i \in L^2(D)$. Define $f := u^i$ and $h := \partial u^i / \partial \nu$ on ∂D.*

1. *If for some $\epsilon > 0$ the incident wave u^i is also defined in $\Omega_{2\epsilon}(z)$ and the restriction of u^i to $\Omega_{2\epsilon}(z)$ is in $H^2(\Omega_{2\epsilon}(z))$, then the solution u to (5.13)–(5.17) satisfies $u \in H^2((\mathbb{R}^2 \setminus \overline{D}) \cap \Omega_\epsilon(z))$ and there is a positive constant C such that*

$$\|u\|_{H^2((\mathbb{R}^2\setminus\overline{D})\cap\Omega_\epsilon(z))} \leq C \left(\|u^i\|_{H^2(\Omega_{2\epsilon}(z))} + \|u^i\|_{H^1(D)}\right).$$

2. *If for some $\epsilon > 0$ the incident wave u^i is also defined in $\Omega_R \setminus \Omega_\epsilon(z)$ and the restriction of u^i to $\Omega_R \setminus \Omega_\epsilon(z)$ is in $H^2(\Omega_R \setminus \Omega_\epsilon(z))$, then the solution u to (5.13)–(5.17) satisfies $u \in H^2(\mathbb{R}^2 \setminus (\overline{D} \cup \Omega_{2\epsilon}(z)))$ and there is a positive constant C such that*

$$\|u\|_{H^2(\mathbb{R}^2\setminus(\overline{D}\cup\Omega_{2\epsilon}(z)))} \leq C \left(\|u^i\|_{H^2(\Omega_R\setminus\Omega_\epsilon(z))} + \|u^i\|_{H^1(D)}\right).$$

This result is proved in Theorem 2 in [81]. The proof employs the interior regularity result stated in Theorem 5.27 and techniques from Theorem 8.8 in [72].

6

Inverse Scattering Problems
for Orthotropic Media

In this chapter we extend the results of Chap. 4 to the case of the inverse scattering problem for an inhomogeneous orthotropic medium. The inverse problem we shall consider in this chapter is to determine the *support* of the orthotropic inhomogeneity given the far-field pattern of the scattered field for many incident directions.

The investigation of the inverse problem is based on the analysis of a nonstandard boundary value problem called the *interior transmission problem*. This problem plays the same role for the inhomogeneous medium problem as the interior impedance problem plays in the solution of the inverse problem for an imperfect conductor, studied in Chap. 4. Having discussed the well-posedness of the interior transmission problem and the existence and countability of transmission eigenvalues, we proceed with a uniqueness result for the inverse problem. We will present here a proof due to Hähner [81] that is based on the use of a regularity result for the solution to the interior transmission problem. We then derive the linear sampling method for finding an approximation to the support of the inhomogeneity. Although the analysis of the justification of the linear sampling method refers to the scattering problem for an orthotropic medium, the implementation of the method does not rely on any a priori knowledge of the physical properties of the scattering object. In particular, we show that the far-field equation we used in Chap. 4 to determine the shape of an imperfect conductor can also be used in the present case where the corresponding far-field pattern is used for the kernel of this equation. Finally, since transmission eigenvalues carry qualitative information about the material properties of the inhomogeneous scattering object (cf. Sect. 6.2), we conclude this chapter by showing how transmission eigenvalues can be determined from the (noisy) far-field data.

F. Cakoni and D. Colton, *A Qualitative Approach to Inverse Scattering Theory*,   111
Applied Mathematical Sciences 188, DOI 10.1007/978-1-4614-8827-9_6,
© Springer Science+Business Media New York 2014

6.1 Formulation of Inverse Problem

Let D be the support and A and n the constitutive parameters of a bounded, orthotropic, inhomogeneous medium in $\mathbb{R}^2$, where D, A, and n satisfy the assumptions given in Sect. 5.1. The scattering of a time-harmonic incident plane wave $u^i := e^{ikx \cdot d}$ by the inhomogeneity D is described by the transmission problem (5.13)–(5.17) with $f := e^{ikx \cdot d}$ and $h := \partial e^{ikx \cdot d}/\partial \nu$, which we recall here for the reader's convenience:

$$\nabla \cdot A\nabla v + k^2 n\, v = 0 \qquad \text{in} \quad D, \tag{6.1}$$

$$\Delta u^s + k^2\, u^s = 0 \qquad \text{in} \quad \mathbb{R}^2 \setminus \bar{D}, \tag{6.2}$$

$$v - u^s = e^{ikx \cdot d} \qquad \text{on} \quad \partial D, \tag{6.3}$$

$$\frac{\partial v}{\partial \nu_A} - \frac{\partial u^s}{\partial \nu} = \frac{\partial e^{ikx \cdot d}}{\partial \nu} \qquad \text{on} \quad \partial D, \tag{6.4}$$

$$\lim_{r \to \infty} \sqrt{r}\left(\frac{\partial u^s}{\partial r} - iku^s \right) = 0, \tag{6.5}$$

where $k > 0$ is the (fixed) wave number, $d := (\cos \phi, \sin \phi)$ is the incident direction, $x = (x_1, x_2) \in \mathbb{R}^2$, and $r = |x|$. In particular, the interior field $v(\cdot) := v(\cdot, \phi)$ and scattered field $u^s(\cdot) := u^s(\cdot, \phi)$ depend on the incident angle ϕ. The radiating scattered field u^s again has the asymptotic behavior

$$u^s(x) = \frac{e^{ikr}}{\sqrt{r}} u_\infty(\theta, \phi) + O(r^{-3/2}), \qquad r \to \infty,$$

where the function $u_\infty(\cdot, \phi)$ defined on $[0, 2\pi]$ is the *far-field pattern* corresponding to the scattering problem (6.1)–(6.5) and the unit vector $\hat{x} := (\cos \theta, \sin \theta)$ is the observation direction. In the same way as in Theorem 4.2 it can be shown that the far-field pattern $u_\infty(\theta, \phi)$ corresponding to (6.1)–(6.5) satisfies the reciprocity relation $u_\infty(\theta, \phi) = u_\infty(\phi + \pi, \theta + \pi)$ and is given by

$$u_\infty(\theta, \phi) = \frac{e^{i\pi/4}}{\sqrt{8\pi k}} \int_{\partial B} \left(u^s(y) \frac{\partial e^{-ik\hat{x} \cdot y}}{\partial \nu} - e^{-ik\hat{x} \cdot y} \frac{\partial u^s(y)}{\partial \nu} \right) ds(y), \tag{6.6}$$

where ∂B is the boundary of a bounded domain containing D (it can also be ∂D).

The following result can be obtained as a consequence of Rellich's lemma (Theorem 4.1).

Theorem 6.1. *Suppose that the far-field pattern u_∞ corresponding to (6.1)–(6.5) satisfies $u_\infty = 0$ for a fixed angle ϕ and all θ in $[0, 2\pi]$. Then $u^s = 0$ in $\mathbb{R}^2 \setminus \bar{D}$.*

Note that by the analyticity of the far-field pattern Theorem 6.1 holds if $u_\infty = 0$ only for a subinterval of $[0, 2\pi]$.

The *inverse scattering problem* we are concerned with is to determine D from a knowledge of the far-field pattern $u_\infty(\theta, \phi)$ for all incident angles $\phi \in [0, 2\pi]$ and all observation angles $\theta \in [0, 2\pi]$. We remark that for an orthotropic medium standard examples [77, 136] show that A and n are not in fact uniquely determined from the far-field pattern $u_\infty(\theta, \phi)$ for all $\phi \in [0, 2\pi]$ and $\theta \in [0, 2\pi]$, but rather what is possible to determine is the support of the inhomogeneity D.

We now consider the *far-field operator* $F : L^2[0, 2\pi] \to L^2[0, 2\pi]$ corresponding to (6.1)–(6.5) defined by

$$(Fg)(\theta) := \int_0^{2\pi} u_\infty(\theta, \phi) g(\phi) d\phi. \tag{6.7}$$

As the reader has already seen (Chap. 4), the far-field operator will play a central role in the solution of the inverse problem. The first problem to resolve is that of injectivity and the denseness of the range of the far-field operator. We recall that a Herglotz function with kernel $g \in L^2[0, 2\pi]$ is given by

$$v_g(x) := \int_0^{2\pi} e^{ikx \cdot d} g(\phi) \, d\phi, \tag{6.8}$$

where $d = (\cos\phi, \sin\phi)$. Note that by superposition, Fg is the far-field pattern of the solution to (6.1)–(6.5), with $e^{ikx \cdot d}$ replaced by v_g. For future reference we note that

$$\tilde{v}_g(x) := \int_0^{2\pi} e^{-ikx \cdot d} g(\phi) \, d\phi \tag{6.9}$$

is also a Herglotz wave function with kernel $g(\phi - \pi)$.

Theorem 6.2. *The far-field operator F corresponding to the scattering problem (6.1)–(6.5) is injective with dense range if and only if there does not exist a Herglotz wave function v_g such that the pair v, v_g is a solution to*

$$\nabla \cdot A\nabla v + k^2 n\, v = 0 \quad \text{and} \quad \Delta v_g + k^2\, v_g = 0 \qquad \text{in} \quad D, \tag{6.10}$$

$$v = v_g \quad \text{and} \quad \frac{\partial v}{\partial \nu_A} = \frac{\partial v_g}{\partial \nu} \qquad \text{on} \quad \partial D. \tag{6.11}$$

Proof. In exactly the same way as in Theorem 4.3, one can show that the far-field operator F is injective if and only if its adjoint operator F^* is injective. Since $\mathrm{N}(F^*)^\perp = \overline{F(L^2[0, 2\pi])}$, to prove the theorem we must only show that F is injective. But $Fg = 0$ with $g \neq 0$ is equivalent to the existence of a nonzero Herglotz wave function v_g with kernel g for which the far-field pattern u_∞ corresponding to (6.1)–(6.5) with $e^{ikx \cdot d}$ replaced by v_g vanishes. By Rellich's lemma we have that $u^s = 0$ in $\mathbb{R}^2 \setminus \bar{D}$, and hence the transmission conditions imply that

$$v = v_g \quad \text{and} \quad \frac{\partial v}{\partial \nu_A} = \frac{\partial v_g}{\partial \nu} \qquad \text{on} \quad \partial D.$$

Since v_g is a solution of the Helmholtz equation, we have that v and v_g satisfy (6.10) as well. This proves the theorem.   □

Motivated by Theorem 6.2, we now define the *interior transmission problem* associated with the transmission problem (5.13)–(5.17).

Interior transmission problem. Given $f \in H^{\frac{1}{2}}(\partial D)$ *and* $h \in H^{-\frac{1}{2}}(\partial D)$, *find two functions* $v \in H^1(D)$ *and* $w \in H^1(D)$ *satisfying*

$$\nabla \cdot A\nabla v + k^2 n\,v = 0 \qquad in \quad D, \tag{6.12}$$

$$\Delta w + k^2\,w = 0 \qquad in \quad D, \tag{6.13}$$

$$v - w = f \qquad on \quad \partial D, \tag{6.14}$$

$$\frac{\partial v}{\partial \nu_A} - \frac{\partial w}{\partial \nu} = h \qquad on \quad \partial D. \tag{6.15}$$

The boundary value problem (6.12)–(6.13) with $f = 0$ and $h = 0$ is called the *homogeneous interior transmission problem* or the *transmission eigenvalue problem*.

Definition 6.3. Values of k for which the homogeneous interior transmission problem has a nontrivial solution are called transmission eigenvalues.

In particular, Theorem 6.2 states that if k is not a transmission eigenvalue, then the range of the far-field operator is dense.

6.2 Interior Transmission Problem

As seen earlier, the interior transmission problem appears naturally in scattering problems for an inhomogeneous medium. Of particular concern to us in this section are the countability and the existence of real transmission eigenvalues, and the approach to studying the interior transmission problem depends on whether or not $n \equiv 1$. In our analysis of the interior transmission problem we exclude the case of $A = I$ and refer the reader to Chap. 8 in [54], which deals with (6.12)–(6.15) when $A = I$.

We begin by establishing the uniqueness of a solution to the interior transmission problem for complex-valued refractive indexes.

Theorem 6.4. *If either* $Im(n) > 0$ *or* $Im\left(\bar{\xi} \cdot A\,\xi\right) < 0$ *at a point* $x_0 \in D$, *then the interior transmission problem (6.12)–(6.15) has at most one solution.*

Proof. Let v and w be a solution of the homogeneous interior transmission problem (i.e., $f = h = 0$). Applying the divergence theorem to $\bar{v}$ and $A\nabla v$ (Corollary 5.8), using the boundary condition and applying Green's first identity to $\bar{w}$ and w (Remark 6.29) we obtain

$$\int\limits_D \nabla \overline{v} \cdot A \nabla v \, dy - \int\limits_D k^2 n |v|^2 \, dy = \int\limits_{\partial D} \overline{v} \cdot \frac{\partial v}{\partial \nu_A} \, dy = \int\limits_D |\nabla w|^2 \, dy - \int\limits_D k^2 |w|^2 \, dy.$$

Hence

$$\mathrm{Im}\left(\int\limits_D \nabla \overline{v} \cdot A \nabla v \, dy \right) = 0 \quad \text{and} \quad \mathrm{Im}\left(\int\limits_D n |v|^2 \, dy \right) = 0. \qquad (6.16)$$

If $\mathrm{Im}(n) > 0$ at a point $x_0 \in D$, and hence by continuity in a small disk $\Omega_\epsilon(x_0)$, then the second equality of (6.16) and the unique continuation principle (Theorem 17.2.6 in [89]) imply that $v \equiv 0$ in D. In the case where $\mathrm{Im}\left(\overline{\xi} \cdot A \xi \right) < 0$ at a point $x_0 \in D$ for all $\xi \in \mathbb{C}^2$, and hence by continuity in a small ball $\Omega_\epsilon(x_0)$, from the first equality of (6.16) we obtain that $\nabla v \equiv 0$ in $\Omega_\epsilon(x_0)$ and from (6.12) $v \equiv 0$ in $\Omega_\epsilon(x_0)$, whence again from the unique continuation principle $v \equiv 0$ in D. From the boundary conditions (6.13) and (6.14), and the integral representation formula, w also vanishes in D. □

We now proceed to the solvability of the interior transmission problem following the approach in [20] and [34]. In the following analysis we assume without loss of generality that D is simply connected. We first study an intermediate problem called the *modified interior transmission problem*, which turns out to be a compact perturbation of our original transmission problem.

The modified interior transmission problem is as follows: given $f \in H^{\frac{1}{2}}(\partial D)$, $h \in H^{-\frac{1}{2}}(\partial D)$, a real-valued function $m \in C(\bar{D})$, and two functions $\rho_1 \in L^2(D)$ and $\rho_2 \in L^2(D)$, find $v \in H^1(D)$ and $w \in H^1(D)$ satisfying

$$\nabla \cdot A \nabla v - m \, v = \rho_1 \qquad \text{in} \quad D, \qquad (6.17)$$

$$\Delta w - w = \rho_2 \qquad \text{in} \quad D, \qquad (6.18)$$

$$v - w = f \qquad \text{on} \quad \partial D, \qquad (6.19)$$

$$\frac{\partial v}{\partial \nu_A} - \frac{\partial w}{\partial \nu} = h \qquad \text{on} \quad \partial D. \qquad (6.20)$$

We now reformulate (6.17)–(6.20) as an equivalent variational problem of the form (5.18). To this end, we define the Hilbert space

$$W(D) := \left\{ \mathbf{w} \in \left(L^2(D) \right)^2 : \nabla \cdot \mathbf{w} \in L^2(D) \quad \text{and} \quad \nabla \times \mathbf{w} = 0 \right\}$$

equipped with the inner product

$$(\mathbf{w}_1, \mathbf{w}_2)_W = (\mathbf{w}_1, \mathbf{w}_2)_{L^2(D)} + (\nabla \cdot \mathbf{w}_1, \nabla \cdot \mathbf{w}_2)_{L^2(D)}$$

and the norm

$$\|\mathbf{w}\|_W^2 = \|\mathbf{w}\|_{L^2(D)}^2 + \|\nabla \cdot \mathbf{w}\|_{L^2(D)}^2.$$

We denote by $\langle \cdot, \cdot \rangle$ the duality pairing between $H^{\frac{1}{2}}(\partial D)$ and $H^{-\frac{1}{2}}(\partial D)$. The duality pairing

$$\langle \varphi, \boldsymbol{\psi} \cdot \nu \rangle = \int_D \varphi \, \nabla \cdot \boldsymbol{\psi} \, dx + \int_D \nabla \varphi \cdot \boldsymbol{\psi} \, dx \qquad (6.21)$$

for $(\varphi, \boldsymbol{\psi}) \in H^1(D) \times W(D)$ will be of particular interest in the sequel. We next introduce the sesquilinear form $\mathcal{A}$ defined on $\{H^1(D) \times W(D)\}^2$ by

$$\mathcal{A}(U, V) = \int_D A \nabla v \cdot \nabla \bar{\varphi} \, dx + \int_D m \, v \, \bar{\varphi} \, dx + \int_D \nabla \cdot \mathbf{w} \, \nabla \cdot \bar{\boldsymbol{\psi}} \, dx + \int_D \mathbf{w} \cdot \bar{\boldsymbol{\psi}} \, dx$$

$$- \langle v, \bar{\boldsymbol{\psi}} \cdot \nu \rangle - \langle \bar{\varphi}, \mathbf{w} \cdot \nu \rangle, \qquad (6.22)$$

where $U := (v, \mathbf{w})$ and $V := (\varphi, \boldsymbol{\psi})$ are in $H^1(D) \times W(D)$. We denote by $L : H^1(D) \times W(D) \to \mathbb{C}$ the bounded conjugate linear functional given by

$$L(V) = \int_D (\rho_1 \, \bar{\varphi} + \rho_2 \, \nabla \cdot \bar{\boldsymbol{\psi}}) \, dx + \langle \bar{\varphi}, h \rangle - \langle f, \bar{\boldsymbol{\psi}} \cdot \nu \rangle. \qquad (6.23)$$

Then the variational formulation of the problem (6.17)–(6.20) is to find $U = (v, \mathbf{w}) \in H^1(D) \times W(D)$ such that

$$\mathcal{A}(U, V) = L(V) \qquad \text{for all} \quad V \in H^1(D) \times W(D). \qquad (6.24)$$

The following theorem proves the equivalence between problems (6.17)–(6.20) and (6.24).

Theorem 6.5. *The problem (6.17)–(6.20) has a unique solution $(v, w) \in H^1(D) \times H^1(D)$ if and only if the problem (6.24) has a unique solution $U = (v, \mathbf{w}) \in H^1(D) \times W(D)$. Moreover if (v, w) is the unique solution to (6.17)–(6.20), then $U = (v, \nabla w)$ is the unique solution to (6.24). Conversely, if $U = (v, \mathbf{w})$ is the unique solution to (6.24), then the unique solution (v, w) to (6.17)–(6.20) is such that $\mathbf{w} = \nabla w$.*

Proof. We first prove the equivalence between the existence of a solution (v, w) to (6.17)–(6.20) and the existence of a solution $U = (v, \mathbf{w})$ to (6.24).

1. Assume that (v, w) is a solution to (6.17)–(6.20), and set $\mathbf{w} = \nabla w$. From (6.18) we see that, since $\nabla \mathbf{w} = w + \rho_2 \in L^2(D)$, then $\mathbf{w} \in W(D)$. Taking the L^2 scalar product of (6.18) with $\nabla \cdot \boldsymbol{\psi}$ for some $\boldsymbol{\psi} \in W(D)$ and using (6.21) we see that

$$\int_D \nabla \cdot \mathbf{w} \, \nabla \cdot \bar{\boldsymbol{\psi}} \, dx + \int_D \mathbf{w} \cdot \bar{\boldsymbol{\psi}} \, dx - \langle w, \bar{\boldsymbol{\psi}} \cdot \nu \rangle = \int_D \rho_2 \, \nabla \cdot \bar{\boldsymbol{\psi}} \, dx.$$

Hence, by (6.19),

$$\int_D \nabla \cdot \mathbf{w} \, \nabla \cdot \bar{\boldsymbol{\psi}} \, dx + \int_D \mathbf{w} \cdot \bar{\boldsymbol{\psi}} \, dx - \langle v, \bar{\boldsymbol{\psi}} \cdot \nu \rangle$$

$$= - \langle f, \bar{\boldsymbol{\psi}} \cdot \nu \rangle + \int_D \rho_2 \, \nabla \cdot \bar{\boldsymbol{\psi}} \, dx. \qquad (6.25)$$

We now take the L^2 scalar product of (6.17) with φ in $H^1(D)$ and integrate by parts. Using the boundary condition (6.20) we see that

$$\int_D A\nabla v \cdot \nabla \bar{\varphi}\, dx + \int_D m\, v\, \bar{\varphi}\, dx - \langle \bar{\varphi}, \mathbf{w} \cdot \nu \rangle = \langle \bar{\varphi}, h \rangle + \int_D \rho_1\, \bar{\varphi}\, dx. \quad (6.26)$$

Finally, adding (6.25) and (6.26) we have that $U = (v, \nabla w)$ is a solution to (6.24).

2. Now assume that $U = (v, \mathbf{w}) \in H^1(D) \times W(D)$ is a solution to (6.24). Since $\nabla \times \mathbf{w} = 0$ and D is simply connected, we deduce the existence of a function $w \in H^1(D)$ such that $\mathbf{w} = \nabla w$, where w is determined up to an additive constant. As we shall see later, this constant can be adjusted so that (v, w) is a solution to (6.17)–(6.20). Obviously, if U satisfies (6.24), then $(v, \mathbf{w})$ satisfies (6.25) and (6.26) for all $(\varphi, \psi) \in H^1(D) \times W(D)$. One can easily see from (6.26) that the pair (v, w) satisfies

$$\nabla \cdot A\nabla v - m\, v = \rho_1 \qquad \text{in} \quad D, \qquad\qquad (6.27)$$

$$\frac{\partial v}{\partial \nu_A} - \frac{\partial w}{\partial \nu} = h \qquad \text{on} \quad \partial D. \qquad\qquad (6.28)$$

On the other hand, substituting for $\mathbf{w}$ in (6.25) and using the duality identity (6.21) in the second integral we have that

$$\int_D (\Delta w - w)\, \nabla \cdot \bar{\psi}\, dx + \langle w - v, \bar{\psi} \cdot \nu \rangle \qquad\qquad (6.29)$$

$$= -\langle f, \bar{\psi} \cdot \nu \rangle + \int_D \rho_2\, \nabla \cdot \bar{\psi}\, dx$$

for all ψ in $W(D)$.

Now consider a function $\phi \in L_0^2(D) = \left\{ \phi \in L^2(D) \ : \ \int_D \phi\, dx = 0 \right\}$, and let $\chi \in H^1(D)$ be a solution to

$$\begin{cases} \Delta \chi = \bar{\phi} \ \text{ in } D, \\ \dfrac{\partial \chi}{\partial \nu} = 0 \ \text{ on } \partial D. \end{cases} \qquad\qquad (6.30)$$

The existence of a solution of the preceding Neumann boundary value problem can be established by the variational methods developed in Chap. 5 (Example 5.15). We leave it to the reader as an exercise [127]. Taking $\psi = \nabla \chi$ in (6.29) [note that from (6.30) $\nabla \cdot \bar{\psi} = \phi$ in D and $\bar{\psi} \cdot \nu = 0$ on ∂D] we have that

$$\int_D (\Delta w - w - \rho_2)\, \phi\, dx = 0 \qquad \text{for all} \quad \phi \in L_0^2(D),$$

which implies the existence of a constant c_1 such that

$$\Delta w - w - \rho_2 = c_1 \quad \text{in } D. \tag{6.31}$$

We now take $\phi \in L^2_0(\partial D)$ and let $\sigma \in H^1(D)$ be a solution to

$$\begin{cases} \Delta \sigma = 0 & \text{in } D, \\ \dfrac{\partial \sigma}{\partial \nu} = \bar{\phi} & \text{on } \partial D. \end{cases} \tag{6.32}$$

Taking $\psi = \nabla \sigma$ in (6.25) [note that (6.32) implies that $\nabla \cdot \bar{\psi} = 0$ in D and $\bar{\psi} \cdot \nu = \phi$ on ∂D] we have that

$$\int_{\partial D} (w - v + f)\phi \, ds = 0 \qquad \text{for all } \phi \in L^2_0(\partial D),$$

which implies the existence of a constant c_2 such that

$$w - v + f = c_2 \quad \text{on } \partial D. \tag{6.33}$$

Substituting (6.31) and (6.33) into (6.29) and using (6.21) we see that

$$(c_1 - c_2) \int_D \nabla \cdot \bar{\psi} \, dx = 0 \quad \forall \, \psi \in W(D),$$

which implies $c_1 = c_2 = c$ [take, for instance, $\psi = \nabla \varrho$, where $\varrho \in H^1_0(D)$ and $\Delta \varrho = 1$ in D]. Equations (6.27), (6.31), and (6.33) show that $(v, w - c)$ is a solution to (6.17)–(6.20).

   We next consider the uniqueness equivalence between (6.17)–(6.20) and (6.24).

3. Assume that (6.17)–(6.20) has at most one solution. Let $U_1 = (v_1, \mathbf{w}_1)$ and $U_2 = (v_2, \mathbf{w}_2)$ be two solutions to (6.24). From step 2 earlier we deduce the existence of w_1 and w_2 in $H^1(D)$ such that $\mathbf{w}_1 = \nabla w_1$ and $\mathbf{w}_2 = \nabla w_2$ and (v_1, w_1) and (v_2, w_2) are solutions to (6.17)–(6.20), whence $(v_1, w_1) = (v_2, w_2)$ and $(v_1, \mathbf{w}_1) = (v_2, \mathbf{w}_2)$.

4. Finally, assume that (6.24) has at most one solution, and consider two solutions (v_1, w_1) and (v_2, w_2) to (6.17)–(6.20). We can deduce from step 1 earlier that $(v_1, \nabla w_1)$ and $(v_2, \nabla w_2)$ are two solutions to (6.24). Hence $v_1 = v_2$ and $w = w_1 - w_2$ is a function in $H^1(D)$ that satisfies

$$\begin{cases} \Delta w - w = 0 & \text{in } D, \\ w = \dfrac{\partial w}{\partial \nu} = 0 & \text{on } \partial D, \end{cases}$$

which implies $w = 0$.

$\square$

We now investigate the modified interior transmission problem in the variational formulation (6.24).

Theorem 6.6. *Assume that there exists a constant $\gamma > 1$ such that, for $x \in D$,*

$$Re\left(\bar{\xi} \cdot A(x)\,\xi\right) \geq \gamma |\xi|^2 \quad for~all~~\xi \in \mathbb{C}^2 \quad and \quad m(x) \geq \gamma. \tag{6.34}$$

Then problem (6.24) has a unique solution $U = (v, \mathbf{w}) \in H^1(D) \times W(D)$. This solution satisfies the a priori estimate

$$
\begin{aligned}
\|v\|_{H^1(D)} + \|\mathbf{w}\|_W \leq 2C\frac{\gamma+1}{\gamma-1}\Big(&\|\rho_1\|_{L^2(D)} + \|\rho_2\|_{L^2(D)} \\
&+ \|f\|_{H^{\frac{1}{2}}(\partial D)} + \|h\|_{H^{-\frac{1}{2}}(\partial D)}\Big),
\end{aligned} \tag{6.35}
$$

where the constant $C > 0$ is independent of ρ_1, ρ_2, f, h, and γ.

Proof. The trace theorems (Sect. 5.2) and Schwarz's inequality ensure the continuity of the conjugate linear functional L on $H^1(D) \times W(D)$ and the existence of a constant c independent of ρ_1, ρ_2, f, and h such that

$$\|L\| \leq C\left(\|\rho_1\|_{L^2} + \|\rho_2\|_{L^2} + \|f\|_{H^{\frac{1}{2}}} + \|h\|_{H^{-\frac{1}{2}}}\right). \tag{6.36}$$

On the other hand, if $U = (v, \mathbf{w}) \in H^1(D) \times W(D)$, then, by assumption (6.34),

$$|\mathcal{A}(U,U)| \geq \gamma \|v\|_{H^1}^2 + \|\mathbf{w}\|_W^2 - 2\operatorname{Re}\left(\langle \bar{v},\, \mathbf{w}\rangle\right). \tag{6.37}$$

According to the duality identity (6.21), one has by Schwarz's inequality that

$$|\langle \bar{v},\, \mathbf{w}\rangle| \leq \|v\|_{H^1}\|\mathbf{w}\|_W,$$

and therefore

$$|\mathcal{A}(U,U)| \geq \gamma \|v\|_{H^1}^2 + \|\mathbf{w}\|_W^2 - 2\|v\|_{H^1}\|\mathbf{w}\|_W.$$

Using the identity $\gamma x^2 + y^2 - 2xy = \frac{\gamma+1}{2}\left(x - \frac{2}{\gamma+1}y\right)^2 + \frac{\gamma-1}{2}x^2 + \frac{\gamma-1}{\gamma+1}y^2$ we conclude that

$$|\mathcal{A}(U,U)| \geq \frac{\gamma-1}{\gamma+1}\left(\|\mathbf{w}\|_W^2 + \|v\|_{H^1}^2\right),$$

whence $\mathcal{A}$ is coercive. The continuity of $\mathcal{A}$ follows easily from Schwarz's inequality, the trace theorem, and Theorem 5.7. Theorem 6.6 is now a direct consequence of the Lax–Milgram lemma applied to (6.24). □

Theorem 6.7. *Assume that there exists a constant $\gamma > 1$ such that, for $x \in D$,*

$$Re\left(\bar{\xi} \cdot A(x)\,\xi\right) \geq \gamma |\xi|^2 \quad \text{for all} \quad \xi \in \mathbb{C}^2 \quad \text{and} \quad m(x) \geq \gamma. \tag{6.38}$$

Then the modified interior transmission problem (6.17)–(6.20) has a unique solution (v, w) that satisfies

$$\begin{aligned}
\|v\|_{H^1(D)} + \|w\|_{H^1(D)} &\leq C\frac{\gamma+1}{\gamma-1}\Big(\|\rho_1\|_{L^2(D)} + \|\rho_2\|_{L^2(D)} \\
&\quad + \|f\|_{H^{\frac{1}{2}}(\partial D)} + \|h\|_{H^{-\frac{1}{2}}(\partial D)}\Big),
\end{aligned} \tag{6.39}$$

where the constant $C > 0$ is independent of ρ_1, ρ_2, f, h, and γ.

Proof. The existence and uniqueness of a solution follow from Theorems 6.5 and 6.6. The a priori estimate (6.39) can be obtained directly from (6.17)–(6.20), but it can also be deduced from (6.35) as follows. Theorem 6.5 tells us that $(v, \nabla w)$ is the unique solution to (6.24). Hence, according to (6.35),

$$\|v\|_{H^1} + \|\nabla w\|_{L^2} \leq C_1 \frac{\gamma+1}{\gamma-1}\left(\|\rho_1\|_{L^2} + \|\rho_2\|_{L^2} + \|f\|_{H^{\frac{1}{2}}} + \|h\|_{H^{-\frac{1}{2}}}\right).$$

From Poincaré's inequality in Sect. 5.2 we can write

$$\|w\|_{H^1(D)} \leq C_2 \left(\|\nabla w\|_{L^2(D)} + \|w\|_{L^2(\partial D)}\right).$$

Now, using the boundary condition (6.19) and the trace theorem we obtain that

$$\|w\|_{H^1(D)} \leq C_2 \left(\|\nabla w\|_{L^2(D)} + \|v\|_{H^1(D)} + \|f\|_{L^2(\partial D)}\right)$$

for some positive constant C_2. The constants C_1 and C_2 can then be adjusted so that (6.39) holds. $\qquad\square$

Now we are ready to show the existence of a solution to the interior transmission problem (6.12)–(6.15).

Theorem 6.8. *Assume that either $Im(n) > 0$ or $Im\left(\bar{\xi} \cdot A\,\xi\right) < 0$ at a point $x_0 \in D$ and that there exists a constant $\gamma > 1$ such that, for $x \in D$,*

$$Re\left(\bar{\xi} \cdot A(x)\,\xi\right) \geq \gamma |\xi|^2 \quad \text{for all} \quad \xi \in \mathbb{C}^2. \tag{6.40}$$

Then (6.12)–(6.15) has a unique solution $(v, w) \in H^1(D) \times H^1(D)$. This solution satisfies the a priori estimate

$$\|v\|_{H^1(D)} + \|w\|_{H^1(D)} \leq C \left(\|f\|_{H^{\frac{1}{2}}(\partial D)} + \|h\|_{H^{-\frac{1}{2}}(\partial D)}\right) \tag{6.41}$$

where the constant $C > 0$ is independent of f and h.

Proof. Set

$$\mathcal{X}(D) = \{(v, w) \in H^1(D) \times H^1(D) \; : \; \nabla \cdot A\nabla v \in L^2(D) \text{ and } \Delta w \in L^2(D)\}$$

and consider the operator $\mathcal{G}$ from $\mathcal{X}(D)$ into $L^2(D) \times L^2(D) \times H^{\frac{1}{2}}(\partial D) \times H^{-\frac{1}{2}}(\partial D)$ defined by

$$\mathcal{G}(v, w) = \left(\nabla \cdot A\nabla v - mv, \Delta w - w, (v - w)|_{\partial D}, \left(\frac{\partial v}{\partial \nu} - \frac{\partial w}{\partial \nu} \right)_{|\partial D} \right) \quad (6.42)$$

where $m \in C(\bar{D})$ and $m > 1$. Obviously $\mathcal{G}$ is continuous and from Theorem 6.7 we know that the inverse of $\mathcal{G}$ exists and is continuous. Now consider the operator $\mathcal{T}$ from $\mathcal{X}(D)$ into $L^2(D) \times L^2(D) \times H^{\frac{1}{2}}(\partial D) \times H^{-\frac{1}{2}}(\partial D)$ defined by

$$\mathcal{T}(v, w) = \left((k^2 n + m)v, \; (k^2 + 1)w, \; 0, \; 0 \right)$$

From the compact embedding of $H^1(D)$ into $L^2(D)$ (Sect. 5.2), the operator $\mathcal{T}$ is compact. Theorem 6.4 implies that $\mathcal{G} + \mathcal{T}$ is injective, and therefore, from Theorem 5.16 we can deduce the existence and the continuity of $(\mathcal{G} + \mathcal{T})^{-1}$, which means in particular the existence of a unique solution to the interior transmission problem (6.12)–(6.15) that satisfies the a priori estimate (6.43). $\square$

The foregoing analysis of the interior transmission problem requires that the matrix A satisfy

$$\text{Re} \left(\bar{\xi} \cdot A(x) \xi \right) \geq \gamma |\xi|^2 \text{ for all } \xi \in \mathbb{C}^2, \quad x \in D \text{ and some constant } \gamma > 1,$$

that is, $\|\text{Re}(A)\| > 1$. The case of $\text{Re}(A)$ positive definite such that $\|\text{Re}(A)\| < 1$ is considered in [34]. By modifying the variational approach of Theorems 6.5 and 6.6 one can prove the following result.

Theorem 6.9. *Assume that either $Im(n) > 0$ or $Im\left(\bar{\xi} \cdot A\xi \right) < 0$ at a point $x_0 \in D$ and that there exists a constant $\gamma > 1$ such that, for $x \in D$,*

$$\text{Re} \left(\bar{\xi} \cdot (A(x))^{-1} \xi \right) \geq \gamma |\xi|^2 \quad \text{for all } \xi \in \mathbb{C}^2 \quad \text{and} \quad \gamma^{-1} \leq m < 1.$$

Then (6.12)–(6.15) has a unique solution $(v, w) \in H^1(D) \times H^1(D)$. This solution satisfies the a priori estimate

$$\|v\|_{H^1(D)} + \|w\|_{H^1(D)} \leq C \left(\|f\|_{H^{\frac{1}{2}}(\partial D)} + \|h\|_{H^{-\frac{1}{2}}(\partial D)} \right), \quad (6.43)$$

where the constant $C > 0$ is independent of f and h.

We remark that a solvability result under less restrictive assumptions on A is obtained later in this chapter (Remark 6.29).

In general we cannot conclude the solvability of the interior transmission problem if A and n do not satisfy the assumptions of the previous theorem. In particular, if $\mathrm{Im}(A) = 0$ and $\mathrm{Im}(n) = 0$ in D, then k may be a transmission eigenvalue (Definition 6.3). Do transmission eigenvalues exist and, if so, do they form a discrete set? The approach in [20] and [34] presented earlier is not suitable to handle these questions, and therefore we devote the next section of the book to address these issues. In particular, we will prove that under appropriate assumptions transmission eigenvalues exist and form a discrete set with infinity as the only accumulation point. As mentioned at the beginning of this section, the analysis of the transmission eigenvalue problem for cases where $n = 1$ and $n \neq 1$ are fundamentally different, and hence we consider each of these cases separately. For the study of the transmission eigenvalue problem if $A = I$ we refer the reader to [32] and to Chap. 10 in [54].

6.3 Transmission Eigenvalue Problem

We recall that the transmission eigenvalue problem is formulated as a problem of finding two nonzero functions $v \in H^1(D)$ and $w \in H^1(D)$ satisfying

$$\nabla \cdot A\nabla v + k^2 n v = 0 \qquad \text{in} \quad D, \tag{6.44}$$

$$\Delta w + k^2 w = 0 \qquad \text{in} \quad D, \tag{6.45}$$

$$v = w \qquad \text{on} \quad \partial D, \tag{6.46}$$

$$\frac{\partial v}{\partial \nu_A} = \frac{\partial w}{\partial \nu} \qquad \text{on} \quad \partial D. \tag{6.47}$$

Since transmission eigenvalues do not exist for complex-valued A and n, henceforth we assume that both A and n are real-valued and define

$$a_{min} := \inf_{x \in D} \inf_{\xi \in \mathbb{R}^2, |\xi|=1} (\xi \cdot A(x)\xi) > 0,$$

$$a_{max} := \sup_{x \in D} \sup_{\xi \in \mathbb{R}^2, |\xi|=1} (\xi \cdot A(x)\xi) < \infty, \tag{6.48}$$

$$n_{min} := \inf_{x \in D} n(x) > 0 \quad \text{and} \quad n_{max} := \sup_{x \in D} n(x) < \infty.$$

Example 6.10. In what follows, we will need to consider a particular case of the interior transmission problem where D is a ball B_R of radius R centered at the origin, $A := a_0 I$, and $n := n_0$, where a_0 and n_0 are positive constants not both equal to one. In this case the interior transmission eigenvalue problem reads as

$$\Delta v + k^2 \frac{n_0}{a_0} v = 0 \quad \text{in } B_R, \tag{6.49}$$

$$\Delta w + k^2 w = 0 \quad \text{in } B_R, \tag{6.50}$$

$$v = w \quad \text{on } \partial B_R, \tag{6.51}$$

$$a_0 \frac{\partial v}{\partial r} = \frac{\partial w}{\partial r} \quad \text{on } \partial B_R, \tag{6.52}$$

where $r = |x|$. To solve (6.49)–(6.52) in $\mathbb{R}^2$, we make the ansatz

$$w(r, \hat{x}) = a_\ell J_\ell(k r) e^{i\ell\theta}, \qquad v(r, \hat{x}) = b_\ell J_\ell\left(k\sqrt{\frac{n_0}{a_0}} r\right) e^{i\ell\theta},$$

where J_ℓ are Bessel functions of order ℓ introduced in Chap. 3. Then using separation of variables one sees that the transmission eigenvalues satisfy

$$W(k) = \det \begin{pmatrix} J_\ell(kR) & J_\ell\left(k\sqrt{\frac{n_0}{a_0}} R\right) \\ k J_\ell'(kR) & k\sqrt{n_0 a_0} J_\ell'\left(k\sqrt{\frac{n_0}{a_0}} R\right) \end{pmatrix} = 0. \tag{6.53}$$

6.3.1 The Case $n = 1$

The case where $n = 1$ corresponds to the electromagnetic scattering problem for an orthotropic medium when the magnetic permeability in the medium is constant and the same as the magnetic permeability in the background. The *transmission eigenvalue problem* reads: find two nonzero functions $v \in H^1(D)$ and $w \in H^1(D)$ satisfying

$$\nabla \cdot A\nabla v + k^2 v = 0 \quad \text{in} \quad D, \tag{6.54}$$

$$\Delta w + k^2 w = 0 \quad \text{in} \quad D, \tag{6.55}$$

$$v = w \quad \text{on} \quad \partial D, \tag{6.56}$$

$$\frac{\partial v}{\partial \nu_A} = \frac{\partial w}{\partial \nu} \quad \text{on} \quad \partial D. \tag{6.57}$$

Our approach follows the one introduced in [20] and developed further in [31], which generalizes the first proof of the existence of transmission eigenvalues given in [134].

The proof of the existence of transmission eigenvalues is based on the following abstract analysis. Let X be a separable Hilbert space with scalar product $(\cdot, \cdot)$ and associated norm $\| \cdot \|$, and let $\mathbb{A}$ be a bounded, positive definite, and self-adjoint operator on X. Under these assumptions $\mathbb{A}^{\pm 1/2}$ are well defined (cf. [115]). In particular, $\mathbb{A}^{\pm 1/2}$ are also bounded, positive definite, and self-adjoint operators, $\mathbb{A}^{-1/2}\mathbb{A}^{1/2} = I$ and $\mathbb{A}^{1/2}\mathbb{A}^{1/2} = \mathbb{A}$. We shall consider the spectral decomposition of the operator $\mathbb{A}$ with respect to self-adjoint nonnegative compact operators. The next two theorems indicate the main properties of such a decomposition.

Definition 6.11. A bounded linear operator $\mathbb{A}$ on a Hilbert space X is said to be nonnegative if $(\mathbb{A}u, u) \geq 0$ for every $u \in X$. $\mathbb{A}$ is said to be strictly coercive if $(\mathbb{A}u, u) \geq \beta\|u\|^2$ for some positive constant β.

Theorem 6.12. *Let $\mathbb{A}$ be a bounded, self-adjoint, and strictly coercive operator on a Hilbert space, and let $\mathbb{B}$ be a nonnegative, self-adjoint, and compact linear operator with null space $N(\mathbb{B})$. Then there exists an increasing sequence of positive real numbers $(\lambda_j)_{j \geq 1}$ and a sequence $(u_j)_{j \geq 1}$ of elements of X satisfying*

$$\mathbb{A}u_j = \lambda_j \mathbb{B}u_j$$

and

$$(\mathbb{B}u_j, u_\ell) = \delta_{j\ell}$$

such that each $u \in [\mathbb{A}(N(\mathbb{B}))]^\perp$ can be expanded in a series

$$u = \sum_{j=1}^{\infty} \gamma_j u_j.$$

If $N(\mathbb{B})^\perp$ has infinite dimension, then $\lambda_j \to +\infty$ as $j \to \infty$.

Proof. This theorem is a direct consequence of the Hilbert–Schmidt theorem applied to the nonnegative self-adjoint compact operator $\tilde{\mathbb{B}} = \mathbb{A}^{-1/2}\mathbb{B}\mathbb{A}^{-1/2}$. Let $(\mu_j, v_j)_{j \geq 1}$ be the sequence of positive eigenvalues and corresponding eigenfunctions associated with $\tilde{\mathbb{B}}$ such that $\{v_j\}_{j \geq 1}$ forms an orthonormal basis for $N(\tilde{\mathbb{B}})^\perp$. Note that zero is the only possible accumulation point for the sequence μ_j. Straightforward calculations show that $\lambda_j = 1/\mu_j$ and $u_j = \sqrt{\lambda_k}\,\mathbb{A}^{-1/2}v_j$ satisfy $\mathbb{A}u_j = \lambda_j \mathbb{B}u_j$. Obviously, if $w \in \mathbb{A}N(\mathbb{B})$, then $w = \mathbb{A}z$ for some $z \in N(\mathbb{B})$, and hence $(u_j, w) = \lambda_j(\mathbb{A}^{-1}\mathbb{B}u_j, w) = \lambda_j(\mathbb{A}^{-1}\mathbb{B}u_j, \mathbb{A}z) = \lambda_j(\mathbb{B}u_j, z) = 0$, which means that $u_j \in [\mathbb{A}N(\mathbb{B})]^\perp$. Furthermore, any $u \in [\mathbb{A}N(\mathbb{B})]^\perp$ can be written as $u = \sum_j \gamma_j u_j = \sum_j \gamma_j \sqrt{\lambda_j}\,\mathbb{A}^{-1/2}v_j$ since $\mathbb{A}^{1/2}u \in \left[N(\mathbb{A}^{-1/2}\mathbb{B}\mathbb{A}^{-1/2})\right]^\perp$. This ends the proof of the theorem. $\square$

Theorem 6.13. *Let $\mathbb{A}$, $\mathbb{B}$, and $(\lambda_j)_{j \geq 1}$ be as in Theorem 6.12, and define the Rayleigh quotient as*

$$R(u) = \frac{(\mathbb{A}u, u)}{(\mathbb{B}u, u)}$$

for $u \notin N(\mathbb{B})$, where $(\cdot, \cdot)$ is the inner product on X. Then the following min-max principle holds:

$$\lambda_j = \min_{W \in \mathcal{U}_j^{\mathbb{A}}} \left(\max_{u \in W \setminus \{0\}} R(u) \right) = \max_{W \in \mathcal{U}_{j-1}^{\mathbb{A}}} \left(\min_{u \in (\mathbb{A}(W + N(\mathbb{B})))^\perp \setminus \{0\}} R(u) \right),$$

where $\mathcal{U}_j^{\mathbb{A}}$ denotes the set of all j-dimensional subspaces of $[\mathbb{A}N(\mathbb{B})]^\perp$.

Proof. The proof follows the classical proof of the Courant min-max principle and is given here for the reader's convenience. It is based on the fact that if $u \in [\mathbb{A}N(B)]^{\perp}$, then from Theorem 6.12 we can write $u = \sum_j \gamma_j u_j$ for some coefficients γ_j, where the u_j are defined in Theorem 6.12 (note that the u_j are orthogonal with respect to the inner product induced by the self-adjoint invertible operator $\mathbb{A}$). Then using the facts that $(\mathbb{B}u_j, u_\ell) = \delta_{j\ell}$ and $\mathbb{A}u_j = \lambda_j \mathbb{B}u_j$ it is easy to see that

$$R(u) = \frac{1}{\sum_j |\gamma_j|^2} \sum_j \lambda_j |\gamma_j|^2.$$

Therefore, if $W_j \in \mathcal{U}_j^{\mathbb{A}}$ denotes the space generated by $\{u_1, \ldots, u_j\}$, then we have that

$$\lambda_j = \max_{u \in W_j \setminus \{0\}} R(u) = \min_{u \in [\mathbb{A}(W_{j-1}+N(\mathbb{B}))]^{\perp} \setminus \{0\}} R(u).$$

Next, let W be any element of $\mathcal{U}_j^{\mathbb{A}}$. Since W has dimension j and $W \subset [\mathbb{A}N(\mathbb{B})]^{\perp}$, then $W \cap [\mathbb{A}W_{j-1} + \mathbb{A}N(\mathbb{B})]^{\perp} \neq \{0\}$. Therefore,

$$\max_{u \in W \setminus \{0\}} R(u) \geq \min_{u \in W \cap [\mathbb{A}(W_{j-1}+N(\mathbb{B}))]^{\perp} \setminus \{0\}} R(u)$$
$$\geq \min_{u \in [\mathbb{A}(W_{j-1}+N(\mathbb{B}))]^{\perp} \setminus \{0\}} R(u) = \lambda_j,$$

which proves the first equality of the theorem. Similarly, if W has dimension $j-1$ and $W \subset [\mathbb{A}N(\mathbb{B})]^{\perp}$, then $W_j \cap (\mathbb{A}W)^{\perp} \neq \{0\}$. Therefore,

$$\min_{u \in [\mathbb{A}(W+N(\mathbb{B}))]^{\perp} \setminus \{0\}} R(u) \leq \max_{u \in W_j \cap (\mathbb{A}W)^{\perp} \setminus \{0\}} R(u) \leq \max_{u \in W_j \setminus \{0\}} R(u) = \lambda_j,$$

which proves the second equality of the theorem. $\square$

The following corollary shows that it is possible to remove the dependence on $\mathbb{A}$ in the choice of the subspaces in the min-max principle for the eigenvalues λ_j.

Corollary 6.14. *Let $\mathbb{A}$, $\mathbb{B}$, $(\lambda_j)_{j \geq 1}$, and R be as in Theorem 6.13. Then*

$$\lambda_j = \min_{W \subset \mathcal{U}_j} \left(\max_{u \in W \setminus \{0\}} R(u) \right), \tag{6.58}$$

where $\mathcal{U}_j$ denotes the set of all j-dimensional subspaces W of X such that $W \cap N(\mathbb{B}) = \{0\}$.

Proof. From Theorem 6.13 and the fact that $\mathcal{U}_j^{\mathbb{A}} \subset \mathcal{U}_j$ it suffices to prove that

$$\lambda_j \leq \min_{W \subset \mathcal{U}_j} \left(\max_{u \in W \setminus \{0\}} R(u) \right).$$

Let $W \in \mathcal{U}_j$, and let $v_1, v_2, \ldots, v_k$ be a basis for W. Each vector v_j can be decomposed into a sum $v_j^0 + \tilde{v}_j$, where $\tilde{v}_j \in [\mathbb{A}N(\mathbb{B})]^\perp$ and $v_j^0 \in N(\mathbb{B})$ (which is the orthogonal decomposition with respect to the scalar product induced by $\mathbb{A}$). Since $W \cap N(B) = \{0\}$, the space $\tilde{W}$ generated by $\tilde{v}_1, \tilde{v}_2, \ldots, \tilde{v}_j$ has dimension j. Moreover, $\tilde{W} \subset [\mathbb{A}N(\mathbb{B})]^\perp$. Now let $\tilde{u} \in \tilde{W}$. Obviously, $\tilde{u} = u - u^0$ for some $u \in W$ and $u^0 \in N(\mathbb{B})$. Since $\mathbb{B}u^0 = 0$ and $(\mathbb{A}u_0, \tilde{u}) = 0$, we have that

$$R(u) = \frac{(\mathbb{A}\tilde{u}, \tilde{u}) + (\mathbb{A}u^0, u^0)}{(\mathbb{B}\tilde{u}, \tilde{u})} = R(\tilde{u}) + \frac{(\mathbb{A}u^0, u^0)}{(\mathbb{B}\tilde{u}, \tilde{u})}.$$

Consequently, since $\mathbb{A}$ is positive definite and $\mathbb{B}$ is nonnegative, we obtain

$$R(\tilde{u}) \leq R(u) \leq \max_{u \in W \setminus \{0\}} R(u).$$

Finally, taking the maximum with respect to $\tilde{u} \in \tilde{W} \subset [\mathbb{A}N(\mathbb{B})]^\perp$ in the preceding inequality, we obtain from Theorem 6.13 that

$$\lambda_j \leq \max_{u \in W \setminus \{0\}} R(u),$$

which completes the proof after taking the minimum over all $W \subset \mathcal{U}_j$.   □

The following theorem provides the theoretical basis of our analysis of the existence of transmission eigenvalues. This theorem is a simple consequence of Theorem 6.13 and Corollary 6.14.

Theorem 6.15. *Let* $\tau \longmapsto \mathbb{A}_\tau$ *be a continuous mapping from* $]0, \infty[$ *to the set of bounded, self-adjoint, and strictly coercive operators on the Hilbert space* X, *and let* $\mathbb{B}$ *be a self-adjoint and nonnegative, compact, bounded, linear operator on* X. *We assume that there exist two positive constants* $\tau_0 > 0$ *and* $\tau_1 > 0$ *such that*

1. $\mathbb{A}_{\tau_0} - \tau_0 \mathbb{B}$ *is positive on* X,
2. $\mathbb{A}_{\tau_1} - \tau_1 \mathbb{B}$ *is nonpositive on a* ℓ-*dimensional subspace* W_j *of* X.

Then each of the equations $\lambda_j(\tau) = \tau$ *for* $j = 1, \ldots, \ell$ *has at least one solution in* $[\tau_0, \tau_1]$, *where* $\lambda_j(\tau)$ *is the* j*th eigenvalue (counting multiplicity) of* $\mathbb{A}_\tau$ *with respect to* $\mathbb{B}$, *i.e.,* $N(\mathbb{A}_\tau - \lambda_j(\tau)\mathbb{B}) \neq \{0\}$.

Proof. First we can deduce from (6.58) that for all $j \geq 1$, $\lambda_j(\tau)$ is a continuous function of τ. Assumption 1 shows that $\lambda_j(\tau_0) > \tau_0$ for all $j \geq 1$. Assumption 2 implies in particular that $W_j \cap N(\mathbb{B}) = \{0\}$. Hence, another application of (6.58) implies that $\lambda_j(\tau_1) \leq \tau_1$ for $1 \leq j \leq \ell$. The desired result is now obtained by applying the intermediate value theorem.   □

The main idea in studying the eigenvalue problem (6.54)–(6.57) is to observe that by making an appropriate substitution one can rewrite it as an equivalent eigenvalue problem for a fourth-order differential equation. To this end, let $w \in H^1(D)$ and $v \in H^1(D)$ satisfy (6.54)–(6.57), and make the substitution

$$\mathbf{v} = A\nabla v \in L^2(D)^2, \quad \text{and} \quad \mathbf{w} = \nabla w \in L^2(D)^2.$$

Since from (6.48) A^{-1} exists and is bounded, we have that

$$\nabla v = A^{-1}\mathbf{v}.$$

Taking the gradient of (6.54) and (6.55), we obtain that $\mathbf{v}$ and $\mathbf{w}$ satisfy

$$\nabla(\nabla \cdot \mathbf{v}) + k^2 A^{-1}\mathbf{v} = 0 \tag{6.59}$$

and

$$\nabla(\nabla \cdot \mathbf{w}) + k^2\mathbf{w} = 0, \tag{6.60}$$

respectively, in D. Obviously, (6.57) implies that

$$\nu \cdot \mathbf{v} = \nu \cdot \mathbf{w} \quad \text{on } \partial D. \tag{6.61}$$

Furthermore, from (6.54) and (6.55) we have that

$$-k^2 v = \nabla \cdot \mathbf{v} \quad \text{and} \quad -k^2 w = \nabla \cdot \mathbf{w},$$

and the transmission condition (6.56) yields

$$\nabla \cdot \mathbf{v} = \nabla \cdot \mathbf{w} \quad \text{on } \partial D. \tag{6.62}$$

We now formulate the interior transmission eigenvalue problem in terms of $\mathbf{w}$ and $\mathbf{v}$. In addition to the usual energy spaces

$$H^1(D) := \left\{ u \in L^2(D) : \nabla u \in L^2(D)^2 \right\},$$
$$H_0^1(D) := \left\{ u \in H^1(D) : u = 0 \text{ on } \partial D \right\},$$

we introduce the Sobolev spaces

$$H(\operatorname{div}, D) := \left\{ \mathbf{u} \in L^2(D)^2 : \nabla \cdot \mathbf{u} \in L^2(D) \right\},$$
$$H_0(\operatorname{div}, D) := \left\{ \mathbf{u} \in H(\operatorname{div}, D) : \nu \cdot \mathbf{u} = 0 \text{ on } \partial D \right\}$$

and

$$\mathcal{H}(D) := \left\{ \mathbf{u} \in H(\operatorname{div}, D) : \nabla \cdot \mathbf{u} \in H^1(D) \right\},$$
$$\mathcal{H}_0(D) := \left\{ \mathbf{u} \in H_0(\operatorname{div}, D) : \nabla \cdot \mathbf{u} \in H_0^1(D) \right\},$$

equipped with the scalar product

$$(\mathbf{u}, \mathbf{v})_{\mathcal{H}(D)} := (\mathbf{u}, \mathbf{v})_{L^2(D)} + (\nabla \cdot \mathbf{u}, \nabla \cdot \mathbf{v})_{H^1(D)}.$$

Letting $N := A^{-1}$, in terms of new vector-valued functions $\mathbf{w}$ and $\mathbf{v}$, the transmission eigenvalue problem can be written as

$$\nabla(\nabla \cdot \mathbf{v}) + k^2 N \mathbf{v} = 0 \quad \text{in} \quad D, \tag{6.63}$$

$$\nabla(\nabla \cdot \mathbf{w}) + k^2 \mathbf{w} = 0 \quad \text{in} \quad D, \tag{6.64}$$

$$\nu \cdot \mathbf{w} = \nu \cdot \mathbf{v} \quad \text{on} \quad \partial D, \tag{6.65}$$

$$\nabla \cdot \mathbf{w} = \nabla \cdot \mathbf{v} \quad \text{on} \quad \partial D. \tag{6.66}$$

Definition 6.16. Values of $k \in \mathbb{C}$ for which the homogeneous interior transmission problem (6.63)–(6.66) has nonzero solutions $\mathbf{w} \in (L^2(D))^2$, $\mathbf{v} \in (L^2(D))^2$ such that $\mathbf{w} - \mathbf{v} \in \mathcal{H}_0(D)$ are called transmission eigenvalues. If k is a transmission eigenvalue, then we call $\mathbf{u} := \mathbf{v} - \mathbf{w}$ the corresponding eigenfunction where $\mathbf{v}$ and $\mathbf{w}$ are a nonzero solution of (6.63)–(6.66).

It is possible to write (6.63)–(6.66) as an equivalent eigenvalue problem for $\mathbf{w} - \mathbf{v} \in \mathcal{H}_0(D)$ satisfying the fourth-order equation

$$\left(\nabla\nabla \cdot + k^2 N\right)(N - I)^{-1}\left(\nabla\nabla \cdot \mathbf{u} + k^2 \mathbf{u}\right) = 0 \quad \text{in} \quad D. \tag{6.67}$$

Equation (6.67) can be written in the variational form

$$\int_D (N - I)^{-1}\left(\nabla\nabla \cdot \mathbf{u} + k^2 \mathbf{u}\right) \cdot \left(\nabla\nabla \cdot \bar{\mathbf{v}} + k^2 N \bar{\mathbf{v}}\right) \, dx = 0 \tag{6.68}$$

for all $\mathbf{v} \in \mathcal{H}_0(D)$. The variational equation (6.68) can in turn be written as an operator equation

$$\mathbb{A}_k \mathbf{u} - k^2 \mathbb{B} \mathbf{u} = 0 \quad \text{for} \quad \mathbf{u} \in \mathcal{H}_0(D), \tag{6.69}$$

where the bounded linear operators $\mathbb{A}_k : \mathcal{H}_0(D) \to \mathcal{H}_0(D)$ and $\mathbb{B} : \mathcal{H}_0(D) \to \mathcal{H}_0(D)$ are defined by means of the Riesz representation theorem

$$(\mathbb{A}_k \mathbf{u}, \mathbf{v})_{\mathcal{H}_0(D)} = \mathcal{A}_k \mathbf{u}(\mathbf{u}, \mathbf{v}) \quad \text{and} \quad (\mathbb{B} \mathbf{u}, \mathbf{v})_{\mathcal{H}_0(D)} = \mathcal{B}(\mathbf{u}, \mathbf{v}), \tag{6.70}$$

with the sesquilinear forms $\mathcal{A}_\tau$ and $\mathcal{B}$ given by

$$\mathcal{A}_k(\mathbf{u}, \mathbf{v}) := \left((N - I)^{-1}\left(\nabla\nabla \cdot \mathbf{u} + k^2 \mathbf{u}\right), \left(\nabla\nabla \cdot \mathbf{v} + k^2 \mathbf{v}\right)\right)_D + k^4 (\mathbf{u}, \mathbf{v})_D$$

and

$$\mathcal{B}(\mathbf{u}, \mathbf{v}) := (\nabla \cdot \mathbf{u}, \nabla \cdot \mathbf{v})_D,$$

respectively, where $(\cdot, \cdot)_D$ denotes the $L^2(D)$ inner product.

Lemma 6.17. $\mathbb{B} : \mathcal{H}_0(D) \to \mathcal{H}_0(D)$ *is a compact operator.*

Proof. Let $\mathbf{u}_n$ be a bounded sequence in $\mathcal{H}_0(D)$. Then there exists a subsequence, denoted again by $\mathbf{u}_n$, that converges weakly to $\mathbf{u}$ in $\mathcal{H}_0(D)$. Since $\nabla \cdot \mathbf{u}_n$ is also bounded in $H^1(D)$, from the Rellich compactness theorem we have that $\nabla \cdot \mathbf{u}_n$ converges strongly to $\nabla \cdot \mathbf{u}_0$ in $L^2(D)$. But

$$\|\mathbb{B}(\mathbf{u}_n - \mathbf{u})\|_{\mathcal{H}_0(D)} \leq \|\nabla \cdot (\mathbf{u}_n - \mathbf{u})\|_{L^2(D)},$$

which proves that $\mathbb{B}\mathbf{u}_n$ converges strongly to $\mathbb{B}\mathbf{u}$.   □

In our discussion we must distinguish between the two cases $a_{min} > 1$ and $a_{max} < 1$. To fix our ideas, we consider in detail only the case where $a_{max} < 1$ (similar results can be obtained for $a_{min} > 1$; cf. [21, 31, 33]). If $\lambda_1(x) \leq \lambda_2(x)$ are the eigenvalues of the matrix $A(x)$, then the condition $a_{max} < 1$ means that $\inf_{x \in D} \lambda_1(x) \leq \sup_{x \in D} \lambda_2(x) = a_{max} < 1$. In particular, we have $\sup_D \|A^{-1}\|_2 > 1/a_{max} > 1$, where $\| \cdot \|_2$ is the Euclidean norm of the matrix, and this implies that $\xi \cdot (N(x) - I)^{-1}\xi \geq \alpha|\xi|^2$ for all $\xi \in \mathbb{R}^2$, $x \in D$, and some constant $\alpha > 0$. More specifically,

$$\xi \cdot (A^{-1} - I)^{-1}\xi \geq \frac{1}{\|A^{-1}\|_2 - 1}|\xi|^2 \geq \frac{1}{\sup_D \|A^{-1}\|_2 - 1}|\xi|^2, \quad \xi \in \mathbb{R}^2, x \in D;$$

thus,

$$\alpha := \frac{1}{\sup_D \|A^{-1}\|_2 - 1}. \tag{6.71}$$

Theorem 6.18. *Assume that $a_{max} < 1$. The set of real transmission eigenvalues is discrete. If k is a real transmission eigenvalue, then*

$$k^2 \geq \frac{\lambda_0(D)}{\sup_D \|A^{-1}\|_2}, \tag{6.72}$$

where $\lambda_0(D)$ is the first eigenvalue of $-\Delta$ on D.

Proof. To prove the first part of the theorem, we consider the formulation (6.69). Since our assumption $a_{max} < 1$ implies $\xi \cdot (N(x) - I)^{-1}\xi \geq \alpha|\xi|^2$ for all $\xi \in \mathbb{R}^2$, and $x \in D$ with α given by (6.71), we have that

$$\mathcal{A}_k(\mathbf{u}, \mathbf{u}) \geq \alpha\|\nabla\nabla \cdot \mathbf{u} + k^2\mathbf{u}\|^2_{L^2(D)} + k^4\|\mathbf{u}\|^2_{L^2(D)}.$$

Setting $X = \|\nabla\nabla \cdot \mathbf{u}\|_{L^2(D)}$ and $Y = k^2\|\mathbf{u}\|_{L^2(D)}$ we have that

$$\|\nabla\nabla \cdot \mathbf{u} + k^2\mathbf{u}\|^2_{L^2(D)} \geq X^2 - 2XY + Y^2,$$

and therefore

$$\mathcal{A}_k(\mathbf{u}, \mathbf{u}) \geq \alpha X^2 - 2\alpha XY + (\alpha + 1)Y^2. \tag{6.73}$$

From the identity

$$\alpha X^2 - 2\alpha XY + (\alpha + 1)Y^2 = \epsilon \left(Y - \frac{\alpha}{\epsilon} X \right)^2 + \left(\alpha - \frac{\alpha^2}{\epsilon} \right) X^2 + (1 + \alpha - \epsilon) Y^2 \tag{6.74}$$

for $\alpha < \epsilon < \alpha + 1$, setting $\epsilon = \alpha + 1/2$ we now obtain that

$$\mathcal{B}_k(\mathbf{u}, \mathbf{u}) \geq \frac{\alpha}{1 + 2\alpha} (X^2 + Y^2). \tag{6.75}$$

From (6.21) we have

$$\|\nabla\nabla \cdot \mathbf{u} + k^2 \mathbf{u}\|_{L^2(D)}^2 = \|\nabla\nabla \cdot \mathbf{u}\|_{L^2(D)}^2 - 2k^2 \|\nabla \cdot \mathbf{u}\|_{L^2(D)}^2 + k^4 \|\mathbf{u}\|_{L^2(D)}^2,$$

which implies that

$$2k^2 \|\nabla \cdot \mathbf{u}\|_{L^2(D)}^2 \leq X^2 + Y^2.$$

Finally, combining the preceding estimates yields the existence of a constant $c_k > 0$ (independent of $\mathbf{u}$ and α) such that

$$\mathcal{A}_k(\mathbf{u}, \mathbf{u}) \geq c_k \frac{\alpha}{1 + 2\alpha} \|\mathbf{u}\|_{\mathcal{H}(D)}^2. \tag{6.76}$$

Hence the sesquilinear form $\mathcal{A}_k(\cdot, \cdot)$ is coercive in $\mathcal{H}_0(D) \times \mathcal{H}_0(D)$, and consequently the operator $\mathbb{A}_k : \mathcal{H}_0(D) \to \mathcal{H}_0(D)$ is a bijection for fixed k. Recall that from Lemma 6.17 the operator $\mathbb{B} : \mathcal{H}_0(D) \to \mathcal{H}_0(D)$ is compact. Hence, to prove that the set of real transmission eigenvalues is discrete, we apply the analytic Fredholm theorem (Theorem 1.24) to

$$\mathbb{A}_k - k^2 \mathbb{B} \quad \text{or} \quad \mathbb{I} - k^2 \mathbb{A}_k^{-1} \mathbb{B}. \tag{6.77}$$

To this end, we observe that the sesquilinear form $\mathcal{A}_k(\cdot, \cdot)$ is analytic in k, which means that the mapping $k \to \mathbb{A}_k$ is analytic (cf. Theorem 8.22 in [54]). By the Lax–Milgram theorem we can conclude that $\mathbb{A}_k^{-1}$ also exists in a neighborhood of the positive real axis and the mapping $k \to \mathbb{A}_k^{-1}$ is analytic. Consequently, the mapping $k \to k^2 \mathbb{A}_k^{-1} \mathbb{B}$ is analytic in a neighborhood of the real axis and for each k the operator $k^2 \mathbb{A}_k^{-1} \mathbb{B}$ is compact. Therefore, the analytic Fredholm theorem (Theorem 1.24) implies that the set of transmission eigenvalues is discrete provided that there exists a $k > 0$ that is not a transmission eigenvalue, i.e., $\left[\mathbb{I} - k^2 \mathbb{A}_k^{-1} \mathbb{B} \right]^{-1}$ exists. In what follows, we will show that if $k > 0$ is sufficiently small, then k is not a transmission eigenvalue by showing that the operator $\mathbb{A}_k - \mathbb{B} : \mathcal{H}_0(D) \to \mathcal{H}_0(D)$ is an isomorphism for $k > 0$ small enough. To this end, for $\nabla \cdot u \in H_0^1(D)$, using the Poincaré inequality (Sect. 5.2), we have that

$$\|\nabla \cdot \mathbf{u}\|_{L^2(D)}^2 \leq \frac{1}{\lambda_0(D)} \|\nabla\nabla \cdot \mathbf{u}\|_{L^2(D)}^2, \tag{6.78}$$

where $\lambda_0(D)$ is the first Dirichlet eigenvalue of $-\Delta$ on D. Hence, from (6.74) and (6.78) for $\alpha < \epsilon < \alpha + 1$ we have that

$$\mathcal{A}_k(\mathbf{u}, \mathbf{u}) - k^2 \mathcal{B}(\mathbf{u}, \mathbf{u}) \geq \left(\alpha - \frac{\alpha^2}{\epsilon}\right) \|\nabla \nabla \cdot \mathbf{u}\|_{L^2(D)}^2 + (1 + \alpha - \epsilon)k^2 \|\mathbf{u}\|_{L^2(D)}^2$$
$$- k^2 \frac{1}{\lambda_0(D)} \|\nabla \nabla \cdot \mathbf{u}\|_{L^2(D)}^2.$$

Therefore, if $k^2 < \left(\alpha - \alpha^2/\epsilon\right) \lambda_0(D)$ for every $\alpha < \epsilon < \alpha + 1$, then $\mathbb{A}_k - k^2 \mathbb{B}$ is invertible. In particular, taking ϵ arbitrarily close to $\alpha + 1$ we have that if $k^2 < \frac{\alpha}{1+\alpha}\lambda_0(D)$, then k is not a transmission eigenvalue. This completes the proof of discreteness of real transmission eigenvalues.

In the foregoing discussion, we showed that if $k > 0$ is a transmission eigenvalue, then it must satisfy $k^2 > \frac{\alpha}{1+\alpha}\lambda_0(D)$, and thus, from (6.71) we obtain that $k^2 \geq \frac{\lambda_0(D)}{\sup_D \|A^{-1}\|_2}$, which proves the theorem.   □

In a similar way it is possible to prove a similar result if $a_{min} > 1$ (see [33] for details). In particular, the following theorem holds.

Theorem 6.19. *Assume that $a_{min} > 1$. The set of real transmission eigenvalues is discrete. If k is a real transmission eigenvalue, then*

$$k^2 \geq \lambda_0(D), \tag{6.79}$$

where $\lambda_0(D)$ is the first eigenvalue of $-\Delta$ on D.

Now we turn our attention to prove the existence of positive transmission eigenvalues. We again only consider in detail the case where $a_{max} < 1$.

Theorem 6.20. *Assume that $a_{max} < 1$. Then there exists an infinite number of positive transmission eigenvalues with $+\infty$ as the only accumulation point.*

Proof. As explained earlier, $k > 0$ is a transmission eigenvalue if and only if the kernel of the operator $\mathbb{A}_k - k^2\mathbb{B}$ or $\mathbb{I} - k^2\mathbb{A}_k^{-1}\mathbb{B}$ is not empty, where the bounded, self-adjoint, strictly positive definite operator $\mathbb{A}_k^{-1}$ and the bounded, self-adjoint, nonnegative, compact operator $\mathbb{B}$ are defined by (6.70). Note that

$$N(\mathbb{B}) = \{\mathbf{u} \in \mathcal{H}_0(D) \quad \text{such that } \mathbf{u} := \text{curl}\,\varphi, \ \varphi \in H(\text{curl}, D)\}.$$

We first observe that the multiplicity of each transmission eigenvalue is finite since it coincides with the multiplicity of the eigenvalue 1 of the compact operator $k^2\mathbb{A}_k^{-1}\mathbb{B}$, which is finite. To analyze the kernel of this operator, we consider the auxiliary eigenvalue problems

$$\mathbb{A}_k\mathbf{u} - \lambda(k)\mathbb{B}\mathbf{u} = 0 \qquad \mathbf{u} \in \mathcal{H}_0(D). \tag{6.80}$$

Thus, a transmission eigenvalue $k > 0$ satisfies $\lambda(k) - k^2 = 0$, where $\lambda(k)$ is an eigenvalue corresponding to (6.80). To prove the existence of an infinite set of transmission eigenvalues, we now use Theorem 6.15 for $\mathbb{A}_k^{-1}$ and $\mathbb{B}$ with $X = \mathcal{H}_0(D)$. Theorem 6.18 states that as long as $0 < k_0^2 < \frac{\lambda_0(D)}{\sup_D \|A^{-1}\|_2}$, the operator $\mathbb{A}_{k_0} - k_0^2 \mathbb{B}$ is positive in $\mathcal{H}_0(D)$, whence assumption 1 of Theorem 6.15 is satisfied for $\tau_0 := k_0^2$. Next, let $k_{1,a_{max}}$ be the first transmission eigenvalue for the disk B of radius $R = 1$ and constant index of refraction $n := a_{max}^{-1}$ [i.e., (6.63)–(6.66) for $D := B$ and $N(x) := nI$ or (6.49)–(6.52) with $R = 1$, $n_0 = 1$, and $a_0 = a_{max}$]. This transmission eigenvalue is the first zero of

$$W(k) = \det \begin{pmatrix} J_0(k) & J_0\left(k\sqrt{\frac{1}{a_{max}}}\right) \\ k\,J_0'(k) & k\sqrt{a_{max}}\,J_0'\left(k\sqrt{\frac{1}{a_{max}}}\right) \end{pmatrix} \qquad (6.81)$$

[if the first zero of the preceding determinant is not the first transmission eigenvalue, then the latter will be a zero of (6.53) for $\ell \geq 1$]. By a scaling argument, it is obvious that $k_\epsilon := k_{1,a_{max}}/\epsilon$ is the first transmission eigenvalue corresponding to a disk of radius $\epsilon > 0$ with index of refraction a_{max}^{-1}. Now take $\epsilon > 0$ small enough such that D contains $m := m(\epsilon) \geq 1$ disjoint disks $B_\epsilon^1, B_\epsilon^2 \ldots B_\epsilon^m$ of radius ϵ, i.e., $\overline{B_\epsilon^j} \subset D$, $j = 1 \ldots m$, and $\overline{B_\epsilon^j} \cap \overline{B_\epsilon^i} = \emptyset$ for $j \neq i$. Then $k_\epsilon = k_{1,a_{max}}/\epsilon$ is the first transmission eigenvalue for each of these disks with index of refraction a_{max}^{-1}, and let $\mathbf{u}^j := \mathbf{u}^{B_\epsilon^j, a_{min}} \in \mathcal{H}_0(B_\epsilon^j)$, $j = 1 \ldots m$, be the corresponding eigenfunctions. We have that $u^j \in \mathcal{H}_0(B_\epsilon^j)$ and

$$\int_{B_\epsilon^j} \frac{1}{n-1}(\nabla\nabla \cdot \mathbf{u}^j + k_\epsilon^2 \mathbf{u}^j) \cdot (\nabla\nabla \cdot \overline{\mathbf{u}}^j + k_\epsilon^2 n \overline{\mathbf{u}}^j)\,dx = 0. \qquad (6.82)$$

By definition, the vectors $\tilde{u}^j$ are not in the kernel of $\mathbb{B}$. The extension by zero $\tilde{u}^j$ of u^j to the whole D is obviously in $\mathcal{H}_0(D)$ due to the boundary conditions on ∂B^j. Furthermore, the functions $\{\tilde{u}^1, \tilde{u}^2, \ldots \tilde{u}^m\}$ are linearly independent and orthogonal in $\mathcal{H}_0(D)$ since they have disjoint supports, and from (6.82) we have that

$$0 = \int_{B_\epsilon^j} \frac{1}{n-1}(\nabla\nabla \cdot \mathbf{u}^j + k_\epsilon^2 \mathbf{u}^j) \cdot (\nabla\nabla \cdot \overline{\mathbf{u}}^j + k_\epsilon^2 n \overline{\mathbf{u}}^j)\,dx \qquad (6.83)$$

$$= \int_D \frac{1}{n-1}|\nabla\nabla \cdot \tilde{\mathbf{u}}^j + k_\epsilon^2 \tilde{\mathbf{u}}^j|^2\,dx + k_\epsilon^4 \int_D |\tilde{\mathbf{u}}^j|^2\,dx - k_\epsilon^2 \int_D |\nabla \cdot \tilde{\mathbf{u}}|^2\,dx$$

for $j = 1 \ldots m$. Denote by W_m the m-dimensional subspace of $\mathcal{H}_0(D)$ spanned by $\{\tilde{u}^1, \tilde{u}^2, \ldots \tilde{u}^m\}$. Since each $\tilde{u}^j$, $j = 1, \ldots, m$, satisfies (6.83) and they have disjoint supports, we have that for k_ϵ and for every $\tilde{u} \in W_m$

$$\left(\mathbb{A}_{k_\epsilon}\tilde{u} - k_\epsilon^2 \mathbb{B}\tilde{u},\ \tilde{u}\right)_{\mathcal{H}_0(D)} \tag{6.84}$$

$$= \int_D (N-I)^{-1}|\nabla\nabla\cdot\tilde{u} + k_\epsilon^2\tilde{u}|^2\,dx + k_\epsilon^4\int_D |\tilde{u}|^2\,dx - k_\epsilon^2\int_D |\nabla\cdot\tilde{u}|^2\,dx$$

$$\leq \int_D \frac{1}{n-1}|\nabla\nabla\cdot\tilde{u} + k_\epsilon^2\tilde{u}|^2\,dx + k_\epsilon^4\int_D |\tilde{u}|^2\,dx - k_\epsilon^2\int_D |\nabla\cdot\tilde{u}|^2\,dx = 0.$$

This means that assumption 2 of Theorem 6.15 is also satisfied, and therefore we can conclude that there are $m(\epsilon)$ transmission eigenvalues (counting multiplicity) inside $[\frac{\lambda_0(D)}{\sup_D \|A^{-1}\|_2},\ \frac{k_{1,\,a_{max}}}{\epsilon}]$. Note that $m(\epsilon)$ and k_ϵ both go to $+\infty$ as $\epsilon \to 0$. Since the multiplicity of each eigenvalue is finite, we have shown, by letting $\epsilon \to 0$, that there exists an infinite countable set of transmission eigenvalues that accumulate at $+\infty$. □

In a similar way it is possible to prove an analogous result if $a_{min} > 1$ (see [33] for details). In particular, the following theorem holds.

Theorem 6.21. *Assume that $a_{min} > 1$. Then there exists an infinite number of positive transmission eigenvalues with $+\infty$ as the only accumulation point.*

The foregoing proof of the existence of transmission eigenvalues provides a framework in which to obtain lower and upper bounds for the first transmission eigenvalue. To this end, denote by $k_{0,A} > 0$ the first positive transmission eigenvalue corresponding to A and D (we omit the dependence on D in our notation since D is assumed to be known). Assume again that $a_{max} < 1$.

Theorem 6.22. *Assume that the index of refraction $A(x)$ satisfies $a_{max} < 1$, where a_{max} and a_{min} are given by (6.48). Then*

$$0 < k_{0,a_{min}} \leq k_{0,A(x)} \leq k_{0,a_{max}}. \tag{6.85}$$

Proof. From the proof of Theorem 6.20 we have that $k_{0,A}^2$ is the smallest zero of

$$\lambda(k,A) - k^2 = 0, \tag{6.86}$$

where

$$\lambda(k,A) = \inf_{\substack{u \in \mathcal{H}_0(D) \\ \|\nabla\cdot u\|_D = 1}} \int_D (A^{-1}-I)^{-1}|\nabla\nabla\cdot u + k^2 u|^2\,dx + k^4\int_D |u|^2\,dx \tag{6.87}$$

and u not in the kernel of $\mathbb{B}$. [Note that any zero of $\lambda(k,A) - k^2 = 0$ leads to a transmission eigenvalue.] Obviously, the mapping $k \to \lambda(k,A)$ is continuous on $(0, +\infty)$. We first note that (6.87) yields

$$\lambda(k, a_{min}) \leq \lambda(k, A(x)) \leq \lambda(k, a_{max}) \tag{6.88}$$

for all $k > 0$. In particular, for $k := k_{0,a_{min}}$ we have that

$$0 = \lambda(k_{0,a_{min}}, a_{min}) - k_{0,a_{min}}^2 \leq \lambda(k_{0,a_{min}}, A(x)) - k_{0,a_{min}}^2,$$

and for $k := k_{0,a_{max}}$ we have that

$$\lambda(k_{0,a_{min}}, A(x)) - k_{0,a_{min}}^2 \leq \lambda(k_{0,a_{max}}, a_{max}) - k_{0,a_{max}}^2 = 0.$$

By continuity of $k \to \lambda(\tau, A) - k^2$, we have that there is a zero $\tilde{k}$ of $\lambda(k, A) - k^2 = 0$ such that $k_{0,a_{min}} \leq \tilde{k} \leq k_{0,a_{max}}$. In particular, the smallest zero $k_{0,A(x)}$ of $\lambda(k, A) - k^2 = 0$ is such that $k_{0,A(x)} \leq \tilde{k} \leq k_{0,a_{max}}$. To end the proof, we need to show that $k_{0,a_{min}} \leq k_{0,A(x)}$, i.e., all the positive zeros of $\lambda(k, A) - k^2 = 0$ are greater than or equal to $k_{0,a_{min}}$. Assume by contradiction that $k_{0,A(x)} < k_{0,a_{min}}$. Then, from (6.88), on the one hand, we have

$$\lambda(k_{0,A(x)}, a_{min}) - k_{0,A(x)}^2 \leq \lambda(k_{0,A(x)}, A(x)) - k_{0,A(x)}^2 = 0.$$

On the other hand, from the proof of Theorem 6.18 we have that for a sufficiently small $k' > 0$, $\lambda(k', a_{min}) - k'^2 > 0$. Hence there exists a zero of $\lambda(k, a_{min}) - k^2 = 0$ between k' and $k_{0,A(x)}$ smaller than $k_{0,a_{min}}$, which contradicts the fact that $k_{0,a_{min}}$ is the smallest zero. Thus we have proven that $k_{0,a_{min}} \leq k_{0,A(x)} \leq k_{0,a_{max}}$, and this completes the proof.   □

In a similar way [31, 33], one can prove the following theorem.

Theorem 6.23. *Assume that the index of refraction $A(x)$ satisfies $a_{min} > 1$, where a_{max} and a_{min} are given by (6.48). Then*

$$0 < k_{0,a_{max}} \leq k_{0,A(x)} \leq k_{0,a_{min}}. \tag{6.89}$$

Theorems 6.22 and 6.23 show in particular that for constant index of refraction $A = aI$ the first transmission eigenvalue $k_{0,a}$ is monotonically increasing if $0 < a < 1$ and is monotonically decreasing if $a > 1$. If fact we can show that this monotonicity is strict, which leads to the following uniqueness result for a constant index of refraction in terms of the first transmission eigenvalue.

Theorem 6.24. *The constant index of refraction $A := aI$ is uniquely determined from a knowledge of the corresponding smallest transmission eigenvalue $k_{0,a} > 0$, provided that it is known a priori that either $a > 1$ or $0 < a < 1$.*

Proof. We show the proof for the case $0 < a < 1$ (a similar proof works for the case $a > 1$). Consider two homogeneous media with constant indexes of refraction a_1 and a_2 such that $a_2 < a_1 < 1$, and let $\mathbf{u}_1 := \mathbf{w}_1 - \mathbf{v}_1$, where $\mathbf{w}_1, \mathbf{v}_1$ is the nonzero solution of (6.63)–(6.66), with $A(x) := a_1 I$ corresponding to the first transmission eigenvalue k_{0,a_1}. Now, setting $k_0 := k_{0,a_1}$ and after normalizing $\mathbf{u}_1$ such that $\nabla \cdot \mathbf{u}_1 = 1$, we have

$$\frac{1}{1/a_1 - 1} \|\nabla\nabla \cdot \mathbf{u}_1 + k_0^2 \mathbf{u}_1\|_{L^2(D)}^2 + k_0^4 \|\mathbf{u}_1\|_{L^2(D)}^2 = k_0^2 = \lambda(k_1, a_1).$$

Furthermore, we have

$$\frac{1}{1/a_2 - 1}\|\nabla\nabla \cdot \mathbf{u} + k^2\mathbf{u}\|^2_{L^2(D)} + k^4\|\mathbf{u}_1\|^2_{L^2(D)}$$

$$\leq \frac{1}{1/a_1 - 1}\|\nabla\nabla \cdot \mathbf{u} + k^2\mathbf{u}\|^2_{L^2(D)} + k^4\|\mathbf{u}_1\|^2_{L^2(D)}$$

for all $\mathbf{u} \in \mathcal{H}_0(D)$ such that $\|\nabla \cdot \mathbf{u}\|_D = 1$, $\mathbf{u}$ not in the kernel of $\mathbb{B}$, and all $k > 0$. In particular, for $\mathbf{u} = \mathbf{u}_1$ and $k = k_0$

$$\frac{1}{1/a_2 - 1}\|\nabla\nabla \cdot \mathbf{u}_1 + k_0^2\mathbf{u}_1\|^2_{L^2(D)} + k_0^4\|\mathbf{u}_1\|^2_{L^2(D)}$$

$$< \frac{1}{1/a_2 - 1}\|\nabla\nabla \cdot \mathbf{u}_1 + k_0^2\mathbf{u}_1\|^2_{L^2(D)} + k_0^4\|\mathbf{u}_1\|^2_{L^2(D)} = \lambda(k_0, a_1).$$

But

$$\lambda(k_0, a_2) \leq \frac{1}{1/a_2 - 1}\|\nabla\nabla \cdot \mathbf{u}_1 + k_0^2\mathbf{u}_1\|^2_{L^2(D)} + k_0^4\|\mathbf{u}_1\|^2_{L^2(D)} < \lambda(k_0, a_1),$$

and hence for this k_0 we have a strict inequality, i.e.,

$$\lambda(k_0, a_2) < \lambda(k_0, a_1). \qquad (6.90)$$

Hence, (6.90) implies the first zero k_{0,a_2} of $\lambda(k, a_2) - k^2 = 0$ is such that $k_{0,a_2} < k_{0,a_1}$ for the first transmission eigenvalues k_{0,a_1} and k_{0,a_2} corresponding to a_1 and a_2, respectively. Hence we have shown that if $0 < a_1 < 1$ and $0 < a_2 < 1$ are such that $a_1 \neq a_2$, then $k_{0,a_1} \neq k_{0,a_2}$, which proves the desired strict monotonicity. The uniqueness result now follows immediately from Theorem 6.22. $\qquad\square$

From the proof of Theorems 6.22 and 6.23, one can see that the following more general monotonicity property of the first transmission eigenvalue with respect to the support of inhomogeneity and the refractive index holds true.

Corollary 6.25. *Let $D_1 \subset D \subset D_2$ and $A_1 < A < A_2$, where A_1, A, A_2 all satisfy the assumptions of either Theorem 6.22 or Theorem 6.23. If $k_{0,A,D}$ denotes the first transmission eigenvalue corresponding to D and A, then*

$$0 < k_{0,A_2,D_2} \leq k_{0,A_2,D} \leq k_{0,A,D} \leq k_{0,A_1,D} \leq k_{0,A_1,D_1}$$

if the assumptions of Theorem 6.22 are satisfied and

$$0 < k_{0,A_1,D_2} \leq k_{0,A_1,D} \leq k_{0,A,D} \leq k_{0,A_2,D} \leq k_{0,A_2,D_1}$$

if the assumptions of Theorem 6.22 are satisfied. Here $A_1 < A$ means that the matrix $A - A_1$ is positive definite uniformly in D, with a similar definition for $A < A_2$.

Remark 6.26. The existence and discreteness of transmission eigenvalues for the problem (6.54)–(6.57) are also considered in [105] using a different approach. In particular, in [105] (see also [54] for the case where $A = I$ and $n \neq 1$) the transmission eigenvalue problem (6.54)–(6.57) is shown to be an eigenvalue problem for a quadratic pencil operator $I - k^2\mathbb{C} + k^4\mathbb{D}$, where $\mathbb{C}$ and $\mathbb{D}$ are self-adjoint compact operators and $\mathbb{D}$ is nonnegative. The latter becomes a linear eigenvalue problem for the non-self-adjoint, matrix-valued operator

$$\begin{pmatrix} \mathbb{C} & \mathbb{D}^{\frac{1}{2}} \\ -\mathbb{D}^{\frac{1}{2}} & 0 \end{pmatrix}.$$

We note that interesting analytical results for this type of non-self-adjoint eigenvalue problems were obtained in [36] and [145]. For more results on a transmission eigenvalue problem as an eigenvalue problem for a quadratic pencil operator see [84, 85, 86].

6.3.2 The Case $n \neq 1$

We now turn our attention to the general case where both $A \neq 1$ and $n \neq 1$. We recall that the *transmission eigenvalue problem* is the problem of finding two nonzero functions $v \in H^1(D)$ and $w \in H^1(D)$ satisfying

$$\nabla \cdot A\nabla v + k^2 n\, v = 0 \qquad \text{in} \quad D, \tag{6.91}$$

$$\Delta w + k^2\, w = 0 \qquad \text{in} \quad D, \tag{6.92}$$

$$v = w \qquad \text{on} \quad \partial D, \tag{6.93}$$

$$\frac{\partial v}{\partial \nu_A} = \frac{\partial w}{\partial \nu} \qquad \text{on} \quad \partial D. \tag{6.94}$$

We already discussed at the beginning of Chap. 6.2 the Fredholm property of the foregoing problem under the assumption that $A - I > 0$ or $I - A > 0$ in D. In fact, we can show that the interior transmission problem satisfies the Fredholm property if the preceding assumptions on the contrast are satisfied only in a neighborhood of the boundary, but in this case we need to impose the same assumptions on the contrast $n - 1$. In addition, the approach we are about to discuss also proves that the set of transmission eigenvalues is discrete. Note that for this general case the existence of transmission eigenvalues can be proven under much more restrictive assumptions using a different approach.

6.3.3 Discreteness of Transmission Eigenvalues

Let $\mathcal{N}$ be a δ-neighborhood of the boundary ∂D in D i.e.,

$$\mathcal{N} := \{x \in D : \text{dist}(x, \partial D) < \delta\},$$

and introduce the following notations:

$$a_* := \inf_{x \in \mathcal{N}} \inf_{\xi \in \mathbb{R}^2, |\xi|=1} (\xi \cdot A(x)\xi) > 0,$$

$$a^* := \sup_{x \in \mathcal{N}} \sup_{\xi \in \mathbb{R}^2, |\xi|=1} (\xi \cdot A(x)\xi) < \infty, \tag{6.95}$$

$$n_* := \inf_{x \in \mathcal{N}} n(x) > 0 \quad \text{and} \quad n^* := \sup_{x \in \mathcal{N}} n(x) < \infty.$$

Note that in (6.95) the infimum and supremum are only taken over a neighborhood of the boundary ∂D as opposed to over the entire domain D as in (6.48).

We consider the Sobolev space

$$\mathcal{H}(D) := \left\{ (v, w) \in H^1(D) \times H^1(D) : v - w \in H_0^1(D) \right\}.$$

Our first observation is that $(v, w) \in H^1(D) \times H^1(D)$ is a solution to (6.91)–(6.94) if and only if

$$a_k((v, w), (v', w')) = 0 \quad \text{for all} \quad (v', w') \in \mathcal{H}(D), \tag{6.96}$$

where the sesquilinear form $a_k(\cdot, \cdot) : \mathcal{H}(D) \to \mathbb{C}$ is defined by

$$a_k((v, w), (v', w')) := \int_D A\nabla v \cdot \nabla \bar{v}' \, dx - \int_D \nabla w \cdot \nabla \bar{w}' \, dx$$
$$- k^2 \int_D n v \, \bar{v}' \, dx + k^2 \int_D w \, \bar{w}' \, dx.$$

Let $\mathbf{A}_k : \mathcal{H}(D) \to \mathcal{H}(D)$ be the bounded linear operator defined by means of the Riesz representation theorem

$$(\mathbf{A}_k(v, w), (v', w'))_{\mathcal{H}(D)} = a_k((v, w), (v', w')). \tag{6.97}$$

Obviously, $\mathbf{A}_k$ depends analytically on $k \in \mathbb{C}$, and furthermore, for any two k and k' the operator $\mathbf{A}_k - \mathbf{A}_{k'}$ is compact, which is a simple consequence of the compact embedding of $\mathcal{H}(D)$ into $L^2(D) \times L^2(D)$. Therefore, to prove the discreteness of transmission eigenvalues, it suffices to prove that $\mathbf{A}_{k'}$ is invertible for some $k' \in \mathbb{C}$ since then we can write $A_k = A_{k'} + (A_k - A_{k'})$ and appeal to the analytic Fredholm theorem (Theorem 1.24). The difficulty in obtaining this result is that the sesquilinear form $a_k((v, w), (v', w'))$ is not coercive for any $k \in \mathbb{C}$ due to the opposite signs in the terms containing the gradients. To show the invertibility of the $\mathbf{A}_k$, we follow the arguments in [9] and [39], which rely on proving that $a_k(\cdot, \cdot)$ is T-coercive (as it is called in [10]) for some k. More specifically, the idea behind T-coercivity is to consider an equivalent formulation of (6.96), where a_k is replaced by a_k^T defined by

$$a_k^T((v,w),(v',w')) := a_k((v,w),\mathbf{T}(v',w')) \tag{6.98}$$

for all $((v,w),(v',w')) \in \mathcal{H}(D) \times \mathcal{H}(D)$, with the operator $\mathbf{T} : \mathcal{H}(D) \to \mathcal{H}(D)$ being an isomorphism. Obviously, $(v,w) \in \mathcal{H}(D)$ satisfies

$$a_k((v,w),(v',w')) = 0 \qquad \text{for all} \qquad (v',w') \in \mathcal{H}(D)$$

if and only if it satisfies

$$a_k^T((v,w),(v',w')) = 0 \qquad \text{for all} \qquad (v',w') \in \mathcal{H}(D).$$

If we can choose $\mathbf{T}$ and k such that a_k^T is coercive, then using the Lax–Milgram theorem and the fact that $\mathbf{T}$ is an isomorphism we can deduce that $\mathbf{A}_k : \mathcal{H}(D) \to \mathcal{H}(D)$ defined by (6.97) is invertible.

Lemma 6.27. *Assume that either* $0 < a^* < 1$ *and* $0 < n^* < 1$, *or* $a_* > 1$ *and* $n_* > 1$. *Then there exists* $k = i\kappa$, *with* $\kappa \in \mathbb{R}$, *such that* $\mathbf{A}_{i\kappa} : \mathcal{H}(D) \to \mathcal{H}(D)$ *is invertible.*

Proof. Let us first consider the case where $0 < a^* < 1$ and $0 < n^* < 1$ and introduce $\chi \in \mathcal{C}^\infty(\overline{D})$, a cutoff function equal to 1 in a neighborhood of ∂D supported in $\mathcal{N}$ such that $0 \leq \chi \leq 1$. We define the isomorphism $\mathbf{T} : \mathcal{H}(D) \to \mathcal{H}(D)$ by

$$\mathbf{T} : (v,w) \mapsto (v - 2\chi w, -w).$$

(Note that $\mathbf{T}$ is an isomorphism since $\mathbf{T}^2 = I$.) We then have that for all $(v,w) \in \mathcal{H}(D)$

$$
\begin{aligned}
\left| a_{i\kappa}^T((v,w),(v,w)) \right| = & \left| (A\nabla v, \nabla v)_D + (\nabla w, \nabla w)_D - 2(A\nabla v, \nabla(\chi w))_D \right. \\
& \left. + \kappa^2 \left((nv,v)_D + (w,w)_D - 2(nv, \chi w)_D \right) \right|, \quad (6.99)
\end{aligned}
$$

where $(\cdot,\cdot)_\mathcal{O}$ for a generic bounded region $\mathcal{O} \subset \mathbb{R}^2$ denotes the $L^2(\mathcal{O})$ inner product. Using Young's inequality

$$|ab| \leq \epsilon a^2 + \frac{1}{\epsilon} b^2, \qquad \epsilon > 0,$$

we can write

$$
\begin{aligned}
2 \left| (A\nabla v, \nabla(\chi w))_D \right| \leq & \ 2 \left| (\chi A\nabla v, \nabla w)_\mathcal{N} \right| + 2 \left| (A\nabla v, \nabla(\chi)w)_\mathcal{N} \right| \\
\leq & \ \eta (A\nabla v, \nabla v)_\mathcal{N} + \eta^{-1}(A\nabla w, \nabla w)_\mathcal{N} \qquad (6.100) \\
& + \alpha(A\nabla v, \nabla v)_\mathcal{N} + \alpha^{-1}(A\nabla(\chi)w, \nabla(\chi)w)_\mathcal{N}
\end{aligned}
$$

and

$$2 \left| (nv, \chi w)_D \right| \leq \beta(nv,v)_\mathcal{N} + \beta^{-1}(nw,w)_\mathcal{N} \tag{6.101}$$

for arbitrary constants $\alpha > 0$, $\beta > 0$, and $\eta > 0$. Substituting (6.100) and (6.101) into (6.99), we now obtain

$$
\begin{aligned}
\left| a_{i\kappa}^T((v,w),(v,w)) \right| &\geq (A\nabla v, \nabla v)_{D\backslash\overline{\mathcal{N}}} + (\nabla w, \nabla w)_{D\backslash\overline{\mathcal{N}}} \\
&\quad + \kappa^2\left((nv,v)_{D\backslash\overline{\mathcal{N}}} + (w,w)_{D\backslash\overline{\mathcal{N}}} \right) \\
&\quad + ((1-\eta-\alpha)A\nabla v, \nabla v)_{\mathcal{N}} + ((I - \eta^{-1}A)\nabla w, \nabla w)_{\mathcal{N}} \\
&\quad + \kappa^2((1-\beta)nv,v)_{\mathcal{N}} + ((\kappa^2(1-\beta^{-1}n) - \sup_{\mathcal{N}}|\nabla\chi|^2 a^*\alpha^{-1})w,w)_{\mathcal{N}}.
\end{aligned}
$$

Taking η, α, and β such that $a^* < \eta < 1$, $n^* < \beta < 1$, and $0 < \alpha < 1 - \eta$, we obtain the coercivity of $a_{i\kappa}^T$ for κ large enough, which proves the lemma. The case where $a_* > 1$ and $n_* > 1$ can be handled in a similar way using $\mathbf{T}(v,w) := (v, -w + 2\chi v)$. $\qquad\square$

Lemma 6.27, combined with the fact that $\mathbf{A}_k - \mathbf{A}_{i\kappa}$ is compact, and an application of the analytic Fredholm theorem (Theorem 1.24) implies the following theorem.

Theorem 6.28. *Assume that either $0 < a^* < 1$ and $0 < n^* < 1$, or $a_* > 1$ and $n_* > 1$. Then the set of transmission eigenvalues is discrete in $\mathbb{C}$.*

Remark 6.29. As a consequence of the proof of Lemma 6.27 we can conclude that the operator $\mathbf{A}_k$ is Fredholm with index zero (cf. [127]). This implies that under the assumptions that either $0 < a^* < 1$ and $0 < n^* < 1$, or $a_* > 1$ and $n_* > 1$, the interior transmission problem (6.91)–(6.94) with boundary data $f \in H^{\frac{1}{2}}(\partial D)$ and $h \in H^{-\frac{1}{2}}(\partial D)$ has a unique solution $(v,w) \in H^1(D) \times H^1(D)$, provided $k \in \mathbb{C}$ is not a transmission eigenvalue. Furthermore, the solution depends continuously on the data f, h.

We conclude this section by showing that if we require that the contrast keep the same sign in D, i.e., $0 < a_{max} < 1$ or $a_{min} > 1$, the T-coercivity approach allows us to prove the discreteness of transmission eigenvalues under more relaxed assumptions on $n - 1$. To this end, taking $v' = w' = 1$ in (6.96) we first notice that the transmission eigenfunctions (v,w) [i.e., the solution to (6.91)–(6.91) corresponding to an eigenvalue k] satisfy $k^2 \int_D (nv-w)dx = 0$. This suggests introducing the subspace of $\mathcal{H}(D)$

$$
\mathcal{Y}(D) := \left\{ (v,w) \in \mathcal{H}(D) \,\Big|\, \int_D (nv - w)dx = 0 \right\}.
$$

Now, suppose $\int_D (n-1)dx \neq 0$. Arguing by contradiction, one can prove the existence of a constant $C_P > 0$ (which depends on D and on n) such that

$$
\|v\|_D^2 + \|w\|_D^2 \leq C_P(\|\nabla v\|_D^2 + \|\nabla w\|_D^2), \quad \forall (v,w) \in \mathcal{Y}(D). \tag{6.102}
$$

Furthermore, we observe that $k \neq 0$ is a transmission eigenvalue if and only if there exists a nontrivial element $(v,w) \in \mathcal{Y}(D)$ such that

$$
a_k((v,w),(v',w')) = 0 \text{ for all } (v',w') \in \mathcal{Y}(D).
$$

Using the variational formulation in this new subspace and (6.102) we can now prove the following theorem, which completes the analysis of the solvability of the interior transmission problem discussed at the beginning of Sect. 6.2.

Theorem 6.30. *Assume that either* $0 < a_{max} < 1$ *or* $a_{min} > 1$, *and* $\int_D (n - 1)dx \neq 0$. *Then the set of transmission eigenvalues is discrete in* $\mathbb{C}$.

Proof. For the sake of simplicity we only consider in detail the case where $0 < a_{max} < 1$. Letting $\lambda(w) := 2\int_D (n - 1)w / \int_D (n - 1)$ we consider the mapping $\mathbf{T} : \mathcal{Y}(D) \to \mathcal{Y}(D)$ defined by

$$\mathbf{T} : (v, w) \mapsto (v - 2w + \lambda(w), -w + \lambda(w)).$$

Note that $\lambda(\lambda(w)) = 2\lambda(w)$, which implies that $\mathbf{T}^2 = I$, and hence $\mathbf{T}$ is an isomorphism in $\mathcal{Y}(D)$. Then for all $(v, w) \in \mathcal{Y}(D)$ we have that

$$\begin{aligned}
&\left| a_k^T((v, w), (v, w)) \right| \\
&= \left| (A\nabla v, \nabla v)_D + (\nabla w, \nabla w)_D - 2(A\nabla v, \nabla w)_D \right. \\
&\quad \left. - k^2 \left((nv, v)_D + (w, w)_D - 2(nv, w)_D \right) \right| \\
&\geq (A\nabla v, \nabla v)_D + (\nabla w, \nabla w)_D - 2 \left| (A\nabla v, \nabla w)_D \right| \\
&\quad - |k|^2 \left((nv, v)_D + (w, w)_D + 2 \left| (nv, w)_D \right| \right) \\
&\geq (1 - \sqrt{a_{max}})((A\nabla v, \nabla v)_D + (\nabla w, \nabla w)_D) \\
&\quad - |k|^2 (1 + \sqrt{n_{max}})((nv, v)_D + (w, w)_D).
\end{aligned}$$

Consequently, for $k \in \mathbb{C}$ such that

$$|k|^2 < (a_{min}(1 - \sqrt{a_{max}}))/(C_P \max(n_{max}, 1)(1 + \sqrt{n_{max}}))$$

a_k^T is coercive on $\mathcal{Y}(D)$. The claim of the theorem follows from the analytic Fredholm theorem.   □

The case $a_{min} > 1$ can be handled in a similar way using the isomorphism $\mathbf{T} : \mathcal{Y}(D) \to \mathcal{Y}(D)$ defined by

$$\mathbf{T} : (v, w) \mapsto (v - \lambda(v), -w + 2v - \lambda(v)).$$

We refer the reader to [9] for estimates on transmission eigenvalues following from the foregoing analysis. For the discreteness of complex transmission eigenvalues in the case where $A = I$ see [154].

6.3.4 Existence of Transmission Eigenvalues for $n \neq 1$

We finally come to the discussion of the existence of positive transmission eigenvalues in the general case of anisotropic media with $n \neq 1$. Unfortunately, the existence of transmission eigenvalues for this case can only be shown under restrictive assumptions on $A - I$ and $n - 1$. The approach presented here follows the lines of [35], where, motivated by the case of $n = 1$, the transmission eigenvalue problem is formulated in terms of the difference $u :=$ $v - w$. However, due to the lack of symmetry, the problem for u is no longer a quadratic eigenvalue problem but takes the form of a more complicated nonlinear eigenvalue problem, as will become clear in what follows.

To simplify the expressions, we set $\tau := k^2$ in (6.91)–(6.94) and observe that, if (w, v) satisfies (6.91)–(6.94), then subtracting the equation for v from the equation for w we obtain

$$
\begin{aligned}
\nabla \cdot A\nabla u + \tau n u &= \nabla \cdot (A - I)\nabla w + \tau(n - 1)\, w \quad \text{in } D, \\
\nu \cdot A\nabla u &= \nu \cdot (A - I)\nabla w \quad \text{on } \partial D,
\end{aligned}
\tag{6.103}
$$

where $u := v - w$. In addition, we also have $u = 0$ on ∂D and

$$
\Delta w + \tau w = 0 \quad \text{in } D. \tag{6.104}
$$

It is easy to verify that (v, w) in $H^1(D) \times H^1(D)$ satisfies (6.91)–(6.94) if and only if (u, w) in $H_0^1(D) \times H^1(D)$ satisfies (6.103)–(6.104). The main idea of the proof of the existence of transmission eigenvalues consists in expressing w in terms of u, using (6.103), and substituting the resulting expression into (6.104) in order to formulate the eigenvalue problem only in terms of u. In the case where $A = I$, this substitution is simple and leads to an explicit expression for the equation satisfied by u (see [54], Sect. 10.5, and [105]). In the current case the substitution requires the inversion of the operator $\nabla \cdot [(A - I)\nabla \cdot] + \tau(n - 1)$ with a Neumann boundary condition. It is then obvious that the case where $(A - I)$ and $(n - 1)$ have the same sign is more problematic since in that case the operator may not be invertible for special values of τ. This is why we only consider in detail the simpler case where $(A - I)$ and $(n - 1)$ have opposite signs almost everywhere in D.

Note that for given $u \in H_0^1(D)$, the problem (6.103) for $w \in H^1(D)$ is equivalent to the variational formulation

$$
\int_D \left[(A - I)\nabla w \cdot \nabla \overline{\psi} - \tau\,(n - 1)\, w\, \overline{\psi} \right] dx = \int_D \left[A\, \nabla u \cdot \nabla \overline{\psi} - \tau n\, u\, \overline{\psi} \right] dx \tag{6.105}
$$

for all $\psi \in H^1(D)$. The following result concerning the invertibility of the operator associated with (6.105) can be proven in a standard way using the Lax–Milgram lemma. We skip the proof here and refer the reader to [35].

Lemma 6.31. *Assume that either $a_{min} > 1$ and $0 < n_{max} < 1$, or $0 < a_{max} < 1$ and $n_{min} > 1$. Then there exists $\delta > 0$ such that for every $u \in H_0^1(D)$ and $\tau \in \mathbb{C}$ with $Re(\tau) > -\delta$ there exists a unique solution $w := w_u \in H^1(D)$ of (6.105). The operator $A_\tau : H_0^1(D) \to H^1(D)$, defined by $u \mapsto w_u$, is bounded and depends analytically on $\tau \in \{z \in \mathbb{C} : Re(z) > -\delta\}$.*

We now set $w_u := A_\tau u$ and denote by $\mathbb{L}_\tau u \in H_0^1(D)$ the unique Riesz representation of the bounded conjugate-linear functional

$$\psi \mapsto \int_D \left[\nabla w_u \cdot \nabla \overline{\psi} - \tau\, w_u\, \overline{\psi} \right] dx \quad \text{for } \psi \in H_0^1(D),$$

i.e.,

$$(\mathbb{L}_\tau u, \psi)_{H^1(D)} = \int_D \left[\nabla w_u \cdot \nabla \overline{\psi} - \tau\, w_u\, \overline{\psi} \right] dx \quad \text{for } \psi \in H_0^1(D). \qquad (6.106)$$

Obviously, $\mathbb{L}_\tau$ also depends analytically on $\tau \in \{ z \in \mathbb{C} : \text{Re}(z) > -\delta \}$. Now we are able to connect a transmission eigenfunction, i.e., a nontrivial solution (v, w) of (6.91)–(6.94), to the kernel of the operator $\mathbb{L}_\tau$.

Theorem 6.32. *The following statements are true:*

1. *Let $(w, v) \in H^1(D) \times H^1(D)$ be a transmission eigenfunction corresponding to some eigenvalue $\tau > 0$. Then $u = w - v \in H_0^1(D)$ satisfies $\mathbb{L}_\tau u = 0$.*
2. *Let $u \in H_0^1(D)$ satisfy $\mathbb{L}_\tau u = 0$ for some $\tau > 0$. Furthermore, let $w := w_u = A_\tau u \in H^1(D)$ be as in Lemma 6.31, i.e., the solution of (6.105). Then τ is a transmission eigenvalue with $(v, w) \in H^1(D) \times H^1(D)$ the corresponding transmission eigenfunction where $v = w - u$.*

Proof. Formula (6.106) implies that $(\mathbb{L}_\tau u, \psi)_{H^1(D)}$ for all $\psi \in H_0^1(D)$, which means that $L_\lambda u = 0$.

The proof of the second part of the theorem is a simple consequence of the observation that (6.104) is equivalent to

$$\int_D \left[\nabla w \cdot \nabla \overline{\psi} - \tau\, w\, \overline{\psi} \right] dx = 0 \quad \text{for all } \psi \in H_0^1(D). \qquad (6.107)$$

Hence $L_\lambda u = 0$ implies that w_u solves the Helmholtz equation in D. Since $v := w - u$, we have that the Cauchy data of w and v coincide. The equation for v follows from (6.105). $\qquad \square$

The operator $\mathbb{L}_\tau$ plays a similar role as the operator $\mathbb{A}_k - k^2 \mathbb{B}$ in (6.77) for the case of $n = 1$.

Theorem 6.33. *The bounded linear operator $\mathbb{L}_\tau : H_0^1(D) \to H_0^1(D)$ satisfies the following statements holds:*

1. *$\mathbb{L}_\tau$ is self-adjoint for all $\tau > 0$.*
2. *$(\sigma \mathbb{L}_0 u, u)_{H^1(D)} \geq c \|u\|_{H^1(D)}^2$ for all $u \in H_0^1(D)$ and $c > 0$ independent of u, where $\sigma = 1$ if $a_{min} > 1$ and $0 < n_{max} < 1$, and $\sigma = -1$ if $0 < a_{max} < 1$ and $n_{min} > 1$.*
3. *$\mathbb{L}_\tau - \mathbb{L}_0$ is compact.*

Proof. 1. Let $u_1, u_2 \in H_0^1(D)$ and $w_1 := w_{u_1}$ and $w_2 := w_{u_2}$ be the corresponding solution of (6.105). Then we have that

$$(\mathbb{L}_\tau u_1, u_2)_{H^1(D)} = \int_D \left[\nabla w_1 \cdot \nabla \overline{u_2} - \tau\, w_1 \overline{u_2}\right] dx$$

$$= \int_D \left[A \nabla w_1 \cdot \nabla \overline{u_2} - \tau n\, w_1 \overline{u_2}\right] dx$$

$$- \int_D \left[(A - I)\nabla w_1 \cdot \nabla \overline{u_2} - \tau\,(n - 1)\, w_1 \overline{u_2}\right] dx\,.$$

Using (6.105) twice, first for $u = u_2$ and the corresponding $w = w_2$ and $\psi = w_1$ and then for $u = u_1$ and the corresponding $w = w_1$ and $\psi = w_2$, yields

$$(\mathbb{L}_\tau u_1, u_2)_{H^1(D)} = \int_D \left[(A - I)\nabla w_1 \cdot \nabla \overline{w_2} - \tau\,(n - 1)\, w_1 \overline{w_2}\right] dx$$

$$- \int_D \left[A \nabla u_1 \cdot \nabla \overline{u_2} - \tau n\, u_1 \overline{u_2}\right] dx, \tag{6.108}$$

which shows that $\mathbb{L}_\tau$ is self-adjoint.

2. To show that $\sigma \mathbb{L}_0 : H_0^1(D) \to H_0^1(D)$ is a strictly coercive operator, we recall the definition (6.106) of $\mathbb{L}_0$ and use the fact that $w = w_u = u + v$ to obtain

$$(\mathbb{L}_0 u, u)_{H^1(D)} = \int_D \nabla w \cdot \nabla \overline{u}\, dx = \int_D |\nabla u|^2\, dx + \int_D \nabla v \cdot \nabla \overline{u}\, dx. \tag{6.109}$$

From (6.105) for $\tau = 0$ and $\psi = v$ we now have that

$$\int_D \nabla v \cdot \nabla \overline{u}\, dx = \int_D (A - I)\nabla v \cdot \nabla \overline{v}\, dx. \tag{6.110}$$

If $a_{min} > 0$, then we have $\int_D (A - I)\nabla w \cdot \nabla \overline{w}\, dx \geq (a_{min} - 1)\|\nabla w\|_{L^2(D)}^2 \geq 0$, and hence

$$(\mathbb{L}_0 u, u)_{H^1(D)} \geq \int_D |\nabla u|^2\, dx.$$

Since from Poincaré's inequality $\|\nabla u\|_{L^2(D)}$ is an equivalent norm in $H_0^1(D)$, this proves the strict coercivity of $\mathbb{L}_0$. Now if $0 < a_{max} < 1$, then from (6.108) with $u_1 = u_2 = u$ and $\tau = 0$ we have

$$- (\mathbb{L}_0 u, u)_{H^1(D)} = - \int_D (A - I)\nabla v \cdot \nabla \overline{v} \, dx + \int_D A \nabla u \cdot \nabla \overline{u} \, dx$$

$$\geq a_{min} \int_D |\nabla u|^2 \, dx,$$

which proves the strict coercivity of $-\mathbb{L}_0$ since $a_{min} > 0$.

3. This follows from the compact embedding of $H_0^1(D)$ into $L^2(D)$.   □

We are now in a position to establish the existence of infinitely many positive transmission eigenvalues, i.e., the existence of a sequence of $\tau_j > 0$, and corresponding $u_j \in H_0^1(D)$, such that $u_j \neq 0$ and $\mathbb{L}_{\tau_j} u_j = 0$. Obviously, these $\tau > 0$ are such that the kernel of $\mathbb{I} - \mathbb{T}_\tau$ is not trivial, which corresponds to 1 being an eigenvalue of the compact self-adjoint operator $\mathbb{T}_\tau$, where $\mathbb{T}_\lambda : H_0^1(D) \to H_0^1(D)$ is defined by

$$\mathbb{T}_\lambda := -(\sigma \mathbb{L}_0)^{-\frac{1}{2}} (\sigma(\mathbb{L}_\tau - \mathbb{L}_0)) (\sigma \mathbb{L}_0)^{-\frac{1}{2}}.$$

Thus, we can conclude that real transmission eigenvalues have finite multiplicity and are such that $\tau := k^2$ are solutions to $\mu_j(\tau) = 1$, where $\{\mu_j(\tau)\}_1^{+\infty}$ is the increasing sequence of the eigenvalues of $\mathbb{T}_\tau$. To prove the existence of positive transmission eigenvalues, we again apply Theorem 6.15 to the continuous operator-valued mapping $\tau \mapsto \mathbb{L}_\tau$, which in our case takes the following form.

Theorem 6.34. *Let $\sigma = 1$ if $a_{min} > 1$ and $0 < n_{max} < 1$, and $\sigma = -1$ if $0 < a_{max} < 1$ and $n_{min} > 1$, and make the following assumptions:*

1. *There is a $\tau_0 \geq 0$ such that $\sigma \mathbb{L}_{\tau_0}$ is positive on $H_0^1(D)$.*
2. *There is a $\tau_1 > \tau_0$ such that $\sigma \mathbb{L}_{\tau_1}$ is nonpositive on some m-dimensional subspace W_m of $H_0^1(D)$.*

Then there are m values of τ in $[\tau_0, \tau_1]$ counting their multiplicity for which $\mathbb{L}_\tau$ fails to be injective.

Using Theorem 6.34 we can now prove the main result of this section.

Theorem 6.35. *Assume that either $a_{min} > 1$ and $0 < n_{max} < 1$, or $0 < a_{max} < 1$ and $n_{min} > 1$. Then there exists an infinite sequence of positive transmission eigenvalues $k_j > 0$ ($\tau_j := k_j^2$) with $+\infty$ as the only accumulation point.*

Proof. We sketch the proof only for the case of $a_{min} > 1$ and $0 < n_{max} < 1$ (i.e., $\sigma = 1$ in Theorem 6.34). First, we recall that assumption 1 of Theorem 6.34 is satisfied with $\tau_0 = 0$ from Theorem 6.33 (2.). Next, from the definition of $\mathbb{L}_\tau$ and the fact that $w = v + u$, we have

$$(\mathbb{L}_\tau u, u)_{H^1(D)} \tag{6.111}$$

$$= \int_D \left[\nabla w \cdot \nabla \overline{u} - \tau w \overline{u} \right] dx = \int_D \left[\nabla v \cdot \nabla \overline{u} - \tau v \overline{u} + |\nabla u|^2 - \tau |u|^2 \right] dx.$$

We also have that v satisfies

$$\int_D \left[(A - I)\nabla v \cdot \nabla \overline{\psi} - \tau(n - 1)\, v\, \overline{\psi}\right] dx \;=\; \int_D \left[\nabla u \cdot \nabla \overline{\psi} - \tau\, u\, \overline{\psi}\right] dx \quad (6.112)$$

for all $\psi \in H^1(D)$. Now taking $\psi = v$ in (6.112) and substituting the result into (6.111) yields

$$(\mathbb{L}_\tau u,\, u)_{H^1(D)} \tag{6.113}$$
$$= \int_D \left[(A - I)\nabla v \cdot \nabla \overline{v} - \tau\, (n - 1)\, |v|^2 + |\nabla u|^2 - \tau\, |u|^2\right] dx.$$

Now let $\hat{\tau}$ be such that $\hat{\tau} := k_1^2$, where k_1 is the first transmission eigenvalue corresponding to (6.49)–(6.52) for the disk B_R with $a_0 := a_{min}$ and $n_0 := n_{max}$. We denote by $\hat{v}$, $\hat{w}$ the corresponding nonzero solutions and set $\hat{u} := \hat{v} - \hat{w} \in H_0^1(B_R)$. We denote the corresponding operator by $\hat{\mathbb{L}}_\tau$. Of course, by construction, we have that (6.113) still holds, i.e., since $\hat{\mathbb{L}}_{\hat{\tau}} \hat{u} = 0$,

$$0 = \left(\hat{\mathbb{L}}_{\hat{\tau}} \hat{u},\, \hat{u}\right)_{H^1(B_R)}, \tag{6.114}$$
$$= \int_{B_R} \left[(a_{min} - 1)|\nabla \hat{v}|^2 - \hat{\tau}\, (n_{max} - 1)|\hat{v}|^2 + |\nabla \hat{u}|^2 - \hat{\tau}\, |\hat{u}|^2\right] dx.$$

Next we denote by $\tilde{u} \in H_0^1(D)$ the extension of $\hat{u} \in H_0^1(B_R)$ by zero to the whole of D and let $\tilde{w} := w_{\tilde{u}}$ be the corresponding solution to (6.105) and $\tilde{v} := \tilde{w} - \tilde{u}$. In particular, $\tilde{v} \in H^1(D)$ satisfies

$$\int_D \left[(A - I)\nabla \tilde{v} \cdot \nabla \overline{\psi} - \hat{\tau}\, p\, \tilde{v}\, \overline{\psi}\right] dx \;=\; \int_D \left[\nabla \tilde{u} \cdot \nabla \overline{\psi} - \hat{\tau}\, \tilde{u}\, \overline{\psi}\right] dx$$
$$= \int_{B_R} \left[\nabla \hat{u} \cdot \nabla \overline{\psi} - \hat{\tau}\, \hat{u}\, \overline{\psi}\right] dx \;=\; \int_{B_R} \left[(a_{min} - 1)\nabla \hat{v} \cdot \nabla \overline{\psi} - \hat{\tau}\, (n_{max} - 1)\, \hat{v}\, \overline{\psi}\right] dx$$

for all $\psi \in H^1(D)$. Therefore, for $\psi = \tilde{v}$ we have

$$\int_D (A - I)\nabla \tilde{v} \cdot \nabla \overline{\tilde{v}} - \hat{\tau}\, (n - 1)\, |\tilde{v}|^2\, dx$$
$$= \int_{B_R} (a_{min} - 1)\, \nabla \hat{v} \cdot \nabla \overline{\tilde{v}} + \hat{\tau}\, |n_{max} - 1|\, \hat{v}\, \overline{\tilde{v}}\, dx.$$

Using the Cauchy–Schwarz inequality we obtain

$$\int_D (A - I)\nabla\tilde{v} \cdot \nabla\overline{\tilde{v}} - \hat{\tau}\,(n - 1)\,|\tilde{v}|^2\,dx$$

$$\leq \left[\int_{B_R} (a_{min} - 1)\,|\nabla\hat{v}|^2 + \hat{\tau}\,|n_{max} - 1|\,|\hat{v}|^2\,dx \right]^{\frac{1}{2}}$$

$$\cdot \left[\int_{B_R} (a_{min} - 1)\,|\nabla\tilde{v}|^2 + \hat{\tau}\,|n_{max} - 1|\,|\tilde{v}|^2\,dx \right]^{\frac{1}{2}}$$

$$\leq \left[\int_{B_R} (a_{min} - 1)\,|\nabla\hat{v}|^2 - \hat{\tau}\,(n_{max} - 1)\,|\hat{v}|^2\,dx \right]^{\frac{1}{2}}$$

$$\cdot \left[\int_D (A - I)\nabla\tilde{v} \cdot \nabla\overline{\tilde{v}} - \hat{\tau}\,(n - 1)\,|\tilde{v}|^2\,dx \right]^{\frac{1}{2}}$$

since $|n - 1| = 1 - n \geq 1 - n_{max} = |n_{max} - 1|$. Hence we have

$$\int_D \left[(A - I)\nabla\tilde{v} \cdot \nabla\overline{\tilde{v}} - \hat{\tau}\,(n - 1)\,|\tilde{v}|^2 \right] dx$$

$$\leq \int_{B_R} \left[(a_{min} - 1)\,|\nabla\hat{v}|^2 - \hat{\tau}\,(n_{max} - 1)\,|\hat{v}|^2 \right] dx\,.$$

Substituting this into (6.113) for $\tau = \hat{\tau}$ and $u = \tilde{u}$ yields

$$\left(\mathbb{L}_{\hat{\tau}}\tilde{u}, \tilde{u} \right)_{H^1(D)} = \int_D \left[(A - I)\nabla\tilde{v} \cdot \nabla\overline{\tilde{v}} - \hat{\tau}\,(n - 1)\,|\tilde{v}|^2 + |\nabla\tilde{u}|^2 - \hat{\tau}\,|\tilde{u}|^2 \right] dx$$

$$\leq \int_{B_R} \left[(a_{min} - 1)|\nabla\hat{v}|^2 - \hat{\tau}\,(n_{max} - 1)\,|\hat{v}|^2 + |\nabla\hat{u}|^2 - \hat{\tau}\,|\hat{u}|^2 \right] dx \;=\; 0$$

by (6.114). Hence from Theorem 6.34 we have that there is a transmission eigenvalue $k > 0$ such that in $k^2 \in (0, \hat{\tau}]$. Finally, repeating this argument for disks of arbitrarily small radius we can show the existence of infinitely many transmission eigenvalues exactly in the same way as in the proof of Theorem 6.18. In a similar way we can prove the same result for the case where $0 < a_{max} < 1$ and $n_{min} > 1$.   □

From the preceding analysis it is possible to obtain bounds for the first transmission eigenvalue stated in the following theorem (here we omit the proof and refer the reader to [35]).

Theorem 6.36. *Let $B_R \subset D$ be the largest disk contained in D and $\lambda_0(D)$ the first Dirichlet eigenvalue of $-\Delta$ in D. Furthermore, let $k_0(A, n, D)$ be the first transmission eigenvalue corresponding to (6.91)–(6.94).*

1. If $a_{min} > 1$ and $0 < n_{max} < 1$ then

$$\lambda_0(D) \leq k_0^2(A, n, D) \leq k_0^2(a_{min}, n_{max}, B_R).$$

2. If $0 < a_{max} < 1$ and $n_{min} > 1$, then

$$\frac{a_{min}}{n_{max}} \lambda_0(D) \leq k_0^2(A, n, D) \leq k_0^2(a_{max}, n_{min}, B_R).$$

For other estimates of the same type we refer the reader to [9].

We end our discussion in this section with a few comments on the case where $(A - I)$ and $(n - 1)$ have the same sign. As indicated earlier, if we follow a similar procedure, then we are faced with the problem that (6.105) is not solvable for all τ. For this reason it is only possible to prove the existence of a finite number of transmission eigenvalues under the restrictive assumption that $n_{max} - 1$ is small enough (for more details we refer the reader to [35]).

In a series of interesting papers [118, 119] and [120] Lakshtanov and Vainberg introduced an alternative approach to showing the discreteness and existence of transmission eigenvalues as well as initiating a studying of the counting function for transmission eigenvalues.

6.4 Uniqueness

The proof of uniqueness for the inverse medium scattering problem is more complicated than for the case of scattering by an imperfect conductor considered in Chap. 4. The idea of the uniqueness proof for the inverse medium scattering problem originates from [93, 94], in which it is shown that the shape of a penetrable, inhomogeneous, isotropic medium is uniquely determined by its far-field pattern for all incident plane waves. The case of an orthotropic medium is due to Hähner [81] (see also [57]), the proof of which is based on the existence of a solution to the modified interior transmission problem. We begin with a simple lemma.

Lemma 6.37. *Assume that either $\bar{\xi} \cdot Re(A) \xi \geq \gamma |\xi|^2$ or $\bar{\xi} \cdot Re(A^{-1}) \xi \geq \gamma |\xi|^2$ for some $\gamma > 1$. Let $\{v_n, w_n\} \in H^1(D) \times H^1(D)$, $n \in \mathbb{N}$, be a sequence of solutions to the interior transmission problem (6.12)–(6.15) with boundary data $f_n \in H^{\frac{1}{2}}(\partial D)$, $h_n \in H^{-\frac{1}{2}}(\partial D)$. If the sequences $\{f_n\}$ and $\{h_n\}$ converge in $H^{\frac{1}{2}}(\partial D)$ and $H^{-\frac{1}{2}}(\partial D)$ respectively, and if the sequences $\{v_n\}$ and $\{w_n\}$ are bounded in $H^1(D)$, then there exists a subsequence $\{w_{n_k}\}$ that converges in $H^1(D)$.*

Proof. Assume first that $\bar{\xi} \cdot \mathrm{Re}(A)\,\xi \geq \gamma |\xi|^2$, $\gamma > 1$, and let $\{v_n, w_n\}$ be as in the statement of the lemma. Due to the compact embedding of $H^1(D)$ into $L^2(D)$, we can select L^2-convergent subsequences $\{v_{n_k}\}$ and $\{w_{n_k}\}$. Hence, $\{v_{n_k}\}$ and $\{w_{n_k}\}$ satisfy

$$
\begin{aligned}
\nabla \cdot A\nabla v_{n_k} - \gamma v_{n_k} &= -(\gamma + k^2 n)v_{n_k} && \text{in} \quad D, \\
\Delta w_{n_k} - w_{n_k} &= -(1 + k^2)w_{n_k} && \text{in} \quad D, \\
v_{n_k} - w_{n_k} &= f_{n_k} && \text{on} \quad \partial D, \\
\frac{\partial v_{n_k}}{\partial \nu_A} - \frac{\partial w_{n_k}}{\partial \nu} &= h_{n_k} && \text{on} \quad \partial D.
\end{aligned}
$$

Then the result of the lemma follows from the a priori estimate of Theorem 6.7. In the case where $\bar{\xi} \cdot \mathrm{Re}(A^{-1})\,\xi \geq \gamma |\xi|^2$, $\gamma > 1$, we use Theorem 6.9 and $1/\gamma$ instead of γ in the preceding equation for v_{n_k} to obtain the same result. $\square$

Note that in the proof of Lemma 6.37 we use the a priori estimate for the modified interior transmission problem instead of the a priori estimate for the interior transmission problem. This allows us to obtain the result without assuming that k is not a transmission eigenvalue.

We can prove a result similar to that in Lemma 6.37 under different assumptions about the physical properties of the medium. In particular, assuming that $\mathrm{Im}(A) = 0$ and $\mathrm{Im}(n) = 0$ we recall definition (6.95) of a_*, a^*, n_*, and n^*.

Lemma 6.38. *Assume that either $0 < a^* < 1$ and $0 < n^* < 1$, or $a_* > 1$ and $n_* > 1$. Let $\{v_n, w_n\} \in H^1(D) \times H^1(D)$, $n \in \mathbb{N}$, be a sequence of solutions to the interior transmission problem (6.12)–(6.15) with boundary data $f_n \in H^{\frac{1}{2}}(\partial D)$, $h_n \in H^{-\frac{1}{2}}(\partial D)$. If the sequences $\{f_n\}$ and $\{h_n\}$ converge in $H^{\frac{1}{2}}(\partial D)$ and $H^{-\frac{1}{2}}(\partial D)$, respectively, and if the sequences $\{v_n\}$ and $\{w_n\}$ are bounded in $H^1(D)$, then there exists a subsequence $\{w_{n_k}\}$ that converges in $H^1(D)$.*

Proof. Similarly to the proof of Lemma 6.37, let $\{v_n, w_n\}$ be as in the statement of the lemma. Due to the compact embedding of $H^1(D)$ into $L^2(D)$, we can select L^2-convergent subsequences $\{v_{n_k}\}$ and $\{w_{n_k}\}$. Hence, $\{v_{n_k}\}$ and $\{w_{n_k}\}$ satisfy

$$
\begin{aligned}
\nabla \cdot A\nabla v_{n_k} - \kappa^2 n v_{n_k} &= (\kappa^2 - k^2)n v_{n_k} && \text{in} \quad D, \\
\Delta w_{n_k} - \kappa^2 w_{n_k} &= (\kappa^2 - k^2)w_{n_k} && \text{in} \quad D, \\
v_{n_k} - w_{n_k} &= f_{n_k} && \text{on} \quad \partial D, \\
\frac{\partial v_{n_k}}{\partial \nu_A} - \frac{\partial w_{n_k}}{\partial \nu} &= h_{n_k} && \text{on} \quad \partial D,
\end{aligned}
$$

where $\kappa > 0$ is chosen as in Lemma 6.27 (i.e., for $k := i\kappa$ the interior transmission problem is invertible). Then the result of the lemma follows from the boundedness of the inverse of the operator equivalent to the interior transmission problem for $k := i\kappa$ (Remark 6.29). □

We are now ready to prove the uniqueness theorem.

Theorem 6.39. *Let the domains D_1 and D_2, the matrix-valued functions A_1 and A_2, and the functions n_1 and n_2 satisfy the assumptions in Sect. 5.2 and the assumptions of either Lemma 6.37 or Lemma 6.38. If the far-field patterns $u^1_\infty(\theta, \phi)$ and $u^2_\infty(\theta, \phi)$ corresponding to D_1, A_1, n_1 and D_2, A_2, n_2, respectively, coincide for all $\theta \in [0, 2\pi]$ and $\phi \in [0, 2\pi]$, then $D_1 = D_2$.*

Proof. Denote by G the unbounded connected component of $\mathbb{R}^2 \setminus (\bar{D}_1 \cup \bar{D}_2)$, and define $D^e_1 := \mathbb{R}^2 \setminus \bar{D}_1$, $D^e_2 := \mathbb{R}^2 \setminus \bar{D}_2$. By Rellich's lemma, we conclude that the scattered fields u_1 and u_2, which are the radiating part of the solution to (5.13)–(5.17) with D_1, A_1, n_1 and D_2, A_2, n_2, respectively, and boundary data with $f := e^{ikx \cdot d}$ and $h := \partial e^{ikx \cdot d}/\partial \nu$, $d = (\cos \phi, \sin \phi)$, coincide in G. Let $\Phi(x, z)$ denote the fundamental solution to the Helmholtz equation given by (3.33).

We now show that the scattered solutions $u_1(\cdot, z)$ and $u_2(\cdot, z)$ also coincide for the incident waves $\Phi(\cdot, z)$ with $z \in G$, i.e., for $f := \Phi(\cdot, z)$ and $h := \partial \Phi(\cdot, z)/\partial \nu$. To this end, choose a large disk Ω_R such that $\bar{D}_1 \cup \bar{D}_2 \subset \Omega_R$ and k^2 is not a Dirichlet eigenvalue for Ω_R. Then, for $z \notin \bar{\Omega}_R$, by Lemma 4.4, there exists a sequence $\{u^i_n\}$ in $\text{span}\{e^{ikx \cdot d} : |d| = 1\}$ such that

$$\|u^i_n - \Phi(\cdot, z)\|_{H^{\frac{1}{2}}(\partial \Omega_R)} \to 0, \quad \text{as } n \to \infty.$$

The well-posedness of the Dirichlet problem for the Helmholtz equation in Ω_R (Example 5.15) implies that u^i_n approximates $\Phi(\cdot, z)$ in $H^1(\Omega_R)$. Then the continuous dependence on the data of the scattered field (5.41), together with the fact that the scattered fields corresponding to u^i_n coincide as linear combinations of scattered fields due to plane waves, implies that $u_1(\cdot, z)$ and $u_2(\cdot, z)$ also coincide for a fixed $z \notin \bar{\Omega}_R$. Since $\Phi(\cdot, z)$ and its derivatives are real-analytic in z, we can again conclude from the well-posedness of the transmission problem (5.13)–(5.17) that $u_1(\cdot, z)$ and $u_2(\cdot, z)$ are real-analytic in z and therefore must coincide for all $z \in G$.

Let us now assume that $\bar{D}_1$ is not included in $\bar{D}_2$. Since D^e_2 is connected, we can find a point $z \in \partial D_1$ and $\epsilon > 0$ with the following properties, where $\Omega_\delta(z)$ denotes the ball of radius δ centered at z:

1. $\Omega_{8\epsilon}(z) \cap \bar{D}_2 = \emptyset$;
2. The intersection $\bar{D}_1 \cap \Omega_{8\epsilon}(z)$ is contained in the connected component of $\bar{D}_1$ to which z belongs;
3. There are points from this connected component of $\bar{D}_1$ to which z belongs that are not contained in $\bar{D}_1 \cap \bar{\Omega}_{8\epsilon}(z)$;
4. The points $z_n := z + \dfrac{\epsilon}{n}\nu(z)$ lie in G for all $n \in \mathbb{N}$, where $\nu(z)$ is the unit normal to ∂D_1 at z.

Due to the singular behavior of $\Phi(\cdot, z_n)$ at the point z_n, it is easy to show that $\|\Phi(\cdot, z_n)\|_{H^1(D_1)} \to \infty$ as $n \to \infty$. We now define

$$w^n(x) := \frac{1}{\|\Phi(\cdot, z_n)\|_{H^1(D_1)}} \Phi(x, z_n), \qquad x \in \bar{D}_1 \cup \bar{D}_2$$

and let v_1^n, u_1^n and v_2^n, u_2^n be the solutions of the scattering problem (5.13)–(5.17) with boundary data $f := w^n$ and $h := \partial w^n/\partial \nu$ corresponding to D_1 and D_2, respectively. Note that for each n, w^n is a solution of the Helmholtz equation in D_1 and D_2. Our aim is to prove that if $\bar{D}_1 \not\subset \bar{D}_2$, then the equality $u_1(\cdot, z) = u_2(\cdot, z)$ for $z \in G$ allows the selection of a subsequence $\{w^{n_k}\}$ from $\{w^n\}$ that converges to zero with respect to $H^1(D_1)$. This certainly contradicts the definition of $\{w^n\}$ as a sequence of functions with $H^1(D_1)$ norm equal to one. Note that $u_1(\cdot, z) = u_2(\cdot, z)$ obviously implies that $u_1^n = u_2^n$ in G.

We begin by noting that, since the functions $\Phi(\cdot, z_n)$ together with their derivatives are uniformly bounded in every compact subset of $\mathbb{R}^2 \setminus \Omega_{2\epsilon}(z)$ and $\|\Phi(\cdot, z_n)\|_{H^1(D_1)} \to \infty$ as $n \to \infty$, then $\|w^n\|_{H^1(D_2)} \to 0$ as $n \to \infty$. Hence, if Ω_R is a large ball containing $\bar{D}_1 \cup \bar{D}_2$, then $\|u_2^n\|_{H^1(\Omega_R \cap G)} \to 0$ as $n \to \infty$ from the a priori estimate (5.41). Since $u_1^n = u_2^n$ in G, then $\|u_1^n\|_{H^1(\Omega_R \cap G)} \to 0$ as $n \to \infty$ as well. Now, with the help of a cutoff function $\chi \in C_0^\infty(\Omega_{8\epsilon}(z))$ satisfying $\chi(x) = 1$ in $\Omega_{7\epsilon}(z)$ (Theorem 5.6), we see that $\|u_1^n\|_{H^1(\Omega_R \cap G)} \to 0$ implies that

$$(\chi u_1^n) \to 0, \qquad \frac{\partial(\chi u_1^n)}{\partial \nu} \to 0 \qquad \text{as } n \to \infty \qquad (6.115)$$

with respect to the $H^{\frac{1}{2}}(\partial D_1)$ norm and $H^{-\frac{1}{2}}(\partial D_1)$ norm, respectively. Indeed, for the first convergence we simply apply the trace theorem, while for the convergence of $\partial(\chi u_1^n)/\partial \nu$ we first deduce the convergence of $\Delta(\chi u_1^n)$ in $L^2(\Omega_R \cap D_1^e)$, which follows from $\Delta(\chi u_1^n) = \chi \Delta u_1^n + 2\nabla\chi \cdot \nabla u_1^n + u_1^n \Delta\chi$, and then apply Theorem 5.7. Note here that we need conditions 2 and 4 on z to ensure $\Omega_{8\epsilon}(z) \cap D_1^e = \Omega_{8\epsilon}(z) \cap G$.

We next note that in the exterior of $\Omega_{2\epsilon}(z)$ the $H^2(\Omega_R \setminus \Omega_{2\epsilon}(z))$ norms of w^n remain uniformly bounded. Then the assertion about the boundary regularity of the solution to (5.13)–(5.17) stated in the second part of Theorem 5.28 implies that u_1^n is uniformly bounded with respect to the $H^2((\Omega_R \cap D_1^e) \setminus \Omega_{4\epsilon}(z))$ norm. Therefore, using the compact embedding of $H^2(\Omega_R \cap D_1^e)$ into $H^1(\Omega_R \cap D_1^e)$, we can select a $H^1(\Omega_R \cap D_1^e)$ convergent subsequence $\{(1 - \chi)u_1^{n_k}\}$ from $\{(1 - \chi)u_1^n\}$. Hence, $\{(1 - \chi)u_1^{n_k}\}$ is a convergent sequence in $H^{\frac{1}{2}}(\partial D_1)$, and, similarly to the foregoing reasoning, we also have that $\{\partial((1 - \chi)u_1^{n_k})/\partial \nu\}$ converges in $H^{-\frac{1}{2}}(\partial D_1)$. This, together with (6.115), implies that the sequences

$$\{u_1^{n_k}\} \qquad \text{and} \qquad \left\{\frac{\partial u_1^{n_k}}{\partial \nu}\right\}$$

converge in $H^{\frac{1}{2}}(\partial D_1)$ and $H^{-\frac{1}{2}}(\partial D_1)$, respectively.

Finally, since the functions $v_1^{n_k}$ and w^{n_k} are solutions to the interior transmissionproblem (6.12)–(6.15) for the domain D_1 with boundary data $f = u_1^{n_k}$ and $h = \partial u_1^{n_k}/\partial \nu$, and since the $H^1(D_1)$ norms of $v_1^{n_k}$ and w^{n_k} remain uniformly bounded, then, according to Lemma 6.37, we can select a subsequence of $\{w^{n_k}\}$, denoted again by $\{w^{n_k}\}$, that converges in $H^1(D_1)$ to a function $w \in H^1(D_1)$. As a limit of weak solutions to the Helmholtz equation, $w \in H^1(D_1)$ is a weak solution to the Helmholtz equation. We also have that $w|_{D_1 \setminus \Omega_{2\epsilon}(z)} = 0$ because the functions w^{n_k} converge uniformly to zero in the exterior of $\Omega_{2\epsilon}(z)$. Hence, w must be zero in all of D_1 [here we make use of condition 3, namely, the fact that the connected component of D_1 containing z has points that do not lie in the exterior of $\bar{\Omega}_{2\epsilon}(z)$]. This contradicts the fact that $\|w^{n_k}\|_{H^1(D_1)} = 1$. Hence the assumption $\bar{D}_1 \not\subset \bar{D}_2$ is false.

Since we can derive an analogous contradiction for the assumption $\bar{D}_2 \not\subset \bar{D}_1$, we have proved that $D_1 = D_2$. $\square$

Remark 6.40. We remark that the proof of the uniqueness of the support of an anisotropic media presented in Theorem 6.39 is valid as long as the material properties A and n guaranty that the corresponding interior transmission problem is a compact perturbation of a well-posed problem.

6.5 Linear Sampling Method

Having shown that the support of an inhomogeneity can be uniquely determined from the far-field pattern, we now want to find an approximation to the support. To this end, we will use the linear sampling method previously introduced in Chap. 4 for the inverse scattering problem for an imperfect conductor. In particular, we shall show that, provided k is not a transmission eigenvalue, the boundary ∂D of the inhomogeneity D can be characterized by the solution of the far-field equation (4.33), where the kernel of the far-field operator is the far-field pattern corresponding to (6.1)–(6.5).

Given $(f, h) \in H^{\frac{1}{2}}(\partial D) \times H^{-\frac{1}{2}}(\partial D)$, let $(v, u) \in H^1(D) \times H^1_{loc}(\mathbb{R}^2 \setminus \bar{D})$ be the unique solution to the corresponding transmission problem (5.13)–(5.17). We recall that the radiating part u has the asymptotic behavior

$$u(x) = \frac{e^{ikr}}{\sqrt{r}} u_\infty(\hat{x}) + O(r^{-3/2}), \qquad r \to \infty, \quad \hat{x} = x/|x|,$$

where u_∞ is the far-field pattern corresponding to (v, u).

Definition 6.41. The bounded linear operator $B : H^{\frac{1}{2}}(\partial D) \times H^{-\frac{1}{2}}(\partial D) \to L^2[0, 2\pi]$ maps $(f, h) \in H^{\frac{1}{2}}(\partial D) \times H^{-\frac{1}{2}}(\partial D)$ onto the far-field pattern $u_\infty \in L^2[0, 2\pi]$, where (v, u) is the solution of (5.13)–(5.17) with the boundary data (f, h).

Note that the fact that B is bounded follows directly from the well-posedness of (5.13)–(5.17).

As in the case of the scattering problem for an imperfect conductor, the operator B will play an important role in the solution of the inverse problem. To determine the range of the operator B, it is more convenient to consider its transpose instead of its adjoint. This is because operating with the duality relation between $H^{\frac{1}{2}}(\partial D)$, $H^{-\frac{1}{2}}(\partial D)$ is much simpler than using the corresponding inner products. In what follows we will define the transpose operator and derive some useful properties of this operator.

Let X and Y be two Hilbert spaces, and let X^* and Y^* be their dual spaces. For any linear mapping $A : X \to Y$, the *transpose* $A^\top : Y^* \to X^*$ is the linear mapping defined by

$$\langle A^\top v, u \rangle_{X,X^*} = \langle v, Au \rangle_{Y,Y^*}, \qquad \text{for all } u \in X \text{ and } v \in Y^*,$$

where $\langle \cdot, \cdot \rangle$ denotes the duality pairing between the denoted spaces.

It can be shown (see Lemma 2.9 in [127]) that the transpose $A^\top$ is bounded if and only if A is bounded. To describe the relation between the range and the kernel of A and $A^\top$, we use the following terminology. For any subset $W \subseteq X$, the *annihilator* W^a is the closed subspace of X^* defined by

$$W^a = \{g \in X^* : \langle g, u \rangle = 0 \text{ for all } u \in W\}.$$

Similarly, for $V \subseteq X^*$ the annihilator ^aV is the closed subspace of X defined by

$$^aV = \{u \in X : \langle g, u \rangle = 0 \text{ for all } g \in V\}.$$

Lemma 6.42. *The null space and range of A and $A^\top$ satisfy*

$$\mathrm{N}(A^\top) = A(X)^a \quad and \quad \mathrm{N}(A) = {}^aA^\top(Y^*).$$

Proof. Applying the various definitions we obtain

$$\begin{aligned}
A(X)^a &= \{g \in Y^* : \langle g, v \rangle = 0 \text{ for all } v \in \mathrm{range}\, A\} \\
&= \{g \in Y^* : \langle g, Au \rangle = 0 \text{ for all } u \in X\} \\
&= \{g \in Y^* : \langle A^\top g, u \rangle = 0 \text{ for all } u \in X\} \\
&= \{g \in Y^* : A^\top g = 0\} = \mathrm{N}(A^\top).
\end{aligned}$$

A similar argument shows that $\mathrm{N}(A) = {}^aA^\top(Y^*)$. $\qquad\square$

It is an easy exercise using the Hahn–Banach theorem [115] to show that a subset $W \subseteq X$ is dense if and only if $W^a = \{0\}$. In particular, from Lemma 6.42 we have the following corollary.

Corollary 6.43. *The operator A has a dense range if and only if the transpose $A^\top$ is injective.*

With the help of the preceding lemma and corollary we can now prove the following result for the operator B.

Theorem 6.44. *The range of $B : H^{\frac{1}{2}}(\partial D) \times H^{-\frac{1}{2}}(\partial D) \to L^2[0, 2\pi]$ is dense in $L^2[0, 2\pi]$.*

Proof. We consider the dual operator $B^\top : L^2[0, 2\pi] \longrightarrow H^{-\frac{1}{2}}(\partial D) \times H^{\frac{1}{2}}(\partial D)$, which maps a function g into $(\tilde{f}, \tilde{h})$ such that

$$\langle B(f,h), g \rangle_{L^2 \times L^2} = \left\langle f, \tilde{f} \right\rangle_{H^{\frac{1}{2}} \times H^{-\frac{1}{2}}} + \left\langle h, \tilde{h} \right\rangle_{H^{-\frac{1}{2}} \times H^{\frac{1}{2}}},$$

where $\langle \cdot, \cdot \rangle$ denotes the duality pairing between the denoted spaces. Now let $(\tilde{v}, \tilde{u})$ be the unique solution of (5.13)–(5.17) with $(f, h) := (\tilde{v}_g|_{\partial D}, \partial \tilde{v}_g / \partial \nu|_{\partial D})$, where $\tilde{v}_g$ is the Herglotz wave function defined by (6.9). Then from (6.6) we have

$$\langle B(f,h), g \rangle = \int_0^{2\pi} u_\infty(\theta) g(\theta) \, d\theta = \int_{\partial D} \left(u(y) \frac{\partial \tilde{v}_g(y)}{\partial \nu} - \tilde{v}_g(y) \frac{\partial u(y)}{\partial \nu} \right) ds(y).$$

Since u and $\tilde{u}$ are solutions of the Helmholtz equation in $\mathbb{R}^2 \setminus \bar{D}$ satisfying the Sommerfeld radiation condition, an application of Green's second identity implies that

$$\int_{\partial D} \left[u(y) \frac{\partial \tilde{u}(y)}{\partial \nu} - \tilde{u}(y) \frac{\partial u(y)}{\partial \nu} \right] ds(y) = 0.$$

Using the transmission conditions on the boundary for $\tilde{u}$ and $\tilde{v}$ we obtain

$$\langle B(f,h), g \rangle_{L^2 \times L^2} =$$

$$= \int_{\partial D} \left[u(y) \left(\frac{\partial \tilde{v}_g(y)}{\partial \nu} + \frac{\partial \tilde{u}(y)}{\partial \nu} \right) - (\tilde{v}_g(y) + \tilde{u}(y)) \frac{\partial u(y)}{\partial \nu} \right] ds(y)$$

$$= \int_{\partial D} \left(u(y) \frac{\partial \tilde{v}(y)}{\partial \nu_A} - \tilde{v}(y) \frac{\partial u(y)}{\partial \nu} \right) ds(y)$$

$$= \int_{\partial D} \left[(v(y) - f(y)) \frac{\partial \tilde{v}(y)}{\partial \nu_A} - \tilde{v}(y) \left(\frac{\partial v(y)}{\partial \nu_A} - h(y) \right) \right] ds(y).$$

Finally, applying Green's (generalized) second identity to v and $\tilde{v}$ we have that

$$\langle B(f,h), g \rangle_{L^2 \times L^2} = \int_{\partial D} \left[f(y) \left(-\frac{\partial \tilde{v}(y)}{\partial \nu_A} \right) + \tilde{v}(y) h(y) \right] ds(y).$$

Hence the dual operator $B^\top$ can be characterized as

$$B^\top g = \left(-\frac{\partial \tilde{v}}{\partial \nu_A} \bigg|_{\partial D}, \tilde{v}|_{\partial D} \right).$$

In what follows we want to show that the operator $B^\top$ is injective. To this end, let $B^\top g \equiv 0$, $g \in L^2[0, 2\pi]$. This implies that $\tilde{v} = 0$ and $\partial \tilde{v}/\partial \nu_A = 0$ on the boundary ∂D. Therefore, $\tilde{u}$ satisfies the Helmholtz equation in $\mathbb{R}^2 \setminus \bar{D}$, the Sommerfeld radiation condition, and, from the transmission conditions,

$$\tilde{u} = -\tilde{v}_g \quad \text{and} \quad \frac{\partial \tilde{u}}{\partial \nu} = -\frac{\partial \tilde{v}_g}{\partial \nu} \quad \text{on } \partial D.$$

Thus, setting $\tilde{u} \equiv -\tilde{v}_g$ in D we have that $\tilde{u}$ can be extended to an entire solution to the Helmholtz equation satisfying the radiation condition. This is only possible if $\tilde{u}$ vanishes, which implies that $\tilde{v}_g$ vanishes also and, thus, $g \equiv 0$, whence $B^\top$ is injective. Finally, from Corollary 6.43 we have that the range of B is dense in $L^2[0, 2\pi]$. $\square$

From Lemma 6.42 we also have that

$$N(B) = B^\top (L^2[0, 2\pi])^a := \left\{ (f_0, h_0) : \int_{\partial D} \left(-f_0 \frac{\partial \tilde{v}}{\partial \nu_A} + h_0 \tilde{v} \right) ds = 0 \right\},$$

where $\tilde{v}$ is as in the proof of Theorem 6.44. Hence, using the divergence theorem, we see that the pairs $(v|_{\partial D}, \partial v/\partial \nu_A|_{\partial D})$, where $v \in H^1(D)$ is a solution of $\nabla \cdot A\nabla v + k^2 n\, v = 0$ in D, are in the kernel of B. So B is not injective. We will restrict the operator B in such a way that the restriction is injective and still has a dense range.

To this end, let us denote by $\overline{H}$ the closure in $H^1(D)$ of all Herglotz wave functions with kernel $g \in L^2[0, 2\pi]$. Note that the space $\overline{H}$ coincides with the space of H^1 weak solutions to the Helmholtz equation. In other words, $\overline{H} = \overline{W(D)}$, where $\overline{W(D)}$ is the closure in $H^1(D)$ of $W(D)$ defined by

$$W(D) := \{ u \in C^2(D) \cap C^1(\overline{D}) : \Delta u + k^2 u = 0 \}.$$

Indeed, if $u \in \overline{W(D)}$, then by seeing u as a weak solution of the interior impedance boundary value problem for the Helmholtz equation in D with $\lambda = 1$ we have from Theorem 8.4 in Chap. 8 (set $\partial D_D = \emptyset$) that there exists a positive constant C such that

$$\|u\|_{H^1(D)} \le C \left\| \frac{\partial u}{\partial \nu} + iu \right\|_{H^{-\frac{1}{2}}(\partial D)}.$$

Then the proof of Theorem 4.10 implies that for any $\epsilon > 0$ there exists a Herglotz wave function v_g such that $\|u - v_g\|_{H^1(D)} < \epsilon$, whence $\overline{H} = \overline{W(D)}$. For later use we state this result in the following lemma.

Lemma 6.45. *Any solution to the Helmhotz equation in a bounded domain $D \subset \mathbb{R}^2$ can be approximated in the $H^1(D)$ norm by a Herglotz wave function.*

Next, we define

$$H(\partial D) := \left\{ \left(u|_{\partial D}, \left. \frac{\partial u}{\partial \nu} \right|_{\partial D} \right) : u \in \overline{H} \right\}.$$

Lemma 6.46. $H(\partial D)$ *is a closed subset of* $H^{\frac{1}{2}}(\partial D) \times H^{-\frac{1}{2}}(\partial D)$.

Proof. Consider $(f, h) \in \overline{H(\partial D)}$. There exists a sequence $\{u_n, \partial u_n/\partial \nu\}$ converging to (f, h) in $H^{\frac{1}{2}}(\partial D) \times H^{-\frac{1}{2}}(\partial D)$, where $u_n \in \overline{H}$. Since the sequence $\{u_n, \partial u_n/\partial \nu\}$ is bounded in $H^{\frac{1}{2}}(\partial D) \times H^{-\frac{1}{2}}(\partial D)$, by considering u_n to be the solution of an impedance boundary value problem in D we can deduce that $\{u_n\}$ is bounded in $H^1(D)$. From this it follows that a subsequence (still denoted by $\{u_n\}$) converges weakly in $H^1(D)$ to a function u that is clearly in $\overline{H}$. From the continuity of the trace operators (Theorems 1.38 and 5.7) we deduce that $\{u_n, \partial u_n/\partial \nu\}$ converges weakly in $H^{\frac{1}{2}}(\partial D) \times H^{-\frac{1}{2}}(\partial D)$ to $(u, \partial u/\partial \nu)$ and by the uniqueness of the limit $(f, h) = (u, \partial u/\partial \nu)$. Hence $(f, h) \in H(\partial D)$, which completes the proof. $\qquad \square$

From the preceding lemma, $H(\partial D)$ equipped with the induced norm from $H^{\frac{1}{2}}(\partial D) \times H^{-\frac{1}{2}}(\partial D)$ is a Banach space.

Now, let B_0 denote the restriction of B to $H(\partial D)$.

Theorem 6.47. *Assume that k is not a transmission eigenvalue. Then the bounded linear operator $B_0 : H(\partial D) \longrightarrow L^2[0, 2\pi]$ is injective and has a dense range.*

Proof. Let $B_0(f, h) = 0$ for $(f, h) \in H(\partial D)$, and let (v, u) be the solution to (5.13)–(5.17) corresponding to these boundary data. Then the radiating solution to the Helmholtz equation in the exterior of D has a zero far-field pattern, whence $u = 0$ for $x \in \mathbb{R}^2 \setminus \bar{D}$. This implies that v satisfies

$$\nabla \cdot A\nabla v + k^2 n\, v = 0 \quad \text{in} \quad D, \qquad v = f \quad \text{and} \quad \frac{\partial v}{\partial \nu} = h \quad \text{on} \quad \partial D.$$

From the definition of $H(\partial D)$, f and h are the traces on ∂D of a $H^1(D)$ solution w to the Helmholtz equation and its normal derivative, respectively. Therefore, (v, w) solves the homogeneous interior transmission problem (6.12)–(6.15), and since k is not a transmission eigenvalue, we have that $w \equiv 0$ and $v \equiv 0$ in D, whence $f = h = 0$.

It remains to show that the set $B_0(H(\partial D))$ is dense in $L^2[0, 2\pi]$. To this end, it is sufficient to show that the range of B is contained in the range of B_0 since from Theorem 6.44 the range of B is dense in $L^2[0, 2\pi]$. Let u_∞ be in the range of B, that is, u_∞ is the far-field pattern of the radiating part u of a solution (v, u) to (5.13)–(5.17). Let (v, w) be the unique solution to (6.12)–(6.15) with the boundary data $(u|_{\partial D}, \partial u/\partial \nu|_{\partial D})$. Hence (v, u) is the solution to (5.13)–(5.17) with boundary data $(w|_{\partial D}, \partial w/\partial \nu|_{\partial D}) \in H(\partial D)$ and has a far-field pattern coinciding with u_∞. This means that $B_0(w|_{\partial D}, \partial w/\partial \nu|_{\partial D}) = u_\infty$. $\qquad \square$

Theorem 6.48. *The operator $B_0 : H(\partial D) \longrightarrow L^2[0, 2\pi]$ is compact.*

Proof. Given $w \in \overline{H}$, consider the solution (v, u) of (5.13)–(5.17) with boundary data $f := w|_{\partial D}$ and $h := \partial w/\partial \nu|_{\partial D}$. Let $\partial \Omega_R$ be the boundary of a disk Ω_R centered at the origin containing $\overline{D}$. The continuous dependence estimate (5.41) implies that the operator $G : H(\partial D) \to H^{\frac{1}{2}}(\partial \Omega_R) \times H^{-\frac{1}{2}}(\partial \Omega_R)$, which maps

$$\left(w|_{\partial D}, \frac{\partial w}{\partial \nu}\Big|_{\partial D} \right) \to \left(u|_{\partial \Omega_R}, \frac{\partial u}{\partial \nu}\Big|_{\partial \Omega_R} \right),$$

is bounded. Next we denote by $K : H^{\frac{1}{2}}(\partial \Omega_R) \times H^{-\frac{1}{2}}(\partial \Omega_R) \to L^2[0, 2\pi]$ the operator that takes $(u|_{\partial \Omega_R}, \partial u/\partial \nu|_{\partial \Omega_R})$ to u_∞ given by

$$u_\infty(\hat{x}) = \frac{e^{i\pi/4}}{\sqrt{8\pi k}} \int\limits_{\partial B} \left(u(y) \frac{\partial e^{-ik\hat{x}\cdot y}}{\partial \nu} - e^{-ik\hat{x}\cdot y} \frac{\partial u(y)}{\partial \nu} \right) ds(y)$$

where $\hat{x} = x/|x|$. An argument similar to that in the proof of Theorem 4.8 shows that K is compact. Therefore, $B_0 = KG$ is compact since it is a composition of a bounded operator with a compact operator.   □

For a Herglotz wave function v_g given by (6.8) with kernel $g \in L^2[0, 2\pi]$ we define $H : L^2[0, 2\pi] \to H(\partial D)$ by

$$Hg := \left(v_g|_{\partial D}, \frac{\partial v_g}{\partial \nu}\Big|_{\partial D} \right).$$

Corollary 6.49. *Assume that $u_\infty \in L^2[0, 2\pi]$ is in the range of B_0. Then for every $\epsilon > 0$ there exists a $g_\epsilon \in L^2[0, 2\pi]$ such that Hg_ϵ satisfies*

$$\|B_0(Hg_\epsilon) - u_\infty\|_{L^2[0, 2\pi]} \leq \epsilon.$$

Proof. The proof is a straightforward application of the definition of the space $H(\partial D)$, the continuity of the trace operator, and the operator B_0, together with Lemma 6.45.   □

Turning to our main goal of finding an approximation to the scattering obstacle D we consider the *far-field equation* corresponding to the scattering by an orthotropic medium given by

$$\int\limits_0^{2\pi} u_\infty(\theta, \phi)g(\phi)d\phi = \gamma e^{-ik\hat{x}\cdot z}, \qquad z \in \mathbb{R}^2, \tag{6.116}$$

where $u_\infty(\theta, \phi)$ is the far-field pattern of the radiating part of the solution to the forward problem (6.1)–(6.5) corresponding to the incident plane wave with incident direction $d = (\cos \phi, \sin \phi)$ and observation direction

$\hat{x} = (\cos\theta, \sin\theta)$. As in Chap. 4 the far-field equation can be written in the form

$$(Fg)(\hat{x}) = \Phi_\infty(\hat{x}, z), \qquad z \in \mathbb{R}^2,$$

where Fg is the far-field operator corresponding to the transmission problem (6.1)–(6.5), and $\Phi_\infty(\hat{x}, z)$ is the far-field pattern of the fundamental solution $\Phi(x, z)$ to the Helmholtz equation in $\mathbb{R}^2$. We observe that the far-field operator Fg can be factored as

$$Fg = B_0(Hg).$$

Hence the far-field equation takes the form

$$(B_0(Hg))(\hat{x}) = \Phi_\infty(\hat{x}, z), \qquad z \in \mathbb{R}^2. \tag{6.117}$$

As the reader has already encountered in the case of scattering by an imperfect conductor, the *linear sampling method* is based on the characterization of the domain D by the behavior of a solution to the far-field equation (6.117). By definition, $B_0(Hg)$ is the far-field pattern of the solution (v, u) to the transmission problem (5.13)–(5.17) with boundary data $(f, h) := Hg$. Therefore, for $z \in D$, from Rellich's lemma the far-field equation implies that this u coincides with $\Phi(\cdot, z)$ in $\mathbb{R}^2 \setminus \bar{D}$. In other words, for $z \in D$, $g \in L^2[0, 2\pi]$ is a solution to the far-field equation if and only if v and $w := v_g$ solve the interior transmission problem

$$\nabla \cdot A\nabla v + k^2 n\, v = 0 \qquad \text{in} \quad D, \tag{6.118}$$

$$\Delta w + k^2\, w = 0 \qquad \text{in} \quad D, \tag{6.119}$$

$$v - w = \Phi(\cdot, z) \qquad \text{on} \quad \partial D, \tag{6.120}$$

$$\frac{\partial v}{\partial \nu_A} - \frac{\partial w}{\partial \nu} = \frac{\partial \Phi(\cdot, z)}{\partial \nu} \qquad \text{on} \quad \partial D, \tag{6.121}$$

where v_g is the Herglotz wave function with kernel g. In general, this is not true. However, in what follows, we will show that one can construct an approximate solution to the far-field equation that behaves in a certain manner.

We first assume that $z \in D$ and that k is not a transmission eigenvalue. Then the interior transmission problem (6.118)–(6.121) has a unique solution (v, w). In this case $(v, \Phi(\cdot, z))$ solves the transmission problem (5.13)–(5.17) with transmission conditions $f := w|_{\partial D}, h := \partial w/\partial \nu|_{\partial D}$. Since the preceding solution has the far-field pattern $\Phi_\infty(\cdot, z)$, we can conclude that $\Phi_\infty(\cdot, z)$ is in the range of B_0. From Corollary 6.49 we can find a g_z^ϵ such that

$$\|B_0(Hg_z^\epsilon) - \Phi_\infty(\cdot, z)\|_{L^2[0, 2\pi]} < \epsilon \tag{6.122}$$

for an arbitrarily small ϵ. From the construction of B_0 and Corollary 6.49 we see that the corresponding Herglotz wave function $v_{g_z^\epsilon}$ approximates w in the $H^1(D)$ norm as $\epsilon \to 0$. Furthermore, for a fixed $\epsilon > 0$, the $H^1(D)$ norm $v_{g_z^\epsilon}$

blows up if z approaches the boundary from the interior of D, as does the $L^2[0, 2\pi]$ norm of g_z^ϵ. To see this, we choose a sequence of points $\{z_j\}$, $z_j \in D$, such that

$$z_j = z^* - \frac{R}{j}\nu(z^*), \qquad j = 1, 2, \ldots,$$

with sufficiently small R, where $z^* \in \partial D$ and $\nu(z^*)$ is the unit outward normal at z^*. We denote by (v_j, w_j) the solution to (6.118)–(6.121) corresponding to $z = z_j$. As $j \to \infty$ the points z_j approach the boundary point z^* and, therefore, $\|\Phi(\cdot, z_j)\|_{H^{\frac{1}{2}}(\partial D)} \to \infty$. From the trace theorem and by using the boundary conditions we can write

$$\|v_j\|_{H^1(D)} + \|w_j\|_{H^1(D)} \geq \|v_j - w_j\|_{H^{\frac{1}{2}}(\partial D)} = \|\Phi(\cdot, z_j)\|_{H^{\frac{1}{2}}(\partial D)}. \qquad (6.123)$$

In particular, we show that the relation (6.123) implies that

$$\lim_{j \to \infty} \|w_j\|_{H^1(D)} = \infty.$$

To this end, we assume, in contrast, that

$$\|w_j\|_{H^1(D)} \leq \bar{C}, \qquad j = 1, 2, \ldots,$$

for some positive constant $\bar{C}$. From the trace theorem we have

$$\|w_j\|_{H^{\frac{1}{2}}(\partial D)} \leq \bar{C} \qquad \text{and} \qquad \|\frac{\partial w_j}{\partial \nu}\|_{H^{\frac{1}{2}}(\partial D)} \leq \bar{C}, \qquad j = 1, 2, \ldots.$$

Recall that for every j the pair $(v_j, \Phi(\cdot, z_j))$ is the solution of (5.13)–(5.17) with $(f, g) := (w_j|_{\partial D}, \partial w_j/\partial \nu|_{\partial D})$. The a priori estimate (5.41) implies that

$$\|v_j\|_{H^1(D)} + \|\Phi(\cdot, z_j)\|_{H^1(\Omega_R \setminus \bar{D})}$$
$$\leq C\left(\|w_j\|_{H^{\frac{1}{2}}(\partial D)} + \|\frac{\partial w_j}{\partial \nu}\|_{H^{-\frac{1}{2}}(\partial D)}\right) \leq 2C\bar{C},$$

which contradicts the fact that $\|\Phi(\cdot, z_j)\|_{H^1(\Omega_R \setminus \bar{D})}$ does not remain bounded as $z_j \to z^* \in \partial D$. So we have that

$$\lim_{j \to \infty} \|w_j\|_{H^1(D)} = \infty.$$

Since for every $j = 1, 2, \ldots$ the corresponding Herglotz wave functions $v_{g_{z_j}^\epsilon}$ satisfying (6.122) approximate the solution w_j in the $H^1(D)$ norm, we conclude that

$$\lim_{j \to \infty} \|v_{g_{z_j}^\epsilon}\|_{H^1(D)} = \infty,$$

and hence

$$\lim_{j \to \infty} \|g_{z_j}^\epsilon\|_{L^2[0, 2\pi]} = \infty.$$

Next we consider $z \in \mathbb{R}^2 \setminus \bar{D}$, and again we assume that k is not a transmission eigenvalue. We would like to show that if g_z^ϵ is such that

$$\|Fg_z^\epsilon - \Phi_\infty(\cdot, z)\|_{L^2[0, 2\pi]} < \epsilon$$

for a given arbitrary $\epsilon > 0$, then the $H^1(D)$ norm of the corresponding Herglotz wave functions $v_{g_z^\epsilon}$ is not bounded as $\epsilon \to 0$. Assume, to the contrary, that there exists a null sequence $\{\epsilon_n\}$ such that $\|v_n\|_{H^1(D)}$ remain bounded as $n \to \infty$, where $v_n := v_{g_z^{\epsilon_n}}$. From the trace theorem $\|Hg_n\|_{H(\partial D)}$ also remain bounded. Then without loss of generality we may assume weak convergence $Hg_n \rightharpoonup h$, where $h := \left(w|_{\partial D}, \dfrac{\partial w}{\partial \nu}\Big|_{\partial D}\right)$ for some $w \in \bar{H}$, i.e., that w is a $H^1(D)$ weak solution to the Helmholtz equation. Since $B_0 : H(\partial D) \to L^2[0, 2\pi]$ is bounded, we also have that $B_0 Hg_n \rightharpoonup B_0 h$ in $L^2[0, 2\pi]$. But by construction, $B_0 Hg_n \to \Phi_\infty(\cdot, z)$, which means that $B_0 h = \Phi_\infty(\cdot, z)$. This contradicts the fact that $\Phi_\infty(\cdot, z)$ does not belong to the range of the operator B_0 because this would mean that $\Phi(\cdot, z)$ solves the Helmholtz equation in the exterior of D.

We summarize the foregoing analysis in the following theorem. To this end, we state the following assumptions on the symmetric matrix-valued function $A = (a_{j,k})_{j,k=1,2}$, $a_{j,k} \in C^1(\bar{D})$ and $n \in C(\bar{D})$:

- *Assumption 1*: $\bar{\xi} \cdot \mathrm{Im}(A)\,\xi = 0$, $\mathcal{I}m\,(n) = 0$, and

$$\text{either } 0 < a^* < 1 \text{ and } 0 < n^* < 1, \text{ or } a_* > 1 \text{ and } n_* > 1,$$

where a_*, a^*, n_*, and n^* are defined by (6.95).
- *Assumption 2*: $\bar{\xi} \cdot \mathrm{Im}(A)\,\xi \leq 0$, $\mathrm{Im}(n) \geq 0$, and

$$\text{either } \bar{\xi} \cdot \mathrm{Re}(A)\,\xi \geq \gamma|\xi|^2 \text{ or } \bar{\xi} \cdot \mathrm{Re}(A^{-1})\,\xi \geq \gamma|\xi|^2$$

for all $\xi \in \mathbb{C}^2$ and $x \in \bar{D}$ with a constant $\gamma > 1$.

Theorem 6.50. *Assume that D is a bounded domain having a C^2 boundary ∂D such that $\mathbb{R}^2 \setminus \bar{D}$ is connected, and A and n satisfy either Assumption 1 or Assumption 2. Furthermore, assume that k is not a transmission eigenvalue. Then if F is the far-field operator (6.7) corresponding to the transmission problem (6.1)–(6.5), we have that*

1. *For $z \in D$ and a given $\epsilon > 0$ there exists a function $g_z^\epsilon \in L^2[0, 2\pi]$ such that*

$$\|Fg_z^\epsilon - \Phi_\infty(\cdot, z)\|_{L^2[0, 2\pi]} < \epsilon$$

and the Herglotz wave function $v_{g_z^\epsilon}$ with kernel g_z^ϵ converges in $H^1(D)$ to w as $\epsilon \to 0$, where (v, w) is the unique solution of (6.118)–(6.121);
2. *For $z \notin D$ and a given $\epsilon > 0$ every function $g_z^\epsilon \in L^2[0, 2\pi]$ that satisfies*

$$\|Fg_z^\epsilon - \Phi_\infty(\cdot, z)\|_{L^2[0, 2\pi]} < \epsilon$$

is such that

$$\lim_{\epsilon \to 0} \|v_{g_z^\epsilon}\|_{H^1(D)} = \infty.$$

The importance of Theorem 6.50 in solving the inverse scattering problem of determining the support D of an orthotropic inhomogeneity from the far-field pattern is clear from our discussion in Chap. 4. In particular, using regularization methods to solve the far-field equation $Fg = \Phi_\infty(\cdot, z)$ for z on an appropriate grid containing D, an approximation to g_z can be obtained, and hence ∂D can be determined by those points where $\|g_z\|_{L^2[0, 2\pi]}$ becomes unbounded. More discussion on the numerical implementation is presented in Chap. 8.

6.6 Determination of Transmission Eigenvalues from Far-Field Data

In the previous section we showed how the linear sampling method could be used to determine the support of the inhomogeneous scattering object provided k is not a transmission eigenvalue. At the same time we showed that the transmission eigenvalues carried qualitative information about the material properties of the scatterer (cf. Theorems 6.22, 6.23, 6.25, and 6.36). To exploit the possibility of using this qualitative information, we are no longer interested in avoiding transmission eigenvalues as in the case of the linear sampling method but rather now want to be able to determine them from the (noisy) far-field data. This last section of our chapter is devoted to this problem.

At this point we assume that D (or a reconstruction of D using the linear sampling method) is known and fix an arbitrary point $z \in D$. In Theorem 6.50 it was shown that if k is not a transmission eigenvalue, then for a given $\epsilon > 0$ there exists a function $g_z^\epsilon \in L^2[0, 2\pi]$ such that

$$\|Fg_z^\epsilon - \Phi_\infty(\cdot, z)\|_{L^2[0, 2\pi]} < \epsilon \qquad (6.124)$$

and the Herglotz wave function $v_{g_z^\epsilon}$ with kernel g_z^ϵ converges in $H^1(D)$ to w as $\epsilon \to 0$, where (v, w) is the unique solution of (6.118)–(6.121). We will now show that if k is a transmission eigenvalue, then the $H^1(D)$ norm of $v_{g_z^\epsilon}$ blows up as $\epsilon \to 0$. More specifically, we can prove the following theorem.

Theorem 6.51. *Assume that either* $0 < a^* < 1$ *and* $0 < n^* < 1$, *or* $a_* > 1$ *and* $n_* > 1$, *where* a_*, a^*, n_*, *and* n^* *are defined by (6.95). Let* k *be a transmission eigenvalue and* g_z^ϵ *satisfy (6.124). Then for every* $z \in D$, *except for a nowhere dense set,* $\|v_{g_z^\epsilon}\|_{H^1(D)}$ *cannot be bounded as* $\epsilon \to 0$.

Proof. Assume to the contrary that for a set of points $z \in D$ that has an accumulation point, there exists a sequence $\epsilon_n \to 0$ such that $\|v_n\|_{H^1(D)}$ remains bounded as $n \to \infty$, where $v_n := v_{g_z^{\epsilon_n}}$, with $g_z^{\epsilon_n}$ satisfying (6.124). Without loss of generality we may assume that v_n converges weakly to $w \in H^1(D)$. In a similar way as in the proof of Theorem 6.50 it is seen that

$w := w_z$, where v_z and w_z solve the interior transmission problem (6.118)–(6.121). But (6.118)–(6.121) is equivalent to the variational form (see the discreteness of transmission eigenvalues for the case $n \neq 1$ in Sect. 6.3)

$$a_k((v, w), (v', w')) = \ell(v', w') \quad \text{for all} \quad (v', w') \in \mathcal{H}(D), \tag{6.125}$$

where

$$\ell(v', w') = \int_{\partial D} \overline{v'} \frac{\partial \Phi(\cdot, z)}{\partial \nu} \, ds - \int_D (\nabla \phi_z \cdot \nabla \overline{w'} - k^2 \phi_z \overline{w'}) \, dx,$$
$$\mathcal{H}(D) := \{(v, w) \in H^1(D) \times H^1(D) : v - w \in H_0^1(D)\},$$

and $\phi_z \in H^1(D)$ is a lifting function such that $\phi_z = \Phi(\cdot, z)$ on ∂D. As discussed in Remark 6.29, (6.125) satisfies the Fredholm alternative. Hence, noting that the operator determined by $a_k(\cdot, \cdot)$ via the Riesz representation theorem is self-adjoint, we have that $w := w_z$ and v_z solve (6.118)–(6.121) if and only if $\ell(v_k, w_k) = 0$, where (v_k, w_k) is a transmission eigenfunction corresponding to the transmission eigenvalue k. Using integration by parts and the facts that $\Delta w_k + k^2 w_k = 0$ in D and $v_k = w_k$ on ∂D we obtain that the solvability condition takes the form

$$\int_{\partial D} \left(\overline{w_k} \frac{\partial \Phi(\cdot, z)}{\partial \nu} - \frac{\partial \overline{w_k}}{\partial \nu} \Phi(\cdot, z) \right) \, ds = 0.$$

Now Green's representation formula and the analyticity of the solution to the Helmholtz equation imply that $w_k = 0$ in D and, consequently, $v_k = 0$ in D. This contradicts the fact that (v_k, w_k) is a transmission eigenfunction, which proves the theorem. $\qquad\square$

Similarly, we can prove the following theorem, which we leave as an exercise for the reader.

Theorem 6.52. *Assume that $n = 1$ and either $a_{max} < 1$ or $a_{min} > 1$, where a_{max} and a_{min} are defined by (6.48). Let k be a transmission eigenvalue and g_z^ϵ satisfy (6.124). Then for every $z \in D$ except for a nowhere dense set, $\|v_{g_z^\epsilon}\|_{H^1(D)}$ cannot be bounded as $\epsilon \to 0$.*

Theorem 6.50, together with Theorems 6.51 and 6.52, suggests that for $z \in D$, $v_{g_z^\epsilon}$ exhibits different behavior if k is not a transmission eigenvalue and if k is a transmission eigenvalue. Hence the far-field equation can be used to determine the transmission eigenvalues in addition to determining the support of the inhomogeneity if the far-field data are available for a range of frequencies.

In practice, only the noisy far-field operator F_δ given by

$$F_\delta g = \int_0^{2\pi} u_\infty^\delta(\hat{x}, d) g(d) \, ds(d)$$

is available, where u_∞^δ is the noisy far-field data with noise level $\delta > 0$. Then we look for the Tikhonov regularized solution $g_z^{\alpha, \delta}$ of the far-field equation defined as the unique minimizer of the *Tikhonov functional*

$$\|F_\delta g - \Phi_\infty(\cdot, z)\|^2_{L^2[0,2\pi]} + \alpha\|g\|^2_{L^2[0,2\pi]},$$

where the positive number $\alpha > 0$ is known as the *Tikhonov regularization parameter* (cf. Sect. 2.1). This regularization parameter depends on the noise level and can be chosen such that $\alpha(\delta) \to 0$ as $\delta \to 0$. If $g_z^\delta := g_z^{\alpha(\delta),\delta}$, then it can be shown (see [22]) that

$$\lim_{\delta \to 0} \|Fg_z^\delta - \Phi_\infty(\cdot, z)\|_{L^2[0,2\pi]} = 0.$$

Hence Theorems 6.51 and 6.52 hold true for the regularized solution g_z^δ, where ϵ is now replaced by δ.

The first part of Theorem 6.50 also holds true for the regularized solution g_z^δ of the far-field equation, but its justification involves the more elaborate argument developed for the Dirichlet obstacle scattering problem by Arens in [5]. This argument can be carried through for the case of inhomogeneous media with real-valued physical parameters, which is the case where transmission eigenvalues exist. It is essential to this generalization to show that $\Phi(\cdot, z)$ is in the range of $(F^*F)^{1/4}$ if and only if $z \in D$, which constitutes the so-called *factorization method*. More generally, the factorization method provides an analytical framework to justify the linear sampling method (i.e., Theorem 6.50) for the regularized solution of the far-field equation that is obtained in practice. The factorization method holds for a restrictive class of scattering problems and is the subject of the following chapter.

In conclusion, to determine the transmission eigenvalues from the far-field data, we choose a point $z \in D$ and the Tikhonov regularized solution g_z^δ to the far-field equation. The transmission eigenvalues will appear as sharp peaks in the plot of $\|v_{g_z^\delta}\|_{H^1(D)}$ or $\|g_z^\delta\|_{L^2[0,2\pi]}$ against the wave number k for a range of interrogating frequencies.

As an example of the use of transmission eigenvalues to determine information about the material properties of the scattering object from far-field data, we consider the scattering problem (5.8)–(5.12) with $n = 1$ and D the unit square $[-1/2, 1/2] \times [-1/2, 1/2]$. We consider four different possibilities for $A = A(x)$:

$$A_{iso} = \begin{pmatrix} 1/4 & 0 \\ 0 & 1/4 \end{pmatrix}, \qquad A_1 = \begin{pmatrix} 1/2 & 0 \\ 0 & 1/8 \end{pmatrix},$$

$$A_2 = \begin{pmatrix} 1/6 & 0 \\ 0 & 1/8 \end{pmatrix}, \qquad A_{2r} = \begin{pmatrix} 0.1372 & 0.0189 \\ 0.0189 & 0.1545 \end{pmatrix},$$

noting that A_{2r} is obtained by rotating matrix A_2 by 1 radian. For each A the direct scattering problem is then solved using finite-element methods, and the far-field equation with noisy far-field data is then solved for 25 random source points z in the unit square (for details see [28]). It is assumed that D is known (for example, through the use of the linear sampling method). In Fig. 6.1 we show a plot of the average norm of the Herglotz kernel $\|g_z^\delta\|_{L^2[0,2\pi]}$ against

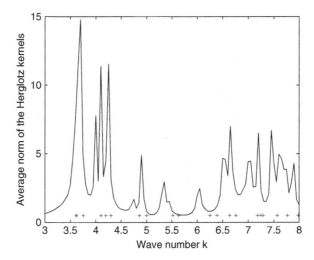

Fig. 6.1. Results for square using anisotropy A_{2r}. We show a plot of the average values of $\|g_z^\delta\|_{L^2(D)}$ against k. We also mark the computed eigenvalues from the finite-element code (shown as + along the bottom of the graph). Good agreement is seen with the lowest computed eigenvalue and the first peak of the norms of g_z^δ][1]

Table 6.1. Our theory implies that the scalar a reconstructed from the first nonzero real transmission eigenvalue should lie between the eigenvalues of matrix A. In the case of an isotropic A, the predicted a should reconstruct the diagonal of A. The table supports both these claims[1]

Domain	Matrix	Eigenvalues	Predicted $k_{1,D,A(x)}$	Predicted a
Square	A_{iso}	1/4,1/4	5.3	0.248
	A_1	1/2,1/8	4.1	0.172
	A_2	1/6,1/8	3.55	0.135
	A_{2r}	1/6,1/8	3.7	0.145

the wave number k corresponding to matrix A_{2r}. Given the first transmission eigenvalue $k_{1,D,A(x)}$ from Fig. 6.1, we can now compute a positive number a such that $k_{1,D,aI} = k_{1,D,A(x)}$. According to Theorem 6.22, a should lie between the smallest and largest eigenvalues of A. Table 6.1 below shows the results of this calculation for each of the preceding cases for A.

Additional numerical examples of the determination of transmission eigenvalues from far-field data and their use to obtain information on the refractive index of the inhomogeneity can be found in [21] and [28].

---

[1]Reprinted from F. Cakoni, D. Colton, P. Monk, and J. Sun, The inverse electromagnetic scattering problem for anisotropic media, Inverse Problems 26 (2010), 074004.

7

Factorization Methods

The linear sampling method introduced in Chaps. 4 and 6 is based on the far-field equation $Fg = \Phi_\infty(\cdot, z)$, where F is the far-field operator corresponding to the scattering problem. In particular, it is shown in Theorem 6.50 that, in the case of noise-free data, for every n there exists an approximate solution $g_n^z \in L^2[0, 2\pi]$ of the far-field equation with discrepancy $1/n$ such that the sequence of Herglotz wave functions $v_{g_n^z}$ with kernel g_n^z converges (in an appropriate norm) if and only if $z \in D$, where D is the support of the scattering object. Unfortunately, since the convergence of $v_{g_n^z}$ is described in terms of a norm depending on D, $v_{g_n^z}$ cannot be used to characterize D. Instead, the linear sampling method characterizes the obstacle by the behavior of g_n^z, and it is not possible to obtain any convergence result for the regularized solution g of the far-field equation if the noise in the data goes to zero. It would be desirable to modify the far-field equation to avoid this difficulty, and this desire motivated Kirsch to introduce in [99] and [100] the *factorization method* for solving both the inverse obstacle scattering problem and the inverse scattering problem for a nonabsorbing inhomogeneous medium. In particular, the factorization method replaces the far-field operator in the far-field equation by the operator $(F^*F)^{1/4}$. One can then show that $(F^*F)^{1/4}g = \Phi_\infty(\cdot, z)$ has a solution if and only if $z \in D$. Despite considerable efforts [101, 103, 75, 76], the factorization method is still limited to a restricted class of scattering problems. In particular, to date the method has not been established for the case of limited aperture data, partially coated obstacles, and many of the basic scattering problems for Maxwell's equations (Chap. 9). On the other hand, when applicable, the factorization method provides a mathematical justification for using the regularized solution of an appropriate far-field equation to determine D, a feature that is in general lacking in the linear sampling method. This is then followed by a derivation of the factorization method for an inhomogeneous anisotropic media. We conclude our chapter by using the factorization method to resolve the aforementioned difficulties with the linear sampling method. For a scholarly and comprehensive discussion of the factorization method we refer the reader to [107].

F. Cakoni and D. Colton, *A Qualitative Approach to Inverse Scattering Theory*, 165
Applied Mathematical Sciences 188, DOI 10.1007/978-1-4614-8827-9_7,
© Springer Science+Business Media New York 2014

7.1 Factorization Method for Obstacle Scattering

7.1.1 Preliminary Results

We begin with some results on single and double layer potentials. In Sects. 3.3 and 4.3 we introduced single and double layer potentials with continuous densities and discussed their continuity properties. In particular, if $D \subset \mathbb{R}^2$ is a bounded domain with C^2 boundary ∂D and ν is the unit outward normal to ∂D, the single layer potential is defined by

$$(\mathcal{S}\psi)(x) := \int_{\partial D} \psi(y)\Phi(x,y)ds_y, \qquad x \in \mathbb{R}^2 \setminus \partial D, \qquad (7.1)$$

and the double layer potential is defined by

$$(\mathcal{D}\psi)(x) := \int_{\partial D} \psi(y)\frac{\partial}{\partial \nu_y}\Phi(x,y)ds_y, \qquad x \in \mathbb{R}^2 \setminus \partial D, \qquad (7.2)$$

where $\Phi(x,y) := i/4H_0^{(1)}(k|x-y|)$ is the fundamental solution to the Helmholtz equation, with $H_0^{(1)}$ being a Hankel function of the first kind of order zero. For $x \in \mathbb{R}^2 \setminus \partial D$, both the single and double layer potentials are solutions to the Helmholtz equation and satisfy the Sommerfeld radiation condition. It can be shown [111, 127] that, for $-1 \leq s \leq 1$, the mapping $\mathcal{S} : H^{s-\frac{1}{2}}(\partial D) \to H_{loc}^{s+1}(\mathbb{R}^2)$ is continuous and the mappings $\mathcal{D} : H^{s+\frac{1}{2}}(\partial D) \to H_{loc}^{s+1}(\mathbb{R}^2 \setminus \bar{D})$ and $\mathcal{D} : H^{s+\frac{1}{2}}(\partial D) \to H^{s+1}(D)$ are continuous.

For smooth densities we define the restriction of $\mathcal{S}$ and $\mathcal{D}$ to the boundary ∂D by

$$(S\psi)(x) := \int_{\partial D} \psi(y)\Phi(x,y)ds_y \qquad x \in \partial D, \qquad (7.3)$$

$$(K\psi)(x) := \int_{\partial D} \psi(y)\frac{\partial}{\partial \nu_y}\Phi(x,y)ds_y \qquad x \in \partial D, \qquad (7.4)$$

and the restriction of the normal derivative of $\mathcal{S}$ and $\mathcal{D}$ to the boundary ∂D by

$$(K'\psi)(x) := \frac{\partial}{\partial \nu_x}\int_{\partial D} \psi(y)\Phi(x,y)ds_y \qquad x \in \partial D, \qquad (7.5)$$

$$(T\psi)(x) := \frac{\partial}{\partial \nu_x}\int_{\partial D} \psi(y)\frac{\partial}{\partial \nu_y}\Phi(x,y)ds_y \qquad x \in \partial D. \qquad (7.6)$$

It can be shown [51, 111] that for smooth densities the single layer potential and the normal derivative of the double layer potential are continuous across ∂D, i.e.,

$$(\mathcal{S}\psi)_+ = (\mathcal{S}\psi)_- = S\psi \qquad \text{on } \partial D, \qquad (7.7)$$

$$\frac{\partial(\mathcal{D}\psi)_+}{\partial \nu} = \frac{\partial(\mathcal{D}\psi)_-}{\partial \nu} = T\psi \qquad \text{on } \partial D, \qquad (7.8)$$

while the normal derivative of the single layer potential and the double layer potential are discontinuous across ∂D and satisfy the following jump relations:

$$\frac{\partial(\mathcal{S}\psi)_\pm}{\partial\nu} = K'\psi \mp \frac{1}{2}\psi \qquad \text{on } \partial D, \tag{7.9}$$

$$(\mathcal{D}\psi)_\pm = K\psi \pm \frac{1}{2}\psi \qquad \text{on } \partial D, \tag{7.10}$$

where the subindexes $+$ and $-$ indicate that x approaches ∂D from outside and from inside D, respectively. It can be shown that for $-1 \leq s \leq 1$ (7.7) and (7.9) remain valid for $\psi \in H^{-\frac{1}{2}+s}(\partial D)$, while (7.8) and (7.10) are valid for $\psi \in H^{\frac{1}{2}+s}(\partial D)$, where $u_\pm$ and $\partial u_\pm(x)/\partial\nu$ are interpreted in the sense of trace theorems for $u \in H^{1+s}(\mathbb{R}^2 \setminus \bar{D})$ and $u \in H^{1+s}(D)$, respectively (see Theorems 1.38 and 5.7 for the case of $s = 0$). Furthermore, the following operators are continuous [90, 127]:

$$S : H^{-\frac{1}{2}+s}(\partial D) \longrightarrow H^{\frac{1}{2}+s}(\partial D), \tag{7.11}$$

$$K : H^{\frac{1}{2}+s}(\partial D) \longrightarrow H^{\frac{1}{2}+s}(\partial D), \tag{7.12}$$

$$K' : H^{-\frac{1}{2}+s}(\partial D) \longrightarrow H^{-\frac{1}{2}+s}(\partial D), \tag{7.13}$$

$$T : H^{\frac{1}{2}+s}(\partial D) \longrightarrow H^{-\frac{1}{2}+s}(\partial D) \tag{7.14}$$

for $-1 \leq s \leq 1$.

Definition 7.1. Let X be a Hilbert space equipped with the operation of conjugation, and let X^* be its dual. If $\langle \cdot, \cdot \rangle$ denotes the duality pairing between X and X^*, then we define

$$(f, u) = \langle f, \bar{u} \rangle \qquad f \in X^*,\ u \in X.$$

Definition 7.2. Let X and Y be Hilbert spaces and $A : X \to Y$ a linear operator. We define the adjoint operator $A^* : Y^* \to X^*$ by

$$(A^*v, u) = (v, Au), \qquad v \in Y^*,\ u \in X,$$

where X^* and Y^* are the duals of X and Y, respectively, and $(\cdot, \cdot)$ is defined by Definition 7.1.

Note that this definition of the adjoint is consistent with that given in Chap. 1. Furthermore, up to conjugation, A^* is the same as the transpose operator $A^\top$ defined in Sect. 6.5.

Theorem 7.3. *Assume that k^2 is not a Dirichlet eigenvalue of $-\Delta$ in D.*

1. *Let S_i be the boundary operator defined by (7.3) with k replaced by i in the fundamental solution. Then S_i satisfies*

$$(S_i\psi, \psi) \geq C\|\psi\|^2_{H^{-\frac{1}{2}}(\partial D)}, \quad \psi \in H^{-\frac{1}{2}}(\partial D),$$

where $(\cdot, \cdot)$ is defined by Definition 7.1.

2. $S - S_i$ is compact from $H^{-\frac{1}{2}}(\partial D)$ to $H^{\frac{1}{2}}(\partial D)$.
3. S is an isomorphism from $H^{-\frac{1}{2}}(\partial D)$ onto $H^{\frac{1}{2}}(\partial D)$.
4. $\text{Im}(S\psi, \psi) = 0$ for some $\psi \in H^{-\frac{1}{2}}(\partial D)$ implies $\psi = 0$.

Proof. Let $v \in H^1_{loc}(\mathbb{R}^2 \setminus \partial D)$ be the single layer potential given by

$$v(x) := \int_{\partial D} \psi(y)\Phi(x, y)\, ds(y), \qquad \psi \in H^{-\frac{1}{2}}(\partial D), \qquad x \in \mathbb{R}^2 \setminus \partial D.$$

In particular, v satisfies the Helmholtz equation in D and $\mathbb{R}^2 \setminus \bar{D}$ and the Sommerfeld radiation condition

$$\lim_{r \to \infty} \sqrt{r}\left(\frac{\partial v}{\partial r} - ikv\right) = 0.$$

1. Set $k = i$ in the definition of v. Applying Green's first identity to v and $\bar{v}$ in D and $\Omega_R \setminus \bar{D}$, where Ω_R is a disk of radius R centered at the origin containing D, and using (7.7) and (7.9), we have that

$$(S_i\psi, \psi) = \left\langle v, \left(\frac{\partial \bar{v}_-}{\partial \nu} - \frac{\partial \bar{v}_+}{\partial \nu}\right)\right\rangle = \int_D \left(|\nabla v|^2 + |v|^2\right) dx$$

$$+ \int_{\Omega_R \setminus \bar{D}} \left(|\nabla v|^2 + |v|^2\right) dx - \int_{|x|=R} v\frac{\partial \bar{v}}{\partial r}ds.$$

From the Sommerfeld radiation condition we obtain

$$(S_i\psi, \psi) = \int_D \left(|\nabla v|^2 + |v|^2\right) dx + \int_{\Omega_R \setminus \bar{D}} \left(|\nabla v|^2 + |v|^2\right) dx$$

$$+ \int_{|x|=R} |v|^2 ds + o(1),$$

and letting $R \to \infty$, noting that v decays exponentially, we have that

$$(S_i\psi, \psi) = \int_{\mathbb{R}^2} \left(|\nabla v|^2 + |v|^2\right) dx. \tag{7.15}$$

Furthermore, from the jump properties of v across the boundary and the trace Theorem 5.7, we can write

$$\|\psi\|_{H^{-\frac{1}{2}}(\partial D)} = \left\|\frac{\partial v_-}{\partial \nu} - \frac{\partial v_+}{\partial \nu}\right\|_{H^{-\frac{1}{2}}(\partial D)} \leq \tilde{C}\|v\|_{H^1(\mathbb{R}^2)}, \tag{7.16}$$

where $\tilde{C} > 0$, and hence combining (7.15) and (7.16) we have that

$$(S_i\psi, \psi) \geq C\|\psi\|_{H^{-\frac{1}{2}}(\partial D)}, \qquad C > 0.$$

2. The kernel of $S - S_i$ is a C^∞ function in a neighborhood of $\partial D \times \partial D$, and hence, as in the first part of Theorem 4.8, we conclude that $S - S_i$ is compact from $H^{-\frac{1}{2}}(\partial D)$ to $H^{\frac{1}{2}}(\partial D)$.

3. Applying the Lax–Milgram lemma to the bounded and coercive sesquilinear form

$$a(\psi, \phi) := (S_i\psi, \phi), \qquad \phi, \psi \in H^{-\frac{1}{2}}(\partial D)$$

we conclude that $S_i^{-1} : H^{\frac{1}{2}}(\partial D) \to H^{-\frac{1}{2}}(\partial D)$ exists and is bounded. From Theorem 5.16 and using part 2, S is an isomorphism if and only if S is injective. To show that S is injective, we consider $\psi \in H^{-\frac{1}{2}}(\partial D)$ such that $S\psi = 0$. Since the single layer potential v is a radiating solution to the homogeneous Dirichlet boundary value problem in $\mathbb{R}^2 \setminus \bar{D}$, $v = 0$ in $\mathbb{R}^2 \setminus \bar{D}$. Similarly, v satisfies the homogeneous Dirichlet boundary value problem in D, and from the assumption that k^2 is not a Dirichlet eigenvalue we conclude that $v = 0$ in D as well. Finally,

$$\psi = \frac{\partial v_-}{\partial \nu} - \frac{\partial v_+}{\partial \nu} = 0,$$

which proves that S is injective.

4. Let $\operatorname{Im}(S\psi, \psi) = 0$ for some $\psi \in H^{-\frac{1}{2}}(\partial D)$. The same argument as in part 1 yields

$$(S\psi, \psi) = \left\langle v, \left(\frac{\partial \bar{v}_-}{\partial \nu} - \frac{\partial \bar{v}_+}{\partial \nu}\right)\right\rangle = \int_D \left(|\nabla v|^2 - k^2|v|^2\right) dx$$

$$+ \int_{\Omega_R \setminus \bar{D}} \left(|\nabla v|^2 - k^2|v|^2\right) dx - \int_{|x|=R} v \frac{\partial \bar{v}}{\partial r} ds$$

$$= \int_{\Omega_R} \left(|\nabla v|^2 - k^2|v|^2\right) dx + ik \int_{|x|=R} |v|^2 ds + o(1), \quad R \to \infty.$$

Taking the imaginary part we see that

$$0 = \operatorname{Im}(S\psi, \psi) = k \lim_{R\to\infty} \int_{|x|=R} |v|^2 ds.$$

Rellich's lemma implies that v vanishes in $\mathbb{R}^2 \setminus \bar{D}$. and thus $S\psi = 0$ on ∂D by the trace theorem (Theorem 1.38). Finally, since S is an isomorphism, we can conclude that $\psi = 0$.

$\square$

Remark 7.4. Property 1 in Theorem 7.3 implies that there exists a square root $S_i^{\frac{1}{2}}$ of S_i and $S_i^{\frac{1}{2}}$ is an isomorphism from $H^{-\frac{1}{2}}(\partial D)$ onto $L^2(\partial D)$ and from $L^2(\partial D)$ onto $H^{\frac{1}{2}}(\partial D)$ (Sect. 9.4 in [90]). Furthermore, $S_i^{\frac{1}{2}}$ is positive definite using the duality defined by Definition 7.1 and self-adjoint, i.e., $S_i^{\frac{1}{2}} = S_i^{\frac{1}{2}*}$, where the adjoint operator is defined by Definition 7.2.

In a similar way as in Theorem 7.3 one can show the following properties for the operator T.

Theorem 7.5. *Assume that k^2 is not a Neumann eigenvalue of $-\Delta$ in D.*

1. *Let T_i be the boundary operator defined by (7.6), with k replaced by i in the fundamental solution. Then T_i satisfies*

$$- (T_i \psi, \psi) \geq C \|\psi\|^2_{H^{\frac{1}{2}}(\partial D)} \qquad \text{for all } \psi \in H^{\frac{1}{2}}(\partial D),$$

where $(\cdot, \cdot)$ is defined by Definition 7.1.

2. $T - T_i$ *is compact from $H^{\frac{1}{2}}(\partial D)$ to $H^{-\frac{1}{2}}(\partial D)$.*
3. T *is an isomorphism from $H^{\frac{1}{2}}(\partial D)$ onto $H^{-\frac{1}{2}}(\partial D)$.*
4. $Im(T\psi, \psi) = 0$ *for some $\psi \in H^{\frac{1}{2}}(\partial D)$ implies $\psi = 0$.*

We now turn our attention to the concept of a Riesz basis in a Hilbert space. Let X be a Hilbert space. A sequence $\{\phi_n\}_1^\infty$ is said to be a *Schauder basis* for X if for each vector $u \in X$ there exists a unique sequence of complex numbers $c_1, c_2, \ldots$ such that $u = \sum_1^\infty c_n \phi_n$, where the convergence is understood as

$$\lim_{k \to \infty} \left\| u - \sum_1^k c_n \phi_n \right\|_X = 0.$$

In particular, a complete orthonormal system is a Schauder basis for X. The simplest way of constructing a new basis from an old is one through an isomorphism. In particular, let $\{\phi_n\}_1^\infty$ be a basis in X and $T : X \to X$ be a bounded linear operator with bounded inverse. Then $\{\psi_n\}_1^\infty$ such that $\psi_n = T\phi_n$, $n = 1, 2, \cdots$ is also a basis for X.

Definition 7.6. Two bases $\{\phi_n\}_1^\infty$ and $\{\psi_n\}_1^\infty$ are said to be *equivalent* if $\sum_1^\infty c_n \phi_n$ converges if and only if $\sum_1^\infty c_n \psi_n$ converges.

The following theorem can be shown [161].

Theorem 7.7. *Two bases $\{\phi_n\}_1^\infty$ and $\{\psi_n\}_1^\infty$ are equivalent if and only if there exists a bounded linear operator $T : X \to X$ with bounded inverse such that $\psi_n = T\phi_n$ for every n.*

In Hilbert spaces the most important bases are orthonormal bases thanks to their nice properties (Theorem 1.13). Second in importance are those bases that are equivalent to some orthonormal basis. They will be called *Riesz bases*.

Definition 7.8. A basis for a Hilbert space is a *Riesz basis* if it is equivalent to an orthonormal basis, that is, if it is obtained from an orthonormal basis by means of a bounded invertible linear operator.

Definition 7.9. Two inner products $(\cdot, \cdot)_1$ and $(\cdot, \cdot)_2$ in a Hilbert space X are said to be *equivalent* if $c\|\cdot\|_1 \leq \|\cdot\|_2 \leq C\|\cdot\|_1$ for some positive constants c, C, where $\|\cdot\|_j$, $j = 1, 2$, is the norm generated by $(\cdot, \cdot)_j$.

The next theorem provides some important properties of Riesz bases.

Theorem 7.10. *Let X be a Hilbert space. Then the following statements are equivalent.*

1. *The sequence $\{\phi_n\}_1^\infty$ forms a Riesz basis for X.*
2. *There exists an equivalent inner product on X with respect to which the sequence $\{\phi_n\}_1^\infty$ becomes an orthonormal basis for X.*
3. *The sequence $\{\phi_n\}_1^\infty$ is complete in X, and there exists positive constants c and C such that for an arbitrary positive integer k and arbitrary complex numbers $c_1, \ldots, c_k$ one has*

$$c \sum_1^k |c_n|^2 \leq \left\| \sum_1^k c_n \phi_n \right\|^2 \leq C \sum_1^k |c_n|^2.$$

Proof. $1 \Longrightarrow 2$: since $\{\phi_n\}_1^\infty$ is a Riesz basis for X, there exists a bounded linear operator T with bounded inverse that transforms $\{\phi_n\}_1^\infty$ into some orthonormal basis $\{e_n\}_1^\infty$, i.e., $T\phi_n = e_n$, $n = 1, 2, \cdots$. Define a new inner product $(\cdot, \cdot)_1$ on X by setting

$$(\phi, \psi)_1 = (T\phi, T\psi), \qquad \phi, \psi \in X,$$

and let $\| \cdot \|_1$ be the norm generated by this inner product. Then

$$\frac{\|\phi\|}{\|T^{-1}\|} \leq \|\phi\|_1 \leq \|T\| \|\phi\|$$

for every $\phi \in X$. Hence the new inner product is equivalent to the original one. Clearly,

$$(\phi_n, \phi_m)_1 = (T\phi_n, T\phi_m) = (e_n, e_m) = \delta_{nm}$$

for every n and m, where $\delta_{nm} = 0$ for $n \neq m$ and $\delta_{nm} = 1$ for $n = m$.
$2 \Longrightarrow 3$: suppose that $(\cdot, \cdot)_1$ is an equivalent inner product on X and $\{\phi_n\}_1^\infty$ is an orthonormal basis with respect to $(\cdot, \cdot)_1$. From the relation

$$c\|\phi\|_1 \leq \|\phi\|_2 \leq C\|\phi\|_1,$$

where c and C are positive constants, it follows that for arbitrary complex numbers $c_1, \ldots, c_k$ one has

$$\frac{1}{C^2} \sum_1^k |c_n|^2 \leq \left\| \sum_1^k c_n \phi_n \right\|^2 \leq \frac{1}{c^2} \sum_1^k |c_n|^2.$$

Clearly, from Theorem 1.13, $\{\phi_n\}_1^\infty$ is complete in X.
$3 \Longrightarrow 1$: let $\{e_n\}_1^\infty$ be an arbitrary orthonormal basis for X. We define operators T and S on the subset of linear combinations of $\{e_n\}_1^\infty$ and $\{\phi_n\}_1^\infty$ by

$$T \sum_1^k c_n e_n = \sum_1^k c_n \phi_n, \qquad S \sum_1^k c_n \phi_n = \sum_1^k c_n e_n.$$

It follows by assumption that T and S are bounded on their domain of definition. Since both $\{e_n\}_1^\infty$ and $\{\phi_n\}_1^\infty$ are complete in X (Theorem 1.13), each of the operators T and S can be extended by continuity to bounded linear operators defined on the entire space X. It is easily seen that $ST = TS = I$, whence $T^{-1} = S$. Hence $\{\phi_n\}_1^\infty$ is a Riesz basis for X. $\qquad\square$

For a more comprehensive study of the Riesz basis we refer the reader to [161].

We end this section with a result on the Riesz basis due to Kirsch [99], which will later play an important role in the factorization method.

Theorem 7.11. *Let X be a Hilbert space. Assume that $K : X \to X$ is a compact linear operator with $\text{Im}(K\phi, \phi) \neq 0$ for all $\phi \in X$, $\phi \neq 0$. Let $\{\phi_n\}_1^\infty$ be a linearly independent and complete sequence in X that is orthogonal in the sense that*

$$((I + K)\phi_n, \phi_m) = c_n \delta_{nm}, \qquad (7.17)$$

where $(\cdot, \cdot)$ is the inner product on X and the constants c_n are such that $\text{Im}(c_n) \to 0$ as $n \to \infty$ and there exists a positive constant $r > 0$ independent of n such that $|c_n| = r$ for all $n = 1, 2, \ldots$. Then $\{\phi_n\}_1^\infty$ is a Riesz basis.

Proof. The proof consists of several steps.

1. We first show that the sequence $\{\phi_n\}_1^\infty$ is bounded. Assume, on the contrary, that there exists a subsequence, still denoted by $\{\phi_n\}_1^\infty$, such that $\|\phi_n\| \to \infty$. Set $\hat{\phi}_n = \phi_n/\|\phi_n\|$, and note that

$$1 + (K\hat{\phi}_n, \hat{\phi}_n) = \left((I + K)\hat{\phi}_n, \hat{\phi}_n\right) \to 0 \qquad \text{as } n \to \infty. \qquad (7.18)$$

   Since $\{\hat{\phi}_n\}_1^\infty$ is bounded, there exists a subsequence, still denoted by $\{\hat{\phi}_n\}_1^\infty$, that converges weakly to a $\hat{\phi} \in X$. Since K is compact, we have that $\|K\hat{\phi}_n - K\hat{\phi}\| \to 0$ for a further subsequence, still denoted by $\{\hat{\phi}_n\}_1^\infty$. Hence $(K\hat{\phi}_n, \hat{\phi}_n) = (K\hat{\phi}_n - K\hat{\phi}, \hat{\phi}_n) + (K\hat{\phi}, \hat{\phi}_n) \to (K\hat{\phi}, \hat{\phi})$ as $n \to \infty$. Then (7.18) implies that $1 + (K\hat{\phi}, \hat{\phi}) = 0$. Taking the imaginary part we see that $\text{Im}(K\hat{\phi}, \hat{\phi}) = 0$ and, thus, $\hat{\phi} = 0$, which contradicts the fact that $1 + (K\hat{\phi}, \hat{\phi}) = 0$.

2. We next show that r is the only accumulation point of $\{c_n\}_1^\infty$. To this end we notice that the conditions on c_n imply that $\pm r$ are the only possible accumulation points of the sequence $\{c_n\}_1^\infty$. Assume now that there exists a subsequence, still denoted by $\{c_n\}_1^\infty$, such that $\{c_n\}_1^\infty \to -r$ as $n \to \infty$. Since from the previous step $\{\phi_n\}_1^\infty$ is bounded, there exists a subsequence, still denoted by $\{\phi_n\}_1^\infty$, such that $\phi_n \rightharpoonup \phi$ weakly. As in step 1 we conclude that $(K\phi_n, \phi_n) \to (K\phi, \phi)$ and, thus, from (7.17)

$$\text{Im}(c_n) = \text{Im}(K\phi_n, \phi_n) \to \text{Im}(K\phi, \phi).$$

On the other hand, since $\text{Im}(c_n) \to 0$, we obtain that $\text{Im}(K\phi, \phi) = 0$, and hence $\phi = 0$. Another application of (7.17) implies that $\|\phi_n\|^2 \to -r$, which is impossible since $r > 0$. Thus we have shown that $c_n \to r$. In particular, there exists an integer n_0 such that $\text{Re}(c_n) \geq r/2$ for all $n \geq n_0$.

3. We define the closed subspace $U \subset X$ by

$$U := \{\phi \in X : ((I+K)\phi, \phi_m) = 0 \quad \text{for } m = 1, \ldots, n_0 - 1\}.$$

We will show that the set $\{\phi_n : n \geq n_0\}$ is complete in U. To this end, we first note that from (7.17) $\phi_n \in U$ for $n \geq n_0$. For a given $\phi \in U$, since $\{\phi_n\}_1^\infty$ is complete in X, there exists $\alpha_n^{(k)} \in \mathbb{C}$, $n = 1, \ldots, k$, and $k \in \mathbb{N}$ such that

$$\sum_{n=1}^{n_0-1} \alpha_n^{(k)} \phi_n + \sum_{n=n_0}^{k} \alpha_n^{(k)} \phi_n \to \phi \quad \text{as } k \to \infty.$$

Applying $I + K$ and taking the inner product of the result with ϕ_m, $m = 1, \ldots, n_0 - 1$, from the continuity of K and of the inner product we obtain

$$\sum_{n=1}^{n_0-1} \alpha_n^{(k)} \underbrace{((I+K)\phi_n, \phi_m)}_{=c_n\delta_{nm}} + \sum_{n=n_0}^{k} \alpha_n^{(k)} \underbrace{((I+K)\phi_n, \phi_m)}_{=0} \to \underbrace{((I+K)\phi, \phi_m)}_{=0},$$

and thus $\alpha_n^{(k)} \to 0$ as $k \to 0$ for every $n = 1, \ldots, n_0 - 1$. This implies that

$$\sum_{n=1}^{n_0-1} \alpha_n^{(k)} \phi_n \to 0 \quad \text{as } k \to \infty,$$

whence

$$\sum_{n=n_0}^{k} \alpha_n^{(k)} \phi_n \to \phi \quad \text{as } k \to \infty,$$

and hence span$\{\phi_n : n \geq n_0\}$ is dense in U.

4. In the next step we show that there exists a $C > 0$ such that

$$\text{Re}\,((I + K)\phi, \phi) \geq C\|\phi\|^2 \qquad \text{for all } \phi \in U. \tag{7.19}$$

To this end, we first claim that

$$\text{Re}\,((I + K)\phi, \phi) > 0 \qquad \text{for all } \phi \in U.$$

Indeed, from step 2 we know that

$$\text{Re}\,((I + K)\phi_n, \phi_n) = \text{Re}(c_n) > 0 \qquad \text{for } n \geq n_0.$$

The orthogonality relation (7.17) yields

$$\text{Re}\left((I+K)\sum_1^k \alpha_n\phi_n, \sum_1^k \alpha_n\phi_n\right) = \sum_1^k \text{Re}(c_n)|\alpha_n|^2 > 0,$$

and the completeness of $\{\phi_n : n \geq n_0\}$ in U proves the claim. Having proved that $\mathrm{Re}\,((I + K)\phi, \phi) > 0$, we now suppose on the contrary that (7.19) is not true. Then there exists a sequence $\{\phi^{(j)}\}$, $\phi^{(j)} \in U$, with $\|\phi^{(j)}\| = 1$ satisfying

$$\mathrm{Re}\left((I + K)\phi^{(j)}, \phi^{(j)}\right) \to 0 \quad \text{as} \quad j \to \infty.$$

By the completeness of $\{\phi_n : n \geq n_0\}$ in U, we can assume without loss of generality that $\phi^{(k)}$ is of the form

$$\phi^{(j)} = \sum_{n=n_0}^{k_j} \alpha_n^{(j)} \phi_n, \qquad \alpha_n^{(j)} \in \mathbb{C}.$$

From the orthogonality relation (7.17) we have that

$$\left((I + K)\phi^{(j)}, \phi^{(j)}\right) = \left((I + K)\sum_{n=n_0}^{k_j} \alpha_n^{(j)} \phi_n, \sum_{n=n_0}^{k_j} \alpha_n^{(j)} \phi_n\right)$$

$$= \sum_{n,m=n_0}^{k_j} \alpha_n^{(j)} \overline{\alpha}_m^{(j)} \left((I + K)\phi_n, \phi_m\right) = \sum_{n=n_0}^{k_j} c_n |\alpha_n^{(j)}|^2.$$

Taking the real part we now have that

$$\sum_{n=n_0}^{k_j} \mathrm{Re}(c_n)|\alpha_n^{(j)}|^2 \to 0 \quad \text{as} \quad j \to \infty.$$

Since from step 2 we have that $r/2 \leq \mathrm{Re}(c_n) \leq r$, this implies that

$$\sum_{n=n_0}^{k_j} |\alpha_n^{(j)}|^2 \to 0 \quad \text{as} \quad j \to \infty,$$

whence

$$\left((I + K)\phi^{(j)}, \phi^{(j)}\right) \to 0 \quad \text{as} \quad j \to \infty. \tag{7.20}$$

Now we proceed as in step 1, where we replace $\hat{\phi}_n$ by $\phi^{(j)}$, to conclude that a subsequence of $\phi^{(j)}$, still denoted by $\phi^{(j)}$, converges weakly to an element ϕ, and consequently, $(K\phi^{(j)}, \phi^{(j)}) \to (K\phi, \phi)$. From (7.20) we conclude that $\mathrm{Im}(K\phi, \phi) = 0$, which implies that $\phi = 0$. From (7.20) again we have that $\|\phi^{(j)}\| \to 0$, which contradicts the fact that $\|\phi^{(j)}\| = 1$.

5. We now define the self-adjoint operator

$$T := I + \frac{1}{2}(K + K^*)$$

and observe that T is strictly coercive in U since

$$
\begin{aligned}
(T\phi, \phi) &= \frac{1}{2}\left((I + K)\phi, \phi\right) + \frac{1}{2}\left((I + K^*)\phi, \phi\right) \\
&= \frac{1}{2}\left((I + K)\phi, \phi\right) + \frac{1}{2}\left(\phi, (I + K)\phi\right) \\
&= \operatorname{Re}\left((I + K)\phi, \phi\right) \geq C\|\phi\| \qquad \text{for all } \phi \in U.
\end{aligned}
$$

Hence from the Lax–Milgram lemma T is an isomorphism on U and the bilinear form

$$
(\phi, \psi)_1 := (T\phi, \psi)
$$

defines an inner product on U, and $(\cdot, \cdot)_1$ is equivalent to the original inner product. Furthermore, the set $\{\phi_n : n \geq n_0\}$ is orthogonal with respect to $(\cdot, \cdot)_1$ since

$$
\begin{aligned}
(\phi_n, \phi_m)_1 &= (T\phi_n, \phi_m) = \frac{1}{2}\left((I + K)\phi_n, \phi_m\right) + \frac{1}{2}\overline{\left((I + K)\phi_n, \phi_m\right)} \\
&= \operatorname{Re}(c_n)\delta_{nm} \qquad \text{for } n, m > n_0.
\end{aligned}
$$

Hence, $\{\phi_n/\sqrt{\operatorname{Re}(c_n)} : n \geq n_0\}$ is a complete orthonormal system in U. Obviously, from Theorem 1.13, for every $\phi \in U$

$$
\phi = \sum_{n_0}^{\infty} \frac{(\phi, \phi_n)_1}{\operatorname{Re}(c_n)}\phi_n = \sum_{n_0}^{\infty} \frac{(T\phi, \phi_n)}{\operatorname{Re}(c_n)}\phi_n,
$$

and Parseval's equality gives

$$
\|\phi\|^2 = \sum_{n_0}^{\infty} \frac{|(T\phi, \phi_n)|^2}{\operatorname{Re}(c_n)}.
$$

In particular, from Theorem 7.10, the set $\{\phi_n : n \geq n_0\}$ forms a Riesz basis for U.

6. Finally, we show that every element $\phi \in X$ can be expanded in a series of the ϕ_n. Let $\phi \in X$, define

$$
\phi^{\{1\}} := \sum_{1}^{n_0-1} \frac{\left((I + K)\phi, \phi_n\right)}{c_n}\phi_n,
$$

and set $\phi^{\{2\}} := \phi - \phi^{\{1\}}$. One can easily see that $\phi^{\{2\}} \in U$ since for $m = 1, \ldots n_0 - 1$

$$
\begin{aligned}
\left((I + K)\phi^{\{2\}}, \phi_m\right) &= \left((I + K)\phi, \phi_m\right) \\
&\quad - \sum_{1}^{n_0-1} \frac{\left((I + K)\phi, \phi_n\right)}{c_n} \underbrace{\left((I + K)\phi_n, \phi_m\right)}_{c_n \delta_{nm}} = 0.
\end{aligned}
$$

Hence, by step 5,

$$\phi = \sum_{n_0}^{\infty} \underbrace{\alpha_n \phi_n}_{=\phi^{\{2\}}} + \sum_{1}^{n_0-1} \underbrace{\alpha_n \phi_n}_{=\phi^{\{1\}}} \, .$$

Thus, $X = U \oplus V$, where V is the finite-dimensional space of linear combinations of ϕ_n for $n = 1, \ldots, n_0 - 1$. From step 5, the fact that V is finite dimensional and the fact that the sum $X = U \oplus V$ is direct (i.e., every $\phi \in X$ can be uniquely written as $\phi = \phi^{\{1\}} + \phi^{\{2\}}$, where $\phi^{\{1\}} \in V$ and $\phi^{\{2\}} \in U$), it is easily seen that $\{\phi_n\}$ forms a Riesz basis for X. The proof is now finished.

$\square$

7.1.2 Properties of Far-Field Operator

We shall now prove some important properties of the far-field operator in the case where the scattering obstacle is a perfect conductor. In particular, consider the direct scattering problem of finding the total field u such that

$$\Delta u + k^2 u = 0 \qquad \text{in } \mathbb{R}^2 \setminus \bar{D}, \tag{7.21}$$

$$u(x) = u^s(x) + u^i(x), \tag{7.22}$$

$$u = 0 \qquad \text{on } \partial D, \tag{7.23}$$

$$\lim_{r \to \infty} \sqrt{r} \left(\frac{\partial u^s}{\partial r} - iku^s \right) = 0, \tag{7.24}$$

where $u^s := u^s(\cdot, \phi)$ is the scattered field due to the incident plane wave $u^i(x) = e^{ikx \cdot d}$ propagating in the incident direction $d = (\cos\phi, \sin\phi)$. This scattering problem is a particular case of the following exterior Dirichlet problem: given $f \in H^{\frac{1}{2}}(\partial D)$ find $u \in H^1_{loc}(\mathbb{R}^2 \setminus \bar{D})$ such that

$$\Delta u + k^2 u = 0 \qquad \text{in } \mathbb{R}^2 \setminus \bar{D}, \tag{7.25}$$

$$u = f \qquad \text{on } \partial D, \tag{7.26}$$

$$\lim_{r \to \infty} \sqrt{r} \left(\frac{\partial u}{\partial r} - iku \right) = 0, \tag{7.27}$$

which is shown in Example 5.23 to be well posed. In particular, the scattered field u^s satisfies (7.25)–(7.27) with $f = -e^{ikx \cdot d}|_{\partial D}$.

The reader has already seen that the Sommerfeld radiation condition implies that a radiating solution u to the Helmholtz equation has the asymptotic behavior

$$u(x) = \frac{e^{ikr}}{\sqrt{r}} u_\infty(\theta) + O(r^{-3/2}) \qquad r = |x| \to \infty \tag{7.28}$$

uniformly in all directions $\hat{x} = (\cos\theta, \sin\theta)$, where $u_\infty(\theta)$ is the far-field pattern given by

$$u_\infty(\theta) = \frac{e^{i\pi/4}}{\sqrt{8\pi k}} \int_{\partial D} \left(u(y) \frac{\partial e^{-iky\cdot\hat{x}}}{\partial \nu} - \frac{\partial u(y)}{\partial \nu} e^{-iky\cdot\hat{x}} \right) ds(y). \qquad (7.29)$$

Now, let $F : L^2[0, 2\pi] \to L^2[0, 2\pi]$ be the far-field operator corresponding to the scattering problem (7.21)–(7.24) given by

$$(Fg)(\theta) := \int_0^{2\pi} u_\infty(\theta, \phi) g(\phi) \, d\phi,$$

where $u_\infty(\theta, \phi)$ is the far-field pattern of $u^s(x, \phi)$.
In the same way as in Theorem 4.2 one can establish the following theorem.

Theorem 7.12. *The far-field pattern $u_\infty(\theta, \phi)$ corresponding to the scattering problem (7.21)–(7.24) satisfies the reciprocity relation*

$$u_\infty(\theta, \phi) = u_\infty(\phi + \pi, \theta + \pi).$$

Using the reciprocity relation, one can now show exactly in the same way as in Theorem 4.3 that the following result is true.

Theorem 7.13. *Assume that k^2 is not a Dirichlet eigenvalue of $-\Delta$ in D. Then the far-field operator corresponding to the scattering problem (7.21)–(7.24) is injective with a dense range.*

We now want to establish the fact that the far-field operator F corresponding to the scattering problem (7.21)–(7.24) is *normal*, i.e., $F^*F = FF^*$, where F^* is the L^2-adjoint of F. To this end, we need the following basic identity [44, 52, 53].

Theorem 7.14. *Let $F : L^2[0, 2\pi] \to L^2[0, 2\pi]$ be the far-field operator corresponding to the scattering problem (7.21)–(7.24). Then for all $g, h \in L^2[0, 2\pi]$ we have*

$$\sqrt{2\pi k}\, e^{-i\pi/4} \,(Fg, h) = \sqrt{2\pi k}\, e^{+i\pi/4} \,(g, Fh) + ik\,(Fg, Fh).$$

Proof. We first note that if u and w are two radiating solutions of the Helmholtz equation with far-field patterns u_∞ and w_∞, then from Green's second identity and the uniformity of the asymptotic relation (7.28) we have that

$$\int_{\partial D} \left(u \frac{\partial \overline{w}}{\partial \nu} - \overline{w} \frac{\partial u}{\partial \nu} \right) ds = -2ik \int_0^{2\pi} u_\infty \overline{w}_\infty d\theta. \qquad (7.30)$$

If v_g is a Herglotz wave function with kernel g given by

$$v_g(x) = \int_0^{2\pi} g(\phi) e^{ikx\cdot d} d\phi, \qquad d := (\cos\phi, \sin\phi),$$

then we have

$$\int_{\partial D} \left(u \frac{\partial \overline{v_g}}{\partial \nu} - \overline{v_g} \frac{\partial u}{\partial \nu} \right) ds = \int_0^{2\pi} \overline{g(\phi)} \int_{\partial D} \left(u \frac{\partial e^{-ikx \cdot d}}{\partial \nu} - \frac{\partial u}{\partial \nu} e^{-ikx \cdot d} \right) ds \, d\phi$$

$$= \sqrt{8\pi k} \, e^{-i\pi/4} \int_0^{2\pi} \overline{g(\phi)} u_\infty(\phi) d\phi. \tag{7.31}$$

Now let v_g and v_h be Herglotz functions with kernels $g, h \in L^2[0, 2\pi]$, respectively. Let u_g^s and u_h^s be the corresponding scattered fields, i.e., u_g^s and u_h^s satisfy (7.21)–(7.24), with u^i replaced by v_g and v_h, respectively, and denote by $u_{g,\infty}$ and $u_{h,\infty}$ the corresponding far-field patterns. Then from (7.30) and (7.31) we have

$$0 = \int_{\partial D} \left((u_g^s + v_g) \frac{\partial \overline{(u_h^s + v_h)}}{\partial \nu} - \overline{(u_h^s + v_h)} \frac{\partial (u_g^s + v_g)}{\partial \nu} \right) ds$$

$$= \int_{\partial D} \left(u_g^s \frac{\partial \overline{u_h^s}}{\partial \nu} - \overline{u_h^s} \frac{\partial u_g^s}{\partial \nu} \right) ds + \int_{\partial D} \left(u_g^s \frac{\partial \overline{v_h}}{\partial \nu} - \overline{v_h} \frac{\partial u_g^s}{\partial \nu} \right) ds$$

$$+ \int_{\partial D} \left(v_g \frac{\partial \overline{u_h^s}}{\partial \nu} - \overline{u_h^s} \frac{\partial v_g}{\partial \nu} \right) ds$$

$$= -2ik \int_0^{2\pi} u_{g,\infty} \overline{u_{h,\infty}} d\phi + \sqrt{8\pi k} e^{-i\pi/4} \int_0^{2\pi} \overline{h} u_{g,\infty} d\phi - \sqrt{8\pi k} e^{i\pi/4} \int_0^{2\pi} g \overline{u_{h,\infty}} d\phi$$

$$= -2ik \, (Fg, Fh) + \sqrt{8\pi k} \, e^{-i\pi/4} \, (Fg, h) - \sqrt{8\pi k} \, e^{i\pi/4} \, (g, Fh),$$

and the proof is complete.   □

Theorem 7.15. *The far-field operator corresponding to the scattering problem (7.21)–(7.24) is normal, i.e., $FF^* = F^*F$.*

Proof. From Theorem 7.14 we have that

$$(g, \, ikF^*Fh) = \sqrt{2\pi k} \left(e^{+i\pi/4} \, (g, Fh) - e^{-i\pi/4} \, (g, F^*h) \right)$$

for all h and g in $L^2[0, 2\pi]$, and hence

$$ikF^*F = \sqrt{2\pi k} \left(e^{-i\pi/4} F - e^{+i\pi/4} F^* \right). \tag{7.32}$$

Using the reciprocity relation as in the proof of the first part of Theorem 4.3 we see that

$$(F^*g)(\theta) = \overline{RFR\overline{g}},$$

where $R : L^2[0, 2\pi] \to L^2[0, 2\pi]$ defines the reflection property $(Rg)(\phi) = g(\phi + \pi)$. From this, observing that $(Rg, Rh) = (g, h) = (\overline{h}, \overline{g})$, we find that

$$(F^*g, F^*h) = (RFR\overline{h}, RFR\overline{g}) = (FR\overline{h}, FR\overline{g}),$$

and hence, using Theorem 7.14 again,

$$
\begin{aligned}
ik\,(F^*g, F^*h) &= \sqrt{2\pi k}\left\{\left(e^{-i\pi/4}FR\overline{h}, R\overline{g}\right) - e^{+i\pi/4}\left(R\overline{h}, FR\overline{g}\right)\right\} \\
&= \sqrt{2\pi k}\left\{e^{-i\pi/4}\,(g, F^*h) - e^{+i\pi/4}\,(F^*g, h)\right\}.
\end{aligned}
$$

If we now proceed as in the derivation of (7.32), then we find that

$$ikFF^* = \sqrt{2\pi k}\left(e^{-i\pi/4}F - e^{+i\pi/4}F^*\right), \tag{7.33}$$

and the proof is complete. □

Assuming that k^2 is not a Dirichlet eigenvalue for $-\Delta$, it can be shown that, since F is normal and injective, there exists a countable number of eigenvalues $\lambda_j \in \mathbb{C}$ of F with $\lambda_j \neq 0$, and the corresponding eigenvectors ψ_j form a complete orthonormal system for $L^2[0, 2\pi]$ [149]. From Theorem 7.14 we see that the eigenvalues of the far-field operator F lie on a circle of radius $\sqrt{2\pi/k}$ with center at $e^{3\pi i/4}\sqrt{2\pi/k}$.

Of importance in studying the far-field operator is the operator $B : H^{\frac{1}{2}}(\partial D) \to L^2[0, 2\pi]$ defined by $Bf = u_\infty$, where u_∞ is the far-field pattern of the radiating solution u to (7.25)–(7.27) with boundary data $f \in H^{\frac{1}{2}}(\partial D)$. We leave to the reader as an exercise to prove, in the same way as Theorem 4.8, the following properties of the operator B.

Theorem 7.16. *Assume that k^2 is not a Dirichlet eigenvalue for $-\Delta$ in D. Then, the operator $B : H^{\frac{1}{2}}(\partial D) \to L^2[0, 2\pi]$ is compact and injective and has dense range in $L^2[0, 2\pi]$.*

We end this section with a factorization formula for the far-field operator F in terms of the operator B and the boundary integral operator S defined by (7.3).

Lemma 7.17. *The far-field operator F can be factored as*

$$F = -\overline{\gamma}^{-1}BS^*B^*,$$

with $B^* : L^2[0, 2\pi] \to H^{-\frac{1}{2}}(\partial D)$ and $S^* : H^{-\frac{1}{2}}(\partial D) \to H^{\frac{1}{2}}(\partial D)$ the adjoints of B and S, respectively (defined by Definition 7.2) and $\gamma = e^{i\pi/4}/\sqrt{8\pi k}$.

Proof. Consider the operator $H : L^2[0, 2\pi] \to H^{\frac{1}{2}}(\partial D)$ defined by $Hg = v_g|_{\partial D}$, where v_g is the Herglotz wave function with kernel g given by

$$v_g(x) := \int_0^{2\pi} g(\theta)e^{ikx\cdot\hat{y}}\,ds \qquad \hat{y} = (\cos\theta,\,\sin\theta).$$

By changing the order of integration it is easy to show that the adjoint (Definition 7.2) $H^* : H^{-\frac{1}{2}}(\partial D) \to L^2[0, 2\pi]$ such that

$$(Hg, \varphi) = (g, H^*\varphi)$$

is given by

$$H^*\varphi(\phi) = \int_{\partial D} \varphi(y)e^{-ik\hat{x}\cdot y}ds(y), \quad \hat{x} = (\cos\phi, \sin\phi). \tag{7.34}$$

By a superposition argument we have that

$$Fg = -BHg. \tag{7.35}$$

On the other hand, from the asymptotic behavior of the fundamental solution (Sect. 4.1) we observe that $\gamma H^*\varphi$ is the far-field pattern of the single layer potential $\mathcal{S}\varphi$ given by (7.1). Since $\mathcal{S}\varphi|_{\partial D} = S\varphi$, where S is given by (7.3), we can write

$$\gamma H^*\varphi = BS\varphi,$$

whence

$$H = \overline{\gamma}^{-1}S^*B^*. \tag{7.36}$$

Substituting H from (7.36) into (7.35) the lemma is proved.   □

7.1.3 Factorization Method

In this section we consider the inverse problem of determining the shape of a perfectly conducting object D from a knowledge of the far-field pattern $u_\infty(\theta, \phi)$ of the scattered field $u^s(x, \phi)$ corresponding to (7.21)–(7.24). In exactly the same way as in Theorem 4.5 one can prove the following uniqueness result.

Theorem 7.18. *Assume that D_1 and D_2 are two obstacles such that the far-field patterns corresponding to the scattering problem (7.21)–(7.24) for D_1 and D_2 coincide for all incident angles $\phi \in [0, 2\pi]$. Then $D_1 = D_2$.*

We shall now use the factorization method introduced by Kirsch in [99] to reconstruct the shape of a perfect conductor from a knowledge of the far-field operator.

We assume that k^2 is not a Dirichlet eigenvalue for D. From the previous section we know that there exists eigenvalues $\lambda_j \neq 0$ of F and that the corresponding eigenvectors form a complete orthonormal system in $L^2[0, 2\pi]$. It is easy to see that $\{|\lambda_j|, \psi_j, \text{sign}(\lambda_j)\psi_j\}_1^\infty$ is a singular system for F (Sect. 2.2), where for $z \in \mathbb{C}$ we define $\text{sign}(z) = z/|z|$. From Lemma 7.17 we can write

$$-\overline{\gamma}^{-1}BS^*B^*\psi_j = \lambda_j\psi_j \qquad j = 1, 2, \cdots.$$

If we define functions $\varphi_j \in H^{-1/2}[0, 2\pi]$ by

$$B^*\psi_j = \sqrt{\lambda_j}\varphi_j, \qquad j = 1, 2, \cdots, \tag{7.37}$$

where the branch of $\sqrt{\lambda_j}$ is chosen such that $\operatorname{Im}\sqrt{\lambda_j}e^{-i\pi/4} > 0$ [note that $\operatorname{Im}(e^{-i\pi/4}\lambda_j) > 0$ since $\lambda_j \neq 0$ lie on a circle of radius $\sqrt{2\pi/k}$ and centered at $e^{3\pi i/4}\sqrt{2\pi/k}$], then we see that

$$BS^*\varphi_j = -\overline{\gamma}\sqrt{\lambda_j}\psi_j. \tag{7.38}$$

Since

$$(S\varphi_j, \varphi_l) = (\varphi_j, S^*\varphi_l) = \frac{1}{\sqrt{\lambda_j}\sqrt{\overline{\lambda_l}}}(B^*\psi_j, S^*B^*\psi_l)$$

$$= \frac{1}{\sqrt{\lambda_j}\sqrt{\overline{\lambda_l}}}(\psi_j, BS^*B^*\psi_l) = -\frac{\gamma\overline{\lambda_l}}{\sqrt{\lambda_j}\sqrt{\overline{\lambda_l}}}(\psi_j, \psi_l),$$

we have that

$$(S\varphi_j, \varphi_l) = c_j\delta_{jl} \qquad \text{where} \quad c_j := -\gamma\frac{\overline{\lambda_j}}{|\lambda_j|}, \quad j, l = 1, 2, \cdots. \tag{7.39}$$

From Sect. 7.1.2 we know that λ_j lies on a circle of radius $\sqrt{2\pi/k}$ and center $e^{3\pi i/4}\sqrt{2\pi/k}$ that passes through the origin. We further know that $\lambda_j \to 0$ as $j \to \infty$. Therefore, we conclude that $|c_j| = 1/\sqrt{8\pi k}$, and $\operatorname{Im}(c_j) \to 0$ as $j \to \infty$.

Let S_i again be the boundary integral operator given by (7.3) corresponding to the wave number $k = i$. Since from Remark 7.4 we have that $S_i^{\frac{1}{2}}$ is well defined and invertible, we can decompose S into

$$S = S_i^{\frac{1}{2}}[I + S_i^{-\frac{1}{2}}(S - S_i)S_i^{-\frac{1}{2}}]S_i^{\frac{1}{2}} = S_i^{\frac{1}{2}}[I + K]S_i^{\frac{1}{2}}, \tag{7.40}$$

where

$$K := S_i^{-\frac{1}{2}}(S - S_i)S_i^{-\frac{1}{2}}. \tag{7.41}$$

Recall from part 2 of Theorem 7.3 that $S - S_i : H^{-\frac{1}{2}}(\partial D) \to H^{\frac{1}{2}}(\partial D)$ is compact. Hence $K : L^2(\partial D) \to L^2(\partial D)$ is compact since it is the composition of bounded operators with a compact operator. Letting

$$\tilde{\varphi}_j := S_i^{\frac{1}{2}}\varphi_j \qquad j = 1, 2, \cdots, \tag{7.42}$$

the orthogonality relation (7.39) takes the form

$$((I + K)\tilde{\varphi}_j, \tilde{\varphi}_l) = c_j\delta_{jl}, \qquad \text{where} \quad c_j := -\gamma\frac{\overline{\lambda_j}}{|\lambda_j|}, \quad j, l = 1, 2, \cdots. \tag{7.43}$$

The main step toward the final result is the following theorem.

Theorem 7.19. *The set* $\{\varphi_j\}_1^\infty$ *defined by (7.37) is a Riesz basis for* $H^{-\frac{1}{2}}(\partial D)$.

Proof. We apply Theorem 7.11 to $X := L^2(\partial D)$, $K = S_i^{-\frac{1}{2}}(S - S_i)S_i^{-\frac{1}{2}}$, and the set $\{\tilde{\varphi}_j\}_1^\infty$ defined by (7.42), which is certainly linearly independent and complete in $L^2(\partial D)$ since B and B^* are injective and S and $S^{\frac{1}{2}}$ are isomorphisms. We need to verify that K satisfies $\mathrm{Im}(K\varphi, \varphi) \neq 0$ for $\varphi \neq 0$. To this end, let $\varphi \in L^2(\partial D)$, and set $\psi = S_i^{-\frac{1}{2}}\varphi$. Then $\psi \in H^{-\frac{1}{2}}(\partial D)$ and

$$(K\varphi, \varphi) = ((S - S_i)\psi, \psi).$$

Since $(S_i\psi, \psi)$ is real-valued, the result follows from part 4 of Theorem 7.3. Hence Theorem 7.11 implies that $\{\tilde{\varphi}_j\}_1^\infty$ is a Riesz basis for $L^2(\partial D)$. Finally, since $S_i^{\frac{1}{2}}$ is an isomorphism from $H^{-\frac{1}{2}}(\partial D)$ onto $L^2(\partial D)$, we obtain that $\{\varphi_j\}_1^\infty$ forms a Riesz basis for $H^{-\frac{1}{2}}(\partial D)$.   □

Remark 7.20. Let $A : X \to X$ be a compact, self-adjoint, positive definite operator in a Hilbert space. It is easy to show that for each $r > 0$ there exists a uniquely defined compact, positive operator $A^r : X \to X$. In particular, this operator is defined in terms of the spectral decomposition

$$A^r\varphi = \sum_1^\infty \lambda_j^r (\varphi, \varphi_j)\varphi_j,$$

where $\lambda_j > 0$ and φ_j, $j = 1, 2, \cdots$, are the eigenvalues and eigenvectors of A, respectively. The inverse of A^r is defined by

$$A^{-r}\varphi = \sum_1^\infty \lambda_j^{-r} (\varphi, \varphi_j)\varphi_j.$$

We are now able to prove the first main result of this section.

Theorem 7.21. *Assume that* k^2 *is not a Dirichlet eigenvalue for* $-\Delta$ *in* D. *Then the range of* $B : H^{\frac{1}{2}}(\partial D) \to L^2[0, 2\pi]$ *is given by*

$$B(H^{\frac{1}{2}}(\partial D)) = \left\{ \sum_1^\infty \rho_j\psi_j : \sum_1^\infty \frac{|\rho_j|^2}{|\lambda_j|} < \infty \right\} = (F^*F)^{\frac{1}{4}}(L^2[0, 2\pi]), \quad (7.44)$$

where $\{|\lambda_j|, \psi_j, \mathrm{sign}(\lambda_j)\psi_j\}_1^\infty$ *is the singular system of the far-field operator* F.

Proof. First, we note that $S^* : H^{-\frac{1}{2}}(\partial D) \to H^{\frac{1}{2}}(\partial D)$ is an isomorphism since $S^*\varphi = \overline{S\overline{\varphi}}$. Suppose that $B\varphi = \psi$ for some $\varphi \in H^{\frac{1}{2}}(\partial D)$. Then $(S^*)^{-1}\varphi \in H^{-\frac{1}{2}}(\partial D)$, and thus $(S^*)^{-1}\varphi = \sum_1^\infty \alpha_j\varphi_j$, with $\sum_1^\infty |\alpha_j|^2 < \infty$, since $\{\varphi_j\}$ forms a Riesz basis for $H^{-\frac{1}{2}}(\partial D)$ (Theorem 7.10). Hence, by (7.38), we have

$$\psi = B\varphi = BS^*\,(S^*)^{-1}\,\varphi = -\overline{\gamma}\sum_{1}^{\infty}\alpha_j\sqrt{\lambda_j}\psi_j = \sum_{1}^{\infty}\rho_j\psi_j,$$

with $\rho_j = -\overline{\gamma}\alpha_j\sqrt{\lambda_j}$, and thus

$$\sum_{1}^{\infty}\frac{|\rho_j|^2}{|\lambda_j|} = \overline{\gamma}^2\sum_{1}^{\infty}|\alpha_j|^2 < \infty. \qquad (7.45)$$

On the other hand, let $\psi = \sum_{1}^{\infty}\rho_j\psi_j$, with the ρ_j satisfying $\sum_{1}^{\infty}\left(|\rho_j|^2/|\lambda_j|\right) < \infty$, and define $\varphi := \sum_{1}^{\infty}\alpha_j\varphi_j$ with $\alpha_j = \overline{\gamma}^{-1}\rho_j/\sqrt{\lambda_j}$. Then $\sum_{1}^{\infty}|\alpha_j|^2 < \infty$, and hence $\varphi \in H^{-\frac{1}{2}}(\partial D)$. But $S^*\varphi \in H^{\frac{1}{2}}(\partial D)$, whence

$$B(S^*\varphi) = -\overline{\gamma}\sum_{1}^{\infty}\alpha_j\sqrt{\lambda_j}\psi_j = \sum_{1}^{\infty}\rho_j\psi_j = \psi.$$

We now observe that $\sqrt{|\lambda_j|}$ and ψ_j are the eigenvalues and eigenfunctions, respectively, of the self-adjoint operator $(F^*F)^{\frac{1}{4}}$ (Remark 7.20). Hence Theorem 2.7 yields

$$(F^*F)^{\frac{1}{4}}(L^2[0, 2\pi]) = \left\{\sum_{1}^{\infty}\rho_j\psi_j : \sum_{1}^{\infty}\frac{|\rho_j|^2}{|\lambda_j|} < \infty\right\} = B(H^{\frac{1}{2}}(\partial D)).$$

$\square$

We recall from Remark 7.20 that $(F^*F)^{-\frac{1}{4}}$ is well defined.

Lemma 7.22. *The operator* $(F^*F)^{-\frac{1}{4}}B$ *is an isomorphism from* $H^{\frac{1}{2}}(\partial D)$ *onto* $L^2[0, 2\pi]$.

Proof. Let $\{\varphi_j\}_{1}^{\infty}$ be defined by (7.37). Then from Theorem 7.10, since $S : H^{-\frac{1}{2}}(\partial D) \to H^{\frac{1}{2}}(\partial D)$ is an isomorphism, we have that $\{S\varphi_j\}_{1}^{\infty}$ is a Riesz basis for $H^{\frac{1}{2}}(\partial D)$. To show that $(F^*F)^{-\frac{1}{4}}B$ is an isomorphism, from Theorem 7.10 it suffices to show that $\left\{(F^*F)^{-\frac{1}{4}}BS\varphi_j\right\}_{1}^{\infty}$ forms a Riesz basis for $L^2[0, 2\pi]$. To this end, using (7.37) and Lemma 7.17, we obtain

$$(F^*F)^{-\frac{1}{4}}BS\varphi_j = \frac{1}{\sqrt{\lambda_j}}(F^*F)^{-\frac{1}{4}}BSB^*\psi_j$$

$$= -\overline{\gamma}\lambda_j(F^*F)^{-\frac{1}{4}}\psi_j = -\overline{\gamma}\sqrt{\frac{\lambda_j}{|\lambda_j|}}\psi_j. \qquad (7.46)$$

The result now follows from the fact that the set $\{\psi_j\}_{1}^{\infty}$ is a complete orthonormal system in $L^2[0, 2\pi]$. $\square$

The following theorem gives examples of functions in the range of B. Recall that $\Phi_\infty(\hat{x}, z)$ denotes the far-field pattern of the fundamental solution $\Phi(x, z)$ of the Helmholtz equation.

Theorem 7.23. $\Phi_\infty(\cdot, z)$ *is in the range of B if and only if $z \in D$.*

Proof. First take $z \in D$ and define $f := \Phi(\cdot, z)|_{\partial D}$. Then, since $\Phi(\cdot, z)$ is a solution to the Helmholtz equation in $\mathbb{R}^2 \setminus \bar{D}$, by definition we have that $Bf = \Phi_\infty(\cdot, z)$.

Next, let $z \in \mathbb{R}^2 \setminus \bar{D}$, and assume that there exists an $f \in H^{\frac{1}{2}}(\partial D)$ such that $Bf = \Phi_\infty(\cdot, z)$. Let u be the solution of the exterior boundary value problem (7.25)–(7.27) with boundary data f. By Rellich's lemma, $u(x) = \Phi(x, z)$ for all x outside of any sphere containing D and z. If $z \notin \bar{D}$, this contradicts the fact that u is analytic in $\mathbb{R}^2 \setminus \bar{D}$, while $\Phi(x, z)$ is singular at $x = z$. If $z \in \partial D$, then we have that $\Phi(x, z) = f(x)$ for $x \in \partial D$, i.e., $\Phi(\cdot, z) \in H^{\frac{1}{2}}(\partial D)$. This is a contradiction since $\nabla \Phi(\cdot, z)$ is in neither $L^2(D)$ nor $L^2_{loc}(\mathbb{R}^2 \setminus \bar{D})$. $\square$

Combining Theorems 7.21 and 7.23 we obtain the main result of this section.

Theorem 7.24. *Assume that k^2 is not a Dirichlet eigenvalue of $-\Delta$ in D, and let F be the far-field operator corresponding to (7.21)–(7.24). Then*

$$D = \left\{ z \in \mathbb{R}^2 : \sum_1^\infty \frac{|\rho_j^{(z)}|^2}{\sigma_j} < \infty \right\}$$

$$= \left\{ z \in \mathbb{R}^2 : \Phi_\infty(\cdot, z) \in (F^*F)^{\frac{1}{4}}(L^2[0, 2\pi]) \right\},$$

where $\left\{ \sigma_j, \psi_j, \tilde{\psi}_j \right\}_1^\infty$ is the singular system of F, and $\rho_j^{(z)} = (\Phi_\infty(\cdot, z), \psi_j)_{L^2}$, $j = 1, 2, \cdots$, are the expansion coefficients of $\Phi_\infty(\hat{x}, z)$ with respect to $\{\psi_j\}_1^\infty$. Moreover, there exists $C > 1$ such that

$$\frac{1}{C^2} \|\Phi(\cdot, z)\|^2_{H^{\frac{1}{2}}(\partial D)} \le \sum_1^\infty \frac{|\rho_j^{(z)}|^2}{\sigma_j} \le C^2 \|\Phi(\cdot, z)\|^2_{H^{\frac{1}{2}}(\partial D)}, \quad z \in D. \quad (7.47)$$

Proof. It only remains to prove the last estimate. From the proof of Theorem 7.21 we have that for $z \in D$

$$g := \sum_1^\infty \frac{\rho_j^{(z)}}{\sqrt{\sigma_j}} \psi_j$$

is the solution of $(F^*F)^{\frac{1}{4}}g = \Phi_\infty(\cdot, z)$. On the other hand, $\Phi_\infty(\cdot, z)$ is the far-field pattern of the fundamental solution $\Phi(\cdot, z)$, i.e., if we define $f := \Phi(\cdot, z)|_{\partial D}$, then $Bf = \Phi_\infty(\cdot, z)$, and hence $g = (F^*F)^{-\frac{1}{4}}Bf$. The estimate (7.47) follows from the fact that $\|g\|^2_{L^2} = \sum_1^\infty |\rho_j^{(z)}|^2/\sigma_j$ and using Lemma 7.22 and Theorem 7.10. $\square$

Remark 7.25. The estimate (7.47) describes how the value of the series blows up when z approaches the boundary ∂D. In particular, it is easily shown that $\|\Phi(\cdot, z)\|^2_{H^{\frac{1}{2}}(\partial D)}$ behaves as $|\ln(d(z, \partial D))|$, where $d(z, \partial D)$ denotes the distance of $z \in D$ from the boundary.

The factorization method looks for a solution to the linear equation

$$(F^*F)^{\frac{1}{4}}g = \Phi_\infty(\cdot, z), \tag{7.48}$$

which is ill posed since $(F^*F)^{\frac{1}{4}} : L^2[0, 2\pi] \to L^2[0, 2\pi]$ is compact. Therefore, a regularization scheme is needed to compute the solution of (7.48). In particular, using Tikhonov regularization, a regularized solution g^α is defined as the solution of the well-posed equation

$$\alpha g^\alpha + (F^*F)^{\frac{1}{2}}g^\alpha = (F^*F)^{\frac{1}{4}}\Phi_\infty(\cdot, z),$$

where $\alpha > 0$ is the regularization parameter, which can be chosen according to the Morozov discrepancy principle (Sect. 2.3) such that

$$\|(F^*F)^{\frac{1}{4}}g^\alpha - \Phi_\infty(\cdot, z)\| = \delta\|g^\alpha\|,$$

with $\delta > 0$ being the error in the measured far-field data. Unlike the far-field equation $Fg = \Phi_\infty(\cdot, z)$ on which the linear sampling method is based, (7.48) is solvable if and only if $z \in D$. Therefore, it is possible to obtain a convergence result for the regularized solution of (7.48) when $\delta \to 0$. This is provided by the following theorem from the theory of ill-posed problems, which we recall for the reader's convenience [54, 99].

Theorem 7.26. *Let $K_\delta : X \to Y$, $\delta \geq 0$, be a family of injective and compact operators with dense range between Hilbert spaces X and Y such that $\|K_0 - K_\delta\| \leq \delta$ for all $\delta > 0$. Furthermore, let $f \in Y$ and $(\alpha_\delta, g_\delta) \in \mathbb{R}^+ \times X$ be the regularized Tikhonov–Morozov solution of the equation $K_\delta g = f$, i.e., the solution of the system*

$$(\alpha_\delta I + K_\delta^* K_\delta)g_\delta = K_\delta^* f \qquad \|K_\delta g - f\| = \delta\|g_\delta\|.$$

Then:

1. *If the noise-free equation $K_0 g = f$ has a unique solution $g \in X$, then $g_\delta \to g$ as $\delta \to 0$.*
2. *If the noise-free equation $K_0 g = f$ has no solution, then $\|g_\delta\| \to \infty$ as $\delta \to 0$.*

7.2 Factorization Method for an Inhomogeneous Medium

We will now develop the factorization method for anisotropic inhomogeneous media, as discussed in Chap. 5, following [100] and [104]. For the sake of simplicity we will assume that $n = 1$, i.e., we consider the following scattering problem:

$$\nabla \cdot A\nabla v + k^2 v = 0 \qquad \text{in} \quad D, \qquad (7.49)$$

$$\Delta u^s + k^2 u^s = 0 \qquad \text{in} \quad \mathbb{R}^2 \setminus \bar{D}, \qquad (7.50)$$

$$v - u^s = u^i \qquad \text{on} \quad \partial D, \qquad (7.51)$$

$$\frac{\partial v}{\partial \nu_A} - \frac{\partial u^s}{\partial \nu} = \frac{\partial u^i}{\partial \nu} \qquad \text{on} \quad \partial D, \qquad (7.52)$$

$$\lim_{r \to \infty} \sqrt{r} \left(\frac{\partial u^s}{\partial r} - iku^s \right) = 0, \qquad (7.53)$$

where, again, $u^s := u^s(\cdot, \phi)$ is the scattered field due to the incident plane wave $u^i(x) = e^{ikx \cdot d}$ propagating in the incident direction $d = (\cos\phi, \sin\phi)$. Note that the matrix-valued function A satisfies the assumptions stated in Chap. 5. Furthermore, we will assume that the medium is nonabsorbing, i.e., $\text{Im}(A) = 0$. This assumption is crucial for the validation of the factorization method since, as will become clear later, it guaranties that the far-field operator is normal.

7.2.1 Preliminary Results

In this section we develop functional analysis tools to justify the factorization method for inhomogeneous media. This new analytical framework can be applied to a larger number of scattering problems and differs from the one in the previous section. Here we follow the discussion of the factorization method given in [54], Chap. 5.

Theorem 7.27. *Let X and H be Hilbert spaces with inner products $(\cdot, \cdot)$, and let X^* be the dual space of X with duality pairing $\langle \cdot, \cdot \rangle$ in X^* and X. Let us assume that $F : H \to H$, $B : X \to H$, and $T : X^* \to X$ are bounded linear operators that satisfy*

$$F = BTB^*, \qquad (7.54)$$

where $B^* : H \to X^*$ is the antilinear adjoint of B defined by

$$\langle \varphi, B^*g \rangle = (B\varphi, g), \qquad g \in H, \varphi \in X.$$

Assume further that

$$|(Tf, f)| \geq c\|f\|_{X^*}^2 \qquad (7.55)$$

for all $f \in B^*(H)$ and some $c > 0$, where (Tf, f) is defined by Definition 7.1 and we have identified X with $(X^*)^*$. Then for any $g \in H$ with $g \neq 0$ we have that $g \in B(H)$ if and only if

$$\inf\{|(F\psi, \psi)| : \psi \in H, (g, \psi) = 1\} > 0. \qquad (7.56)$$

Proof. From (7.60)–(7.56) we obtain that

$$|(F\psi, \psi)| = |(TB^*\psi, B^*\psi)| \geq c\|B^*\psi\|_{X^*}^2, \qquad \text{for all } \psi \in H. \qquad (7.57)$$

Now assume that $g = B\varphi$ for some $\varphi \in X$ and $g \neq 0$. Then for each $\psi \in H$ with $(g, \psi) = 1$ we can estimate

$$c = c\,|(B\varphi, \psi)|^2 = c\,|<\varphi, B^*\psi>|^2 \leq c\|\varphi\|_X^2 \|B^*\psi\|_{X^*}^2 \leq \|\varphi\|_X^2\,|(F\psi, \psi)|\,,$$

and consequently (7.56) is satisfied.

Conversely, let (7.56) be satisfied, and assume that $g \notin B(X)$. We define $V := [\text{span}\{g\}]^\perp$ and show that $B^*(V)$ is dense in $\overline{B^*(H)}$. Via the antilinear isomorphism J from the Riesz representation theorem given by

$$\langle \varphi, f \rangle = (\varphi, Jf)\,, \qquad \varphi \in X, f \in X^*,$$

we can identify $X = J(X^*)$. In particular, $JB^* : H \to X$ is the Hilbert space adjoint of $B : X \to H$, and it suffices to show that $JB^*(V)$ is dense in $\overline{JB^*(H)}$. To this end, let $\varphi = \lim_{n\to\infty} JB^*\psi_n$ with $\psi_n \in H$ orthogonal to $JB^*(V)$. Then

$$(B\varphi, \psi) = (\varphi, JB^*\psi) = 0 \qquad \text{for all } \psi \in V,$$

and hence $B\varphi \in V^\perp = \text{span}\{g\}$. Since $g \notin B(X)$, this implies $B\varphi = 0$. But then

$$\|\varphi\|^2 = \lim_{n\to\infty} (\varphi, JB^*\psi_n) = \lim_{n\to\infty} (B\varphi, \psi_n) = 0,$$

and hence $JB^*(V)$ is dense in $\overline{JB^*(H)}$.

Now we can choose a sequence $(\tilde{\psi})$ in V such that

$$B^*\tilde{\psi}_n \to \frac{1}{\|g\|^2} B^*g, \qquad n \to \infty.$$

Setting

$$\psi_n := \tilde{\psi}_n + \frac{1}{\|g\|^2} g$$

we have $(g, \psi_n) = 1$ for all n and $B^*\psi_n \to 0$ for $n \to \infty$. Then from the first equation in (7.57) we observe that

$$|(F\psi_n, \psi_n)| \leq \|T\| \|B^*\psi_n\|_{X^*}^2 \to 0, \qquad n \to \infty,$$

which contradicts the assumption that (7.56) is satisfied. Hence g must belong to $B(X)$, and this concludes the proof. □

We note that an equivalent formulation of Theorem 7.27 can be stated without referring to the dual space of the Hilbert space X via the Riesz representation theorem as in the foregoing proof. The corresponding formulation is for a factorization $F = B\tilde{T}\tilde{B}^*$, where $\tilde{T} : X \to X$ and $\tilde{B}^* : H \to X$ is the Hilbert space adjoint of $B : X \to H$. Both formulations are connected via $\tilde{B}^* = JB^*$ and $\tilde{T} = TJ^{-1}$. Condition (7.55) becomes

$$\left|(\tilde{T}\varphi, \varphi)\right| \geq c\|\varphi\|_X^2 \qquad (7.58)$$

for all $\varphi \in \tilde{B}^*(H)$ and some $c > 0$. For the special case where $X = H$ in the sequel (for example, subsequently in the proof of Theorem 7.29), we will always refer to this second variant of Theorem 7.27.

The following lemma provides a tool for checking the strong coercivity assumption (7.55) in Theorem 7.27.

Lemma 7.28. *In the setting of Theorem 7.27 let $T : X^* \to X$ satisfy*

$$\mathrm{Im}\,(Tf, f) \neq 0 \qquad (7.59)$$

for all $f \in \overline{B^*(H)}$, with $f \neq 0$. In addition, let us assume that T is of the form $T = T_0 + C$, where C is compact and T_0 strictly coercive, i.e.,

$$(T_0 f, f) \geq c_0 \|f\|_{X^*}^2.$$

for all $f \in \overline{B^*(H)}$ and some $c_0 > 0$. Then T satisfies (7.55).

Proof. Assume to the contrary that (7.55) is not satisfied. Then there exists a sequence $\{f_n\}$ in $B^*(H)$ with $\|f_n\| = 1$ for all n and

$$(Tf_n, f_n) \to 0, \qquad n \to \infty.$$

We assume that $\{f_n\}$ converges weakly to some $f \in \overline{B^*(H)}$. From the compactness of C, writing

$$(Cf_n, f_n) = (Cf_n - Cf, f_n) + (Cf, f_n)$$

we observe that

$$(Cf_n, f_n) \to (Cf, f), \qquad n \to \infty,$$

and consequently

$$(T_0 f_n, f_n) \to -(Cf, f), \qquad n \to \infty.$$

Taking the imaginary part implies $\mathrm{Im}\,(Cf, f) = 0$ because $(T_0 f, f)$ is real. Therefore, $\mathrm{Im}\,(Tf, f) = 0$, whence $f = 0$ by assumption (7.61). This yields $(T_0 f_n, f_n) \to 0$ for $n \to \infty$, which contradicts $\|f_n\| = 1$ for all n and the fact that T_0 is strictly coercive. $\qquad\square$

Now we are ready to formulate the main theorem that will be the basis of the factorization method for scattering from an inhomogeneous medium.

Theorem 7.29. *Let X and H be Hilbert spaces with inner products $(\cdot,\cdot)$, and let X^* be the dual space of X with duality pairing $\langle\cdot,\cdot\rangle$ in X^* and X. Let us assume that $F : H \to H$, $B : X \to H$, and $T : X^* \to X$ are bounded, linear operators that satisfy*

$$F = BTB^*, \tag{7.60}$$

where $B^* : H \to X^*$ is the antilinear adjoint of B. In addition, let the operator F be compact and injective, and assume that $I+\tau iF$ is unitary for some $\tau > 0$. Assume further that

$$\mathrm{Im}\,(Tf, f) \neq 0 \tag{7.61}$$

for all $f \in \overline{B^*(H)}$, with $f \neq 0$, and T is of the form $T = T_0 + C$, where C is compact and T_0 strictly coercive, i.e.,

$$(T_0 f, f) \geq c_0 \|f\|_{X^*}^2$$

for all $f \in \overline{B^*(H)}$ and some $c_0 > 0$. Then the ranges $B(X)$ and $(F^*F)^{1/4}(H)$ coincide.

Proof. First we note that, by Lemma 7.28, the operator T satisfies assumption (7.55) of Theorem 7.27. Since $I + i\tau F$ is unitary, F is normal, i.e., $F^*F = FF^*$. Therefore, by the spectral theorem for compact normal operators ([149], cf. Theorem 1.30 for the special case where F is self-adjoint) there exists a complete set of orthonormal eigenfunctions $\psi_n \in H$ with corresponding eigenvalues λ_n, $n = 1, 2, \cdots$. In particular, the spectral theorem also provides the expansion

$$F\psi = \sum_{n=1}^{\infty} \lambda_n(\psi, \psi_n)\psi_n, \qquad \psi \in H. \tag{7.62}$$

From this we observe that F has a second factorization in the form

$$F = (F^*F)^{1/4}\tilde{F}(F^*F)^{1/4}, \tag{7.63}$$

where $(F^*F)^{1/4} : H \to H$ is given by

$$(F^*F)^{1/4}\psi = \sum_{n=1}^{\infty} \sqrt{|\lambda_n|}(\psi, \psi_n)\psi_n, \qquad \psi_n \in H, \tag{7.64}$$

and $\tilde{F} : H \to H$ is given by

$$\tilde{F}\psi = \sum_{n=1}^{\infty} \frac{\lambda_n}{|\lambda_n|}(\psi, \psi_n)\psi_n, \qquad \psi \in H. \tag{7.65}$$

We will show that $\tilde{F}$ also satisfies assumption (7.55) of Theorem 7.27. Then the statement of the theorem follows by applying Theorem 7.27 to both factorizations of F since both ranges are characterized by the same criterion (7.56).

Since the operator $I + i\tau F$ is unitary, the eigenvalues λ_n lie on a circle of radius $r := 1/\tau$ and center ir. We set

$$s_n := \frac{\lambda_n}{|\lambda_n|}, \qquad n \in \mathbb{N}, \tag{7.66}$$

and from $|\lambda_n - ir| = r$ and the only accumulation point $\lambda_n \to 0$, $n \to \infty$, we conclude that 1 and -1 are the only possible accumulation points of the sequence $\{s_n\}$. We will show that 1 is indeed the only accumulation point. To this end, we define $\varphi_n \in X^*$ by

$$\varphi_n := \frac{1}{\sqrt{\lambda_n}} B^* \psi_n, \qquad n \in \mathbb{N},$$

where the branch of the square root is chosen such that $\mathrm{Im}\sqrt{\lambda_n} > 0$. Then from $BTB^*\psi_n = F\psi_n = \lambda_n \psi_n$ we readily observe that

$$(T\varphi_n, \varphi_n) = s_n, \qquad n \in \mathbb{N}. \tag{7.67}$$

Consequently, since T satisfies assumption (7.55) of Theorem 7.27, we can estimate

$$c\|\varphi_n\|^2 \le |(T\varphi_n, \varphi_n)| = |s_n| = 1$$

for all $n \in \mathbb{N}$ and some positive constant $c > 0$, that is, the sequence $\{\varphi_n\}$ is bounded.

Now we assume that -1 is an accumulation point of the sequence $\{s_n\}$. Then, by the boundedness of the sequence $\{\varphi_n\}$, without loss of generality we may assume that $s_n \to -1$ and $\varphi_n \rightharpoonup \varphi \in X^*$ for $n \to \infty$ (Theorem 2.17). From (7.67) we then have that

$$(T_0\varphi_n, \varphi_n) + (C\varphi_n, \varphi_n) = (T\varphi_n, \varphi_n) \to -1, \qquad n \to \infty, \tag{7.68}$$

and the compactness of C implies that $C\varphi_n \to C\varphi$, $n \to \infty$. Consequently,

$$|(C\varphi_n - C\varphi, \varphi_n)| \le \|C\varphi_n - C\varphi\|\|\varphi_n\| \to 0, \qquad n \to \infty,$$

which yields

$$(C\varphi_n, \varphi_n) \to (C\varphi, \varphi), \qquad n \to \infty.$$

Taking the imaginary part of (7.68) we now obtain that $\mathrm{Im}\,(T\varphi, \varphi) = \mathrm{Im}\,(C\varphi, \varphi) = 0$, and therefore $\varphi = 0$ by the assumption of the theorem. Then (7.68) implies

$$(T_0\varphi_n, \varphi_n) \to -1, \qquad n \to \infty,$$

and this contradicts the coercivity of T_0.

Now we can write $s_n = e^{it_n}$, where $0 \le t_n \le \pi - 2\delta$ for all $n \in \mathbb{N}$ and some $0 < \delta \le \pi/2$. Then

$$\mathrm{Im}\left\{e^{i\delta} s_n\right\} \ge \sin\delta, \qquad n \in \mathbb{N},$$

and using $|(\tilde{F}\psi, \psi)| = |e^{i\delta}(\tilde{F}\psi, \psi)|$ we can estimate

$$|(\tilde{F}\psi, \psi)| \geq \text{Im} \sum_{n=1}^{\infty} e^{i\delta} s_n |(\psi, \psi_n)|^2 \geq \sin \delta \sum_{n=1}^{\infty} |(\psi, \psi_n)|^2 = \sin \delta \|\psi\|^2$$

for all $\psi \in H$, which proves that $\tilde{F}$ also satisfies assumption (7.55) of Theorem 7.27. This concludes the proof.   □

Note that Theorem 7.29 could also be used to justify the factorization method for the scattering problem for a perfect conductor instead of the analytical framework developed in Sect. 7.1, and we refer the reader for such a discussion to Chap. 5 of [54].

7.2.2 Properties of Far-Field Operator

Now consider the far-field operator $F : L^2[0, 2\pi] \to L^2[0, 2\pi]$ given by

$$(Fg)(\theta) := \int_0^{2\pi} u_\infty(\theta, \phi) g(\phi) \, d\phi,$$

where $u_\infty(\theta, \phi)$ is the far-field pattern given by (7.29) corresponding to the scattered field u^s that solves (7.49)–(7.53). We can again establish the following reciprocity relation.

Theorem 7.30. *The far-field pattern $u_\infty(\theta, \phi)$ corresponding to the scattering problem (7.49)–(7.53) satisfies the reciprocity relation*

$$u_\infty(\theta, \phi) = u_\infty(\phi + \pi, \theta + \pi).$$

This result can be proven in the same way as in Theorem 4.2, where using the symmetry of A and with the help of Green's theorem the integral over ∂D in (4.10) is moved to the integral over $|y| = a$.

Furthermore, thanks to Theorem 6.2, we can state the following theorem.

Theorem 7.31. *If k is not a transmission eigenvalue, then the far-field operator corresponding to the scattering problem (7.49)–(7.53) is injective with dense range.*

Similarly to Sect. 7.1, we need to show that for real-valued A the far-field operator corresponding to (7.49)–(7.53) is normal, i.e., $F^*F = FF^*$, where F^* is the L^2-adjoint of F. To this end, we follow the proof in [46].

Theorem 7.32. *If A is real-valued and symmetric, then the far-field operator F corresponding to (7.49)–(7.53) is normal, i.e., $FF^* = F^*F$.*

Proof. Let v_g^i and v_h^i be the Herglotz wave functions with kernel $g, h \in L^2[0, 2\pi]$, respectively, and let (v_g, u_g^s) and (v_h, u_h^s) be the solutions of (7.49)–(7.53), with the incoming wave u^i replaced by v_g^i and v_h^i, respectively. Let us denote the total fields by $u_g = u_g^s + v_g^i$ and $u_h = u_h^s + v_h^i$. Then we have

$$
2i \int_D \mathrm{Im} \left(\nabla v_g \cdot \overline{A \nabla v_h} \right) \, dx = 2i \int_D \mathrm{Im} \left(\nabla \cdot (v_g \overline{A \nabla v_h}) - v_g \nabla \cdot (\overline{A \nabla v_h}) \right) \, dx,
$$

$$
= 2i \int_D \left(\nabla \cdot (v_g \overline{A \nabla v_h}) - \nabla \cdot (\overline{v_g} A \nabla v_h) - v_g \nabla \cdot (\overline{A \nabla v_h}) + \overline{v_g} \nabla \cdot (A \nabla v_h) \right) \, dx,
$$

$$
= \int_{\partial D} \left(\nu \cdot (v_g \overline{A \nabla v_h}) - \nu \cdot (\overline{v_g} A \nabla v_h) \right) \, ds + k^2 \int_D \left(v_g \overline{v_h} - \overline{v_g} v_h \right) \, dy,
$$

$$
= \int_{\partial D} \left(u_g \frac{\partial \overline{u_h}}{\partial \nu} - \overline{u_g} \frac{\partial u_h^s}{\partial \nu} \right) \, ds + k^2 \int_D \left(v_g \overline{v_h} - \overline{v_g} v_h \right) \, dy.
$$

Since A is a real-valued symmetric matrix, we have that

$$
\int_D \mathrm{Im} \left(\nabla v_g \cdot \overline{A \nabla v_h} \right) \, dx = 0.
$$

Hence

$$
\int_{\partial D} \left(u_g \frac{\partial \overline{u_h}}{\partial \nu} - \overline{u_g} \frac{\partial u_h}{\partial \nu} \right) \, ds = k^2 \int_D \left(\overline{v_g} v_h - v_g \overline{v_h} \right) \, dy
$$

for all $g, h \in L^2[0, 2\pi]$. In particular, interchanging g and h and taking conjugates yields

$$
\int_{\partial D} \left(\overline{u_h} \frac{\partial u_g}{\partial \nu} - u_h \frac{\partial \overline{u_g}}{\partial \nu} \right) \, ds = k^2 \int_D \left(\overline{v_g} v_h - v_g \overline{v_h} \right) \, dy
$$

$$
= \int_{\partial D} \left(u_g \frac{\partial \overline{u_h}}{\partial \nu} - \overline{u_g} \frac{\partial u_h}{\partial \nu} \right) \, ds.
$$

From this we conclude that

$$
\int_{\partial D} \left(\overline{u_h} \frac{\partial u_g}{\partial \nu} - u_h \frac{\partial \overline{u_g}}{\partial \nu} \right) \, ds = \int_{\partial D} \left(u_g \frac{\partial \overline{u_h}}{\partial \nu} - \overline{u_g} \frac{\partial u_h}{\partial \nu} \right) \, ds,
$$

i.e., the integral is real for any $g, h \in L^2[0, 2\pi]$. Replacing h by ih, we can conclude that

$$
\int_{\partial D} \left(u_g \frac{\partial \overline{u_h}}{\partial \nu} - \overline{u_g} \frac{\partial u_h}{\partial \nu} \right) \, ds = 0
$$

for all $g, h \in L^2[0, 2\pi]$. Interchanging g and h and using the identities (7.30) and (7.31) (note that u_h^s and u_g^s are radiating solutions to the Helmholtz equation) now shows that

$$
0 = \int_{\partial D} \left(u_g \frac{\partial \overline{u_h}}{\partial \nu} - \overline{u_g} \frac{\partial u_h}{\partial \nu} \right) \, ds
$$

$$
= -2ik \left(Fg, Fh \right) + \sqrt{8\pi k} \, e^{-i \frac{\pi}{4}} \left(Fg, h \right) - \sqrt{8\pi k} \, e^{i \frac{\pi}{4}} \left(g, Fh \right). \quad (7.69)
$$

Then, from (7.69), exactly as in the proof of Theorem 7.15, we can now show that $FF^* = F^*F$, which concludes the proof. $\qquad\square$

Remark 7.33. The normality of the far-field operator F and the identity (7.69) imply that the scattering operator $\mathcal{S}$ defined by $\mathcal{S} = I + i\sqrt{\frac{k}{2\pi}}e^{-\pi i/4}F$ is unitary, i.e., $\mathcal{S}\mathcal{S}^* = \mathcal{S}^*\mathcal{S} = I$.

Assuming that k is not a transmission eigenvalue, it follows that, since F is normal and injective, there exists a countable number of eigenvalues $\lambda_j \in \mathbb{C}$ of F with $\lambda_j \neq 0$, and the corresponding eigenvectors ψ_j form a complete orthonormal system for $L^2[0, 2\pi]$ [149]. From Theorem 7.32 we see that the eigenvalues of the far-field operator F lie on the circle $\sqrt{8\pi k}\,\mathrm{Im}\left(e^{-i\frac{\pi}{4}}\lambda\right) - k|\lambda|^2 = 0$ (which is a circle of radius $\sqrt{2\pi/k}$ with center at $e^{3\pi i/4}\sqrt{2\pi/k}$).

7.2.3 Factorization Method

To fix our ideas, we assume that A is such that $a_{min} > 1$, and let us denote by $Q = A - I$ the contrast in the media. The assumption on A means that $Q - I$ is a positive definite matrix for all $x \in D$, i.e., $\overline{\xi} \cdot A\xi \geq \alpha|\xi|^2$, where $\alpha = a_{min} - 1 > 0$. In particular, the square root $Q^{1/2}$ is well defined for all $x \in D$ and is also positive definite with inverse $Q^{-1/2}$. For later use we need to consider the following problem:

$$\nabla \cdot \tilde{A}\nabla u + k^2 u = \nabla \cdot \left(Q^{1/2}f\right) \quad \text{in} \quad \mathbb{R}^2, \qquad (7.70)$$

$$\lim_{r\to\infty} \sqrt{r}\left(\frac{\partial u}{\partial r} - iku\right) = 0, \qquad (7.71)$$

where $f \in (L^2(D))^2$ and $\tilde{A} := A$ in D and $\tilde{A} := I$ in $\mathbb{R}^2 \setminus \overline{D}$.

Lemma 7.34. *The problem (7.70)–(7.71) has a unique solution $u \in H^1_{loc}(\mathbb{R}^2)$.*

Proof. In a similar way as in Sect. 5.4, we can show that (7.70)–(7.71) is equivalent to

$$\nabla \cdot \tilde{A}\nabla u + k^2 u = \nabla \cdot \left(Q^{1/2}f\right) \quad \text{in} \quad \Omega_R \qquad (7.72)$$

$$\frac{\partial u}{\partial \nu} = Tu \quad \text{on} \quad \partial\Omega_R, \qquad (7.73)$$

where T is the Dirichlet-to-Neumann operator defined in Definition 5.21. Uniqueness follows from Lemma 5.25, whereas existence follows from applying in a similar way as in Sect. 5.4 the Lax–Milgram lemma and Fredholm alternative to the variational equation

$$\int_{\Omega_R} \left(\tilde{A}\nabla u \nabla\overline{\phi} - k^2 u\overline{\phi}\right) dx - \int_{\partial\Omega_R} Tu\,\overline{\phi}\,ds = \int_D Q^{1/2}f \cdot \nabla\overline{\phi}\,dx \qquad (7.74)$$

for all $\phi \in H^1(\Omega_R)$. $\qquad\square$

In the following analysis, two operators $G : (L^2(D))^2 \to L^2[0, 2\pi]$ and $\mathcal{H} : L^2[0, 2\pi] \to (L^2(D))^2$, defined below, will play an important role. The operator $\mathcal{H} : L^2[0, 2\pi] \to (L^2(D))^2$ is defined by

$$(\mathcal{H}g)(x) = Q^{1/2}(x)\nabla \int_0^{2\pi} g(d)e^{ikd\cdot x}\, ds(d), \qquad x \in D, \qquad (7.75)$$

and $G : (L^2(D))^2 \to L^2[0, 2\pi]$ is defined by $Gf = u_\infty$, where u_∞ is the far-field pattern of the radiating solution u to (7.70)–(7.71) given by (4.6). Since we can write

$$(\mathcal{H}g)(x) = ikQ^{1/2}(x) \int_0^{2\pi} d\, g(d)e^{ikd\cdot x}\, ds(d), \qquad x \in D,$$

we can easily see that H is injective. Furthermore, from the fact that the scattered field u^s satisfies

$$\nabla \cdot \tilde{A}\nabla u^s + k^2\, u^s = -\nabla \cdot \left(Q\nabla u^i\right)$$

and using superposition we observe that $F = -G\mathcal{H}$. The adjoint $\mathcal{H}^* : (L^2(D))^2 \to L^2[0, 2\pi]$ is given by

$$(\mathcal{H}^*h)(\hat{x}) = -ik \int_D \hat{x}\, Q^{1/2}(y)h(y)e^{-ik\hat{x}\cdot y}\, dy$$

$$= \int_D \left(\nabla_y e^{-ik\hat{x}\cdot y}\right) Q^{1/2}(y)h(y)\, dy. \qquad (7.76)$$

Therefore, $\gamma\mathcal{H}^*h = p_\infty$, where p is given by the volume potential

$$p(x) = \int_D \nabla_y \Phi(x, y)Q^{1/2}(y)h(y)\, dy$$

$$= -\int \nabla_x \cdot \left[\Phi(x, y)Q^{1/2}(y)h(y)\right]\, dy = -\nabla \cdot \tilde{p}(x),$$

where

$$\tilde{p}(x) = \int_D \Phi(x, y)Q^{1/2}(y)h(y)\, dy, \qquad x \in \mathbb{R}^2,$$

and $\Phi(x, y)$ is the fundamental solution of the Helmholtz equation given by (3.33) and $\gamma = e^{i\pi/4}/\sqrt{8\pi k}$. But since $\Delta\tilde{p} + k^2\tilde{p} = -Q^{1/2}h$, we can conclude that

$$\Delta p + k^2 p = \nabla \cdot (Q^{1/2}h).$$

We recall that $Gf = u_\infty$, where $u \in H^1_{loc}(\mathbb{R}^2)$ is the radiating solution to (7.70)–(7.71), which can be rewritten as

$$\Delta u + k^2 u = \nabla \cdot \left[Q^{1/2}(f - Q^{1/2}\nabla u)\right] \qquad \text{in } \mathbb{R}^2. \qquad (7.77)$$

Now we are ready to obtain a factorization of the far-field operator. To this end, let us define the operator $\mathbb{T}_k : (L^2(D))^2 \to (L^2(D))^2$ by

$$\mathbb{T}_k f = f - Q^{1/2}\nabla u,$$

where $u \in H^1_{loc}(\mathbb{R}^2)$ satisfies (7.70)–(7.71) (here k indicates the dependence of the operator on the wave number). From the preceding discussion we conclude that $G = \gamma \mathcal{H}^* \mathbb{T}_k$, and hence

$$F = -\gamma \mathcal{H}^* \mathbb{T}_k \mathcal{H} \tag{7.78}$$

or

$$\tilde{F} := \gamma^{-1} F = -\mathcal{H}^* \mathbb{T}_k \mathcal{H}, \tag{7.79}$$

with $\mathcal{H}$ and $\mathcal{H}^*$ given by (7.75) and (7.76), respectively.

Lemma 7.35. *The operator $\mathbb{T}_k : (L^2(D))^2 \to (L^2(D))^2$ satisfies the following properties:*

1. If $k > 0$, then

$$Im\,(\mathbb{T}_k f, f)_{L^2(D)} \le 0 \qquad for\ all \quad f \in (L^2(D))^2,$$

   where $(\cdot, \cdot)_{L^2(D)}$ is the $L^2(D)$ inner product.
2. If $k > 0$ is not a transmission eigenvalue, then

$$Im\,(\mathbb{T}_k f, f)_{L^2(D)} < 0 \qquad for\ all \quad f \in \overline{R(\mathcal{H})},\ f \ne 0,$$

   where $\overline{R(\mathcal{H})}$ is the closure of the range of $\mathcal{H}$ in $L^2(D)$.
3. For $k = i$ the operator $\mathbb{T}_i$ is strictly coercive, i.e., there exists $c > 0$ such that

$$(\mathbb{T}_i f, f)_{L^2(D)} \ge c\|f\|^2_{L^2(D)} \qquad for\ all\ f \in (L^2(D))^2.$$

4. The operator $\mathbb{T}_k - \mathbb{T}_i$ is compact.

Proof. Part 1: let $\mathbb{T}_k f = g$, where by definition $g = f - Q^{1/2}\nabla u$ and $u \in H^1_{loc}(\mathbb{R}^2)$ satisfies (7.70)–(7.71). Obviously, from (7.77) we have that u satisfies

$$\int_{\mathbb{R}^2} \left(\nabla u \nabla \overline{\phi} - k^2 u\,\overline{\phi}\right) dx = \int_D Q^{1/2} g \nabla \overline{\phi}\, dx \tag{7.80}$$

for any $\phi \in H^1(\mathbb{R}^2)$ with compact support (note that the integral on the left-hand side is over the support of ϕ). We choose $\chi \in C_0^\infty(\mathbb{R}^2)$ such that $\chi = 1$ for $|x| \le r$, where r is such that $\overline{D}$ is contained in a disk of radius r. Setting $\phi = \chi u$ in (7.80) yields

$$\int_{|x|<r} \left(|\nabla u|^2 - k^2 |u|^2\right) dx + \int_{|x|>r} \left(\nabla u \nabla \overline{\phi} - k^2 u \overline{\phi}\right) dx = \int_D Q^{1/2} g \nabla \overline{u}\, dx.$$

Outside D, u is a smooth solution of the Helmholtz equation $\Delta u + k^2 u = 0$, and hence from Green's theorem we have

$$\int_{|x|>r} \left(\nabla u \nabla \overline{\phi} - k^2 u \overline{\phi} \right) dx = - \int_{|x|=r} \overline{u} \frac{\partial u}{\partial \nu} \, ds.$$

Therefore, we can now write for every $r > 0$ sufficiently large

$$\int_{|x|<r} \left(|\nabla u|^2 - k^2 |u|^2 \right) dx - \int_{|x|=r} \overline{u} \frac{\partial u}{\partial \nu} \, ds = \int_D Q^{1/2} g \nabla \overline{u} \, dx.$$

Letting $r \to \infty$ we obtain

$$\int_{\mathbb{R}^2} \left(|\nabla u|^2 - k^2 |u|^2 \right) dx - ik \int_0^{2\pi} |u_\infty|^2 \, ds = \int_D Q^{1/2} g \nabla \overline{u} \, dx,$$

where u_∞ is the far-field pattern of the radiating solution u. Now we obtain

$$
\begin{aligned}
(\mathbb{T}_k f, f)_{L^2(D)} &= \int_D |g|^2 \, dx + \int_D g \, \overline{Q^{1/2} \nabla u} \, dx \\
&= \|g\|_{L^2(D)}^2 + \int_D Q^{1/2} g \nabla \overline{u} \, dx \quad\quad\quad (7.81) \\
&= \|g\|_{L^2(D)}^2 + \int_{\mathbb{R}^2} \left(|\nabla u|^2 - k^2 |u|^2 \right) dx - ik \int_0^{2\pi} |u_\infty|^2 \, ds.
\end{aligned}
$$

Taking the imaginary part of (7.81) proves the claim.

To prove part 2, we assume that $\mathrm{Im}\,(\mathbb{T}_k f, f)_{L^2(D)} = 0$ for some $f \in \overline{R(\mathcal{H})}$. From (7.81) we conclude that $u_\infty = 0$, and hence by Rellich's lemma and analyticity $u = 0$ outside D. Therefore, u satisfies

$$\nabla \cdot A \nabla u + k^2 u = \nabla \cdot \left(Q^{1/2} f \right) \quad \text{in} \quad D, \quad\quad\quad (7.82)$$

$$u = 0 \quad \text{and} \quad \frac{\partial u}{\partial \nu_A} = 0 \quad \text{on} \quad \partial D. \quad\quad\quad (7.83)$$

Since $f \in \overline{R(\mathcal{H})}$, there exist Herglotz wave functions

$$v_{g_n} = \int_0^{2\pi} g_n(d) e^{ikd \cdot x} \, ds(d), \quad\quad x \in \mathbb{R}^2,$$

such that $f_n = Q^{1/2} \nabla v_{g_n}$ converge to f on $L^2(D)$. From continuous dependence we conclude that u_n converges to u in $H^1(D)$, where $u_n, w \in H^1_{loc}(\mathbb{R}^2)$ satisfy (7.70)–(7.71) for f_n and f, respectively. In addition, it is obvious that v_{g_n} converges in $H^1(D)$ to some solution w to the Helmholtz equation, which implies that $f = Q^{1/2} \nabla w$ or $Q^{1/2} f = Q \nabla w$. Substituting the latter into (7.82) and recalling that $Q = A - I$ yields

$$\nabla \cdot A \nabla u + k^2 u = \nabla \cdot (A - I) \nabla w \quad \text{in } D.$$

Recalling that $\Delta w + k^2 w = 0$ in D we have that $v := w - u$ and w satisfy the transmission eigenvalue problem (6.54)–(6.57), and since k is not a transmission eigenvalue, we conclude that $v = w = 0$, which implies $f = 0$. This proves the claim.

Part *3* follows from (7.81) and the well-posedness of (7.70)–(7.71), which imply

$$(\mathbb{T}_k f, f)_{L^2(D)} \geq C_1 \|u\|^2_{H^1_{loc}(D)} \geq C_2 \|f\|_{L^2(D)}.$$

Finally, we prove part *4*. To this end, we note that $T_k f - T_i f = Q^{1/2}\nabla (u_i - u_k)$, where $\tilde{u} = u_i - u_k$ satisfies

$$\nabla \cdot A\nabla\tilde{u} + k^2\tilde{u} = (k^2 + 1)u_i.$$

The boundedness of $f \mapsto u_i$ from $(L^2(D))^2$ into $H^1(D)$ and $u_i \mapsto \tilde{u}$ from $L^2(D)$ into $H^1(D)$, respectively, and the compactness of the embedding of $H^1(D)$ into $L^2(D)$ imply that $T_k - T_i$ is a compact operator. $\qquad\square$

Using Lemma 7.35, we can now apply Theorem 7.29 to our factorization $\tilde{F} = -\mathcal{H}^*\mathbb{T}_k\mathcal{H}$ to obtain the following result, where $H := L^2[0, 2\pi]$, $X = X^* = (L^2(D))^2$ (the duality pairing coincides with the L^2 inner product). Note that from Remark 7.33 $\tilde{F}$ is such that $(I + i\sqrt{k/2\pi}\tilde{F})$ is unitary and the range of $\tilde{F} := \gamma^{-1}F$ coincides with the range of F.

Theorem 7.36. *Assume that k is not a transmission eigenvalue. Then the range of $\mathcal{H}^*$ and the range of $(F^*F)^{1/4}$ coincide.*

The last step of the factorization method is the characterization of D by the range of $\mathcal{H}^*$. Then this result, combined with Theorem 7.36, yields a characterization of D in terms of the range of $(F^*F)^{1/4}$.

Recall that $\Phi_\infty(\hat{x}, y)$ is the far-field pattern of the fundamental solution $\Phi(x, y)$ of the Helmholtz equation.

Theorem 7.37. $\Phi_\infty(\cdot, z) = \gamma e^{-ik\hat{x}\cdot z}$ *is in the range of $\mathcal{H}$ if and only if $z \in D$.*

Proof. Let $z \in D$. Choose a small disk B centered at z such that $B \subset D$ and a function $\varphi \in C^\infty(\mathbb{R}^2)$ with $\varphi(x) = \Phi(x, z)$ far all $x \notin B$. The function φ can also be chosen such that $k^2 \int_D \varphi\, dx = -\int_{\partial D} \frac{\partial\Phi(\cdot,z)}{\partial\nu}\, ds$. Then, in particular, $\varphi = \Phi(\cdot, z)$ outside D and the Cauchy data of φ and $\Phi(\cdot, z)$ coincide on ∂D. Consider the following interior Neumann boundary value problem for $\rho \in C^1(\overline{D}) \cap C^2(D)$ (e.g., [111]):

$$\Delta\rho = \Delta\varphi + k^2\varphi \quad \text{in } D, \qquad \frac{\partial\rho}{\partial\nu} = 0 \quad \text{on } \partial D.$$

This problem has a solution since by Green's theorem

$$\int_D (\Delta\varphi + k^2\varphi)\, dx = k^2 \int_D \varphi\, dx + \int_{\partial D} \frac{\partial\Phi(\cdot, z)}{\partial\nu}\, ds = 0.$$

Setting $f = Q^{-1/2}\nabla\rho$ (see [111]) we would like to show that $\mathcal{H}^* f = \Phi_\infty(\cdot, z)$. From the characterization (7.76) of $\mathcal{H}^*$ we have that $\mathcal{H}^* f = p_\infty$, where p is given by

$$p(x) = \int_D \nabla_y \Phi(x, y)\nabla\rho \, dy = -\nabla_x \cdot \int_D \Phi(x, y)\nabla\rho \, dy. \qquad (7.84)$$

Using the choice of ρ, it is easy to check, using potential theory [43], that

$$\Delta p + k^2 p = \nabla \cdot \nabla\rho = \Delta\rho = \Delta\varphi + k^2\varphi \qquad \text{in } D,$$

i.e., $\Delta(p - \varphi) + k^2(p - \varphi) = 0$ in D. Outside of D both functions, p and $\varphi = \Phi(\cdot, z)$, satisfy the Helmholtz equation. Furthermore, φ and $\partial\varphi/\partial\nu$, and p and $\partial p/\partial\nu$ are continuous across the boundary ∂D [43] (for the latter we use that $\partial\rho/\partial\nu = 0$ on ∂D.) Hence $p - \varphi$ is an entire solution to the Helmholtz equation, and it satisfies the radiation condition, which implies that $p = \varphi$. This means that $\mathcal{H}^* f = \Phi_\infty(\cdot, z)$.

Let now $z \notin D$, and assume to the contrary that $\mathcal{H}^* f = \Phi_\infty(\cdot, z)$ for some $f \in (L^2(D))^2$. Let p be given by (7.76). Then, by definition, $\mathcal{H}^* f = p_\infty$. By Rellich's lemma and analyticity of the solution to the Helmholtz equation we have that p and $\Phi(\cdot, z)$ coincide in $\mathbb{R}^2 \setminus (D \cup \{z\})$. This is a contradiction since $p \in H^1(B)$ and $\Phi(\cdot, z) \notin H^1(B)$ for any disk B containing z in its interior. $\square$

Combining Theorems 7.36 and 7.37 we can now formulate the main theorem of this section, which constitutes the factorization method [104].

Theorem 7.38. *Assume that k is not a transmission eigenvalue and $a_{min} > 1$. Then $\Phi_\infty(\cdot, z)$ belongs to the range of $(F^*F)^{1/4}$ if and only if $z \in D$. In other words, the equation*

$$(F^*F)^{1/4}g = \Phi_\infty(\cdot, z) \qquad (7.85)$$

is solvable in $L^2[0, 2\pi]$ if and only if $z \in D$.

Recall that F possesses a complete set $\{\psi_j : j \in \mathbb{N}\}$ of eigenfunctions corresponding to eigenvalues λ_j. Then we can write the solvability condition (7.37) as

$$z \in D \Longleftrightarrow \sum_{j=1}^\infty \frac{(\Phi_\infty(\cdot, z), \psi_j)}{|\lambda_j|} < \infty.$$

Thus $W(z)$ defined by

$$W(z) = \left[\sum_{j=1}^\infty \frac{(\Phi_\infty(\cdot, z), \psi_j)}{|\lambda_j|}\right]^{-1}, \qquad z \in \mathbb{R}^2,$$

is the characteristic function of D since it is nonzero only inside D.

Results similar to those obtained earlier can also be obtained if the condition $a_{min} > 1$ is replaced by $a_{max} < 1$.

7.3 Justification of Linear Sampling Method

As explained earlier, the linear sampling method lacks a complete justification when it comes to the regularized solution of the far-field equation. In particular, the theory stipulates that the far-field equation has an approximate solution such that the corresponding Herglotz wave function is bounded in the $H^1(D)$ norm inside the support of the scatterer and becomes arbitrarily large outside. There are two problems associated with this result: (1) the statement about the behavior of the Herglotz wave function depends on D, and unfortunately such a claim cannot be made about the approximate solution (which is the kernel of the Herglotz wave function) of the far-field equation; (2) it is not clear that the Tikhonov regularized solution of the far-field equation inherits the same behavior as the approximate solution of the far-field equation. Both these issues are resolved for the scattering problems for which the support is characterized by the range of $(F^*F)^{1/4}$ thanks to the following theorem due to Arens and Lechleiter [7].

To this end, let H and X be Hilbert spaces and X^* be the dual of X. In what follows, we will assume that the normal operator $F : H \to H$ is factorized as $F = BTB^*$, where the bounded linear operators $B : X \to H$ and $T : X^* \to X$ satisfy the assumptions that guaranty that $B(X) = (F^*F)^{1/4}(H)$ (e.g., the assumptions of Theorem 7.36). The following result holds true.

Theorem 7.39. *For $\alpha > 0$ let g_α denote the Tikhonov regularized solution of the equation $Fg = \varphi$ for $\varphi \in H$, i.e., the solution of*

$$\alpha g_\alpha + F^*Fg_\alpha = F^*\varphi.$$

1. *If φ is in a range of $(F^*F)^{1/4}$, that is, $\varphi = (F^*F)^{1/4}g$ for some $g \in H$, then $\lim_{\alpha \to 0}(g_\alpha, \varphi)_H$ exists and*

$$c\|g\|^2 \leq \lim_{\alpha \to 0} |(g_\alpha, \varphi)| \leq \|g\|^2 \tag{7.86}$$

   for some $c > 0$ depending only on F.
2. *If $\varphi \notin (F^*F)^{1/4}(H)$, then $\lim_{\alpha \to 0}(g_\alpha, \varphi)_H = \infty$.*

Proof. Let $\psi_n \in H$, $n = 1, 2 \cdots$, be a complete set of orthonormal eigenfunctions of the normal operator $F : H \to H$ with corresponding eigenvalues λ_n. Hence the operator F can be written as

$$F\psi = \sum_{n=1}^{\infty} \lambda_n(\psi, \psi_n)\psi_n, \qquad \psi \in H,$$

which implies that

$$F^*\psi = \sum_{n=1}^{\infty} \overline{\lambda}_n(\psi, \psi_n)\psi_n, \qquad \psi \in H,$$

and consequently we have that

$$g_\alpha = \sum_{n=1}^{\infty} \frac{\bar{\lambda}_n}{\alpha + |\lambda_n|^2} (\varphi, \psi_n) \psi_n$$

and

$$(g_\alpha, \varphi) = \sum_{n=1}^{\infty} \frac{\bar{\lambda}_n}{\alpha + |\lambda_n|^2} |(\varphi, \psi_n)|^2. \tag{7.87}$$

If $\varphi = (F^*F)^{1/4} g$ for some $g \in H$, then

$$(\varphi, \psi_n) = \left((F^*F)^{1/4} g, \psi_n \right) = \left(g, (F^*F)^{1/4} \psi_n \right) = \sqrt{|\lambda_n|} (g, \psi_n),$$

whence

$$(g_\alpha, \varphi) = \sum_{n=1}^{\infty} \frac{\bar{\lambda}_n |\lambda_n|}{\alpha + |\lambda_n|^2} |(g, \psi_n)|^2 \tag{7.88}$$

follows. Proceeding as in the proof of Theorem 2.6 we can obtain that

$$\lim_{\alpha \to 0} (g_\alpha, \varphi) = \sum_{n=1}^{\infty} \bar{s}_n |(g, \psi_n)|^2 \tag{7.89}$$

with the complex numbers $s_n = \lambda_n / |\lambda_n|$. By Parseval's equality, (7.88) implies $|(g_\alpha, \varphi)| \leq \|g\|^2$, and the second inequality in (7.86) is obvious. For $g \neq 0$, by Parseval's equality from (2.6), we observe that

$$\frac{1}{\|g\|^2} \lim_{\alpha \to 0} (g_\alpha, \varphi)$$

belongs to the closure M of the convex hull of $\{\bar{s}_n : n \in \mathbb{N}\} \subset \mathbb{C}$. From the proof of Theorem 7.29 we know that the s_n lie on the upper half-circle $\{e^{it} : 0 \leq t \leq \pi - 2\delta\}$ for some $0 < \delta \leq \pi/2$. This implies that the set M has a positive lower bound c depending on the operator F, and this proves the first inequality in (7.86).

Conversely, assume that $\lim_{\alpha \to 0} (g_\alpha, \varphi)$ exists. Then from (7.87) we have that

$$\left| \sum_{n=1}^{\infty} \frac{\lambda_n}{\alpha + |\lambda_n|^2} |(\varphi, \psi_n)|^2 \right| \leq C \tag{7.90}$$

for all $\alpha > 0$ and some $C > 0$. Since 1 is the only accumulation point of the sequence $\{s_n\}$, there exists $n_0 \in \mathbb{N}$ such that $\mathrm{Re}(\lambda_n) \geq 0$ for all $n \geq n_0$. From (7.90) and the triangle inequality it follows that

$$\left| \sum_{n=n_0}^{\infty} \frac{\lambda_n}{\alpha + |\lambda_n|^2} |(\varphi, \psi_n)|^2 \right| \leq C_1$$

for all $\alpha > 0$ and some $C_1 > 0$ because the remaining finite sum is bounded. From this we can estimate

$$\sum_{n=n_0}^{\infty} \frac{\lambda_n}{\alpha + |\lambda_n|^2} |(\varphi, \psi_n)|^2 \leq \sum_{n=n_0}^{\infty} \frac{\text{Re}(\lambda_n) + \text{Im}(\lambda_n)}{\alpha + |\lambda_n|^2} |(\varphi, \psi_n)|^2 \qquad (7.91)$$

$$\leq \sqrt{2} \left| \sum_{n=n_0}^{\infty} \frac{\lambda_n}{\alpha + |\lambda_n|^2} |(\varphi, \psi_n)|^2 \right| \leq \sqrt{2} C_1.$$

Proceeding as in the proof of Theorem 2.6 we can pass to the limit $\alpha \to 0$ and conclude that the series

$$\sum_{n=n_0}^{\infty} \frac{1}{|\lambda_n|} |(\varphi, \psi_n)|^2$$

converges. Therefore, by Picard's Theorem 2.7 the equation $(F^*F)^{1/4}g = \varphi$ has a solution $g \in H$, and this concludes the proof of the second statement. $\square$

Finally, the discussions in Sects. 7.1 and 7.2, combined with Theorem 7.39, imply the following theorem, which provides a rigorous justification of the linear sampling method.

Theorem 7.40. *Let F be the far-field operator corresponding to (7.21)–(7.24) or to (7.49)–(7.53), and k^2 is not a Dirichlet eigenvalue or k is not a transmission eigenvalue, respectively. For $z \in D$ denote by g_z the solution of $(F^*F)^{1/4}g_z = \Phi_\infty(\cdot, z)$, and for $\alpha > 0$ and $z \in \mathbb{R}^2$ let g_z^α denote the solution of the far-field equation $Fg_z^\alpha = \Phi_\infty(\cdot, z)$ obtained by Tikhonov regularization, i.e., the solution of*

$$\alpha g_z^\alpha + F^* F g_z^\alpha = F^* \Phi_\infty(\cdot, z),$$

and let $v_{g_z^\alpha}$ denote the Herglotz wave function with kernel g_z^α. Then:

1. *If $z \in D$, then $\lim_{\alpha \to 0} v_{g_z^\alpha}(z)$ exists and*

$$c\|g_z\|^2 \leq \lim_{\alpha \to 0} |v_{g_z^\alpha}(z)| \leq \|g_z\|^2$$

   for some positive c depending only on D.
2. *If $z \notin D$, then $\lim_{\alpha \to 0} v_{g_z^\alpha}(z) = \infty$.*

Proof. Observing that $v_{g_z^\alpha}(z) = (g_z^\alpha, \Phi_\infty(\cdot, z))_{L^2[0,2\pi]}$ the statement follows from (7.86) and the fact that, as discussed in Sects. 7.1 and 7.2, in both cases F is normal $\Phi_\infty(\cdot, z)$ is in the range of $(F^*F)^{1/4}$ if and only if $z \in D$. $\square$

7.4 Closing Remarks

The factorization method described in the previous section relies in an essential manner on the fact that the far-field operator corresponding to the scattering problem is normal. Unfortunately, this is not always the case. In particular, the far-field operator is not normal in the case of the scattering problem for an imperfect conductor considered in Chap. 3 and the scattering problem for an absorbing inhomogeneous medium. A version of the factorization method that does not need the far-field operator to be normal was introduced by Kirsch in [101, 103].

A drawback of both the linear sampling method and the factorization method is the large amount of data needed for the inversion procedure. In particular, the factorization method has not been established for limited aperture data. Although the linear sampling method is valid for limited-aperture, far-field data (Sect. 4.5), one still needs a multistatic set of data, i.e., the far field measured at all observation directions on a subset of the unit circle with incident directions on a (possibly different) subset of the unit circle.

What happens if the far-field pattern is only known for a finite number of incident waves? In certain cases, it has been shown [41, 62, 146, 151] that only a finite number of incident plane waves is sufficient to uniquely determine the support of the scattering object. Progress has recently been made in the use of qualitative methods that use only a finite number of incident plane waves. In particular, it was shown in [74, 116, 117] and [141] that a single or a few incident waves can determine the *convex scattering support* that provides a lower bound for the convex hull of the scatterer.

8

Mixed Boundary Value Problems

This chapter is devoted to the study of mixed boundary value problems in electromagnetic scattering theory. Mixed boundary value problems typically model scattering by objects that are coated with a thin layer of material on part of the boundary. We shall consider here two main problems: (1) the scattering by a perfect conductor that is partially coated with a thin dielectric layer and (2) scattering by an orthotropic dielectric that is partially coated with a thin layer of highly conducting material. The first problem leads to an exterior mixed boundary value problem for the Helmholtz equation where on the coated part of the boundary the total field satisfies an impedance boundary condition and on the remaining part of the boundary the total field vanishes, while the second problem leads to a transmission problem with mixed transmission-conducting boundary conditions. In this chapter we shall present a mathematical analysis of these two mixed boundary value problems.

In the study of inverse problems for partially coated obstacles, it is important to mentioned that, in general, it is not known a priori whether or not the scattering object is coated and, if so, what the extent of the coating is. Hence the linear sampling method becomes the method of choice for solving inverse problems for mixed boundary value problems since it does not make use of the physical properties of the scattering object. In addition to the reconstruction of the shape of the scatterer, a main question in this chapter will be to determine whether the obstacle is coated and if so what the electrical properties of the coating are. In particular, we will show that the solution of the far-field equation that was used to determine the shape of the scatterer by means of the linear sampling method can also be used in conjunction with a variational method to determine the maximum value of the surface impedance of the coated portion in the case of partially coated perfect conductors and of the surface conductivity in the case of partially coated dielectrics.

Finally, we will extend the linear sampling method to the scattering problem by very thin objects, referred to as cracks, which are modeled by open arcs in $\mathbb{R}^2$.

F. Cakoni and D. Colton, *A Qualitative Approach to Inverse Scattering Theory*, 203
Applied Mathematical Sciences 188, DOI 10.1007/978-1-4614-8827-9_8,
© Springer Science+Business Media New York 2014

8.1 Scattering by a Partially Coated Perfect Conductor

We consider the scattering of an electromagnetic time-harmonic plane wave by a perfectly conducting infinite cylinder in $\mathbb{R}^3$ that is partially coated with a thin dielectric material. In particular, the total electromagnetic field on the uncoated part of the boundary satisfies the perfect conducting boundary condition, that is, the tangential component of the electric field is zero, whereas the boundary condition on the coated part is described by an impedance boundary condition [79].

More precisely, let D denote the cross section of the infinitely long cylinder and assume that $D \subset \mathbb{R}^2$ is an open bounded region with C^2 boundary ∂D such that $\mathbb{R}^2 \setminus \bar{D}$ is connected. The boundary ∂D has the dissection $\partial D = \overline{\partial D_D} \cup \overline{\partial D_I}$, where ∂D_D and ∂D_I are disjoint, relatively open subsets (possibly disconnected) of ∂D. Let ν denote the unit outward normal to ∂D, and assume that the surface impedance $\lambda \in C(\overline{\partial D_I})$ satisfies $\lambda(x) \geq \lambda_0 > 0$ for $x \in \partial D_I$. Then the total field $u = u^s + u^i$, given as the sum of the unknown scattered field u^s and the known incident field u^i, satisfies

$$\Delta u + k^2 u = 0 \quad \text{in} \quad \mathbb{R}^2 \setminus \bar{D}, \tag{8.1}$$

$$u = 0 \quad \text{on} \quad \partial D_D, \tag{8.2}$$

$$\frac{\partial u}{\partial \nu} + i\lambda u = 0 \quad \text{on} \quad \partial D_I, \tag{8.3}$$

where $k > 0$ is the wave number and u^s satisfies the Sommerfeld radiation condition

$$\lim_{r \to \infty} \sqrt{r}\left(\frac{\partial u^s}{\partial r} - iku^s\right) = 0 \tag{8.4}$$

uniformly in $\hat{x} = x/|x|$ with $r = |x|$. Note that here again the incident field u^i is usually an entire solution of the Helmholtz equation. In particular, in the case of incident plane waves, we have $u^i(x) = e^{ikx \cdot d}$, where $d := (\cos \phi, \sin \phi)$ is the incident direction and $x = (x_1, x_2) \in \mathbb{R}^2$.

Due to the boundary condition, the preceding exterior mixed boundary value problem may not have a solution in $C^2(\mathbb{R}^2 \setminus \bar{D}) \cap C^1(\mathbb{R} \setminus D)$, even for incident plane waves and analytic boundary. In particular, the solution fails to be differentiable at the boundary points of $\overline{\partial D}_D \cap \overline{\partial D}_I$. Therefore, looking for a weak solution in the case of mixed boundary value problems is very natural.

To define a weak solution to the mixed boundary value problem in the energy space $H^1(D)$, we need to understand the respective trace spaces on parts of the boundary. To this end, we now present a brief discussion of Sobolev spaces on open arcs. The classic reference for such spaces is [124]. For a systematic treatment of these spaces, we refer the reader to [127].

Let $\partial D_0 \subseteq \partial D$ be an open subset of the boundary. We define

$$H^{\frac{1}{2}}(\partial D_0) := \{u|_{\partial D_0} : u \in H^{\frac{1}{2}}(\partial D),\}$$

i.e., the space of restrictions to ∂D_0 of functions in $H^{\frac{1}{2}}(\partial D)$, and define

$$\tilde{H}^{\frac{1}{2}}(\partial D_0) := \{u \in H^{\frac{1}{2}}(\partial D) : \operatorname{supp} u \subseteq \overline{\partial D_0}, \}$$

where $\operatorname{supp} u$ is the essential support of u, i.e., the largest relatively closed subset of ∂D such that $u = 0$ almost everywhere on $\partial D \setminus \operatorname{supp} u$. We can identify $\tilde{H}^{\frac{1}{2}}(\partial D_0)$ with a trace space of $H_0^1(D, \partial D \setminus \overline{\partial D_0})$, where

$$H_0^1(D, \partial D \setminus \overline{\partial D_0}) = \left\{u \in H^1(D) : u|_{\partial D \setminus \overline{\partial D_0}} = 0 \text{ in the trace sense}\right\}.$$

A very important property of $\tilde{H}^{\frac{1}{2}}(\partial D_0)$ is that the extension by zero of $u \in \tilde{H}^{\frac{1}{2}}(\partial D_0)$ to the whole ∂D is in $H^{\frac{1}{2}}(\partial D)$ and the zero extension operator is bounded from $\tilde{H}^{\frac{1}{2}}(\partial D_0)$ to $H^{\frac{1}{2}}(\partial D)$. It can also be shown (cf. Theorem A4 in [127]) that there exists a bounded extension operator $\tau : H^{\frac{1}{2}}(\partial D_0) \to H^{\frac{1}{2}}(\partial D)$. In other words, for any $u \in H^{\frac{1}{2}}(\partial D_0)$ there exists an extension $\tau u \in H^{\frac{1}{2}}(\partial D)$ such that

$$\|\tau u\|_{H^{\frac{1}{2}}(\partial D)} \leq C\|u\|_{H^{\frac{1}{2}}(\partial D_0)}, \tag{8.5}$$

with C independent of u, where

$$\|u\|_{H^{\frac{1}{2}}(\partial D_0)} := \min\left\{\|U\|_{H^{\frac{1}{2}}(\partial D)} \text{ for } U \in H^{\frac{1}{2}}(\partial D), U|_{\partial D_0} = u\right\}.$$

Example 8.1. Consider the step function

$$u(t) = \begin{cases} 1 & t \in [0, \pi], \\ 0 & t \in (\pi, 2\pi]. \end{cases}$$

Using the definition of Sobolev spaces in terms of the Fourier coefficients (Sect. 1.4) it is easy to show that the step function is not in $H^{\frac{1}{2}}[0, 2\pi]$. In particular, the Fourier coefficients of u are $a_{2k} = 0$ and $a_{2k+1} = 1/(i(2k+1)\pi)$, whence

$$\sum_{-\infty}^{\infty} \left(1 + m^2\right)^{\frac{1}{2}} |a_m|^2 = \sum_{-\infty}^{\infty} \left(1 + (2k+1)^2\right)^{\frac{1}{2}} \frac{1}{\pi^2(2k+1)^2} = +\infty.$$

Now consider the unit circle $\partial \Omega = \{x \in \mathbb{R}^2 : x = (\sin t, \cos t), t \in [0, 2\pi]\}$, and denote by $\partial \Omega_0 = \{x \in \mathbb{R}^2 : x = (\sin t, \cos t), t \in [0, \pi]\}$ the upper half-circle. Let $v : \partial \Omega_0 \to \mathbb{R}$ be the constant function $v = 1$. By definition, $v \in H^{\frac{1}{2}}(\partial \Omega_0)$ since it is the restriction to $\partial \Omega_0$ of the constant function 1 defined on the whole circle $\partial \Omega$ that is in $H^{\frac{1}{2}}(\partial \Omega)$. But $v \notin \tilde{H}^{\frac{1}{2}}(\partial \Omega_0)$ since its extension by zero to the whole circle is not in $H^{\frac{1}{2}}(\partial \Omega)$ [note that the extension $\tilde{v}(\sin t, \cos t)$ is a step function and from the preceding discussion is not in $H^{\frac{1}{2}}[0, 2\pi]$].

The foregoing example shows that if $u \in \tilde{H}^{\frac{1}{2}}(\partial D_0)$, then it has a certain behavior at the boundary of ∂D_0 in ∂D. A better insight into this behavior is given in [124]. In particular, the space $\tilde{H}^{\frac{1}{2}}(\partial D_0)$ coincides with the space

$$H_{00}^{\frac{1}{2}}(\partial D_0) := \{u \in H^{\frac{1}{2}}(\partial D_0) : \ r^{-\frac{1}{2}}u \in L^2(\partial D_0)\},$$

where r is the polar radius.

Both $H^{\frac{1}{2}}(\partial D_0)$ and $\tilde{H}^{\frac{1}{2}}(\partial D_0)$ are Hilbert spaces when equipped with the restriction of the inner product of $H^{\frac{1}{2}}(\partial D)$. Hence, we can define the corresponding dual spaces

$$H^{-\frac{1}{2}}(\partial D_0) := \left(\tilde{H}^{\frac{1}{2}}(\partial D_0)\right)' = \text{the dual space of} \quad \tilde{H}^{\frac{1}{2}}(\partial D_0)$$

and

$$\tilde{H}^{-\frac{1}{2}}(\partial D_0) := \left(H^{\frac{1}{2}}(\partial D_0)\right)' = \text{the dual space of} \quad H^{\frac{1}{2}}(\partial D_0)$$

with respect to the duality pairing explained in what follows.

A bounded linear functional $F \in H^{-\frac{1}{2}}(\partial D_0)$ can in fact be seen as the restriction to ∂D_0 of some $\tilde{F} \in H^{-\frac{1}{2}}(\partial D)$ in the following sense: if $\tilde{u} \in H^{\frac{1}{2}}(\partial D)$ denotes the extension by zero of $u \in \tilde{H}^{\frac{1}{2}}(\partial D_0)$, then the restriction $F := \tilde{F}|_{\partial D_0}$ is defined by

$$F(u) = \tilde{F}(\tilde{u}).$$

With the preceding understanding, to unify the notations, we identify

$$H^{-\frac{1}{2}}(\partial D_0) := \{v|_{\partial D_0} : v \in H^{-\frac{1}{2}}(\partial D)\}$$

and

$$\langle v, u \rangle_{H^{-\frac{1}{2}}(\partial D_0), \tilde{H}^{\frac{1}{2}}(\partial D_0)} = \langle v, \tilde{u} \rangle_{H^{-\frac{1}{2}}(\partial D), H^{\frac{1}{2}}(\partial D)},$$

where $\langle \cdot, \cdot \rangle$ denotes the duality pairing between the denoted spaces and $\tilde{u} \in H^{\frac{1}{2}}(\partial D)$ is the extension by zero of $u \in \tilde{H}^{\frac{1}{2}}(\partial D_0)$.

For a bounded linear functional $F \in H^{-\frac{1}{2}}(\partial D)$, we define $\operatorname{supp} F$ to be the largest relatively closed subset of ∂D such that the restriction of F to $\partial D \setminus \operatorname{supp} F$ is zero. Similarly, for $\tilde{H}^{\frac{1}{2}}(\partial D_0)$ we can now write

$$\tilde{H}^{-\frac{1}{2}}(\partial D_0) := \{v \in H^{-\frac{1}{2}}(\partial D) : \operatorname{supp} v \subseteq \overline{\partial D_0}\}.$$

Therefore, the extension by zero $\tilde{v} \in H^{-\frac{1}{2}}(\partial D)$ of $v \in \tilde{H}^{-\frac{1}{2}}(\partial D_0)$ is well defined and

$$\langle \tilde{v}, u \rangle_{H^{-\frac{1}{2}}(\partial D), H^{\frac{1}{2}}(\partial D)} = \langle v, u \rangle_{\tilde{H}^{-\frac{1}{2}}(\partial D_0), H^{\frac{1}{2}}(\partial D_0)},$$

where $u \in H^{\frac{1}{2}}(\partial D)$.

We can now formulate the following mixed boundary value problems:

Exterior mixed boundary value problem: Let $f \in H^{\frac{1}{2}}(\partial D_D)$ and $h \in H^{-\frac{1}{2}}(\partial D_I)$. Find a function $u \in H^1_{loc}(\mathbb{R}^2 \setminus \bar{D})$ such that

$$\Delta u + k^2 u = 0 \qquad \text{in} \qquad \mathbb{R}^2 \setminus \bar{D}, \tag{8.6}$$

$$u = f \qquad \text{on} \qquad \partial D_D, \tag{8.7}$$

$$\frac{\partial u}{\partial \nu} + i\lambda u = h \qquad \text{on} \qquad \partial D_I, \tag{8.8}$$

$$\lim_{r \to \infty} \sqrt{r} \left(\frac{\partial u}{\partial r} - iku \right) = 0. \tag{8.9}$$

Note that the scattering problem for a partially coated perfect conductor (8.1)–(8.4) is a special case of (8.6)–(8.9). In particular, the scattered field u^s satisfies (8.6)–(8.9) with $f := -u^i|_{\partial D_D}$ and $h := -\partial u^i/\partial \nu - i\lambda u^i|_{\partial D_I}$.

For later use we also consider the corresponding interior mixed boundary value problem.

Interior mixed boundary value problem: Let $f \in H^{\frac{1}{2}}(\partial D_D)$ and $h \in H^{-\frac{1}{2}}(\partial D_I)$. Find a function $u \in H^1(D)$ such that

$$\Delta u + k^2 u = 0 \qquad \text{in} \qquad D, \tag{8.10}$$

$$u = f \qquad \text{on} \qquad \partial D_D, \tag{8.11}$$

$$\frac{\partial u}{\partial \nu} + i\lambda u = h \qquad \text{on} \qquad \partial D_I. \tag{8.12}$$

Theorem 8.2. *Assume that $\partial D_I \neq \emptyset$ and $\lambda \neq 0$. Then the interior mixed boundary value problem (8.10)–(8.12) has at most one solution in $H^1(D)$.*

Proof. Let u be a solution to (8.10)–(8.12), with $f \equiv 0$ and $h \equiv 0$. Then an application of Green's first identity in D yields

$$- k^2 \int_D |u|^2 \, dx + \int_D |\nabla u|^2 \, dx = \int_{\partial D} \frac{\partial u}{\partial \nu} \bar{u} \, ds, \tag{8.13}$$

and making use of homogeneous boundary condition we obtain

$$- k^2 \int_D |u|^2 \, dx + \int_D |\nabla u|^2 \, dx = -i \int_{\partial D_I} \lambda |u|^2 \, ds. \tag{8.14}$$

Since λ is a real-valued function and $\lambda(x) \geq \lambda_0 > 0$, taking the imaginary part of (8.14) we conclude that $u|_{\partial D_I} \equiv 0$ as a function in $H^{\frac{1}{2}}(\partial D_I)$, and consequently $\partial u/\partial \nu|_{\partial D_I} \equiv 0$ as a function in $H^{-\frac{1}{2}}(\partial D_I)$.

Now let Ω_ρ be a disk of radius ρ with center on ∂D_I such that $\bar{\Omega}_\rho \cap \partial D_D = \emptyset$, and define $v = u$ in $D \cap \Omega_\rho$, $v = 0$ in $(\mathbb{R}^2 \setminus \bar{D}) \cap \Omega_\rho$. Then applying Green's first identity in each of these domains to v and a test function $\overline{\varphi} \in C_0^\infty(\Omega_\rho)$ we see that v is a weak solution to the Helmholtz equation in Ω_ρ. Thus v is a real-analytic solution in Ω_ρ. We can now conclude that $u \equiv 0$ in Ω_ρ, and thus $u \equiv 0$ in D. $\qquad\square$

Theorem 8.3. *The exterior mixed boundary value problem (8.6)–(8.9) has at most one solution in $H^1_{loc}(\mathbb{R}^2 \setminus \bar{D})$.*

Proof. The proof of the theorem is essentially the same as the proof of Theorem 3.3. $\qquad\square$

Theorem 8.4. *Assume that $\partial D_I \neq \emptyset$ and $\lambda \neq 0$. Then the interior mixed boundary value problem (8.10)–(8.12) has a solution that satisfies the estimate*

$$\|u\|_{H^1(D)} \le C \left(\|f\|_{H^{\frac{1}{2}}(\partial D_D)} + \|h\|_{H^{-\frac{1}{2}}(\partial D_I)} \right), \tag{8.15}$$

with C a positive constant independent of f and h.

Proof. To prove the theorem, we use the variational approach developed in Sect. 5.3. (For a solution procedure based on integral equations of the first kind we refer the reader to [23]). Let $\tilde{f} \in H^{\frac{1}{2}}(\partial D)$ be the extension of the Dirichlet data $f \in H^{\frac{1}{2}}(\partial D_D)$ that satisfies $\|\tilde{f}\|_{H^{\frac{1}{2}}(\partial D)} \le C\|f\|_{H^{\frac{1}{2}}(\partial D_D)}$ given by (8.5), and let $u_0 \in H^1(D)$ be such that $u_0 = \tilde{f}$ on ∂D and $\|u_0\|_{H^1(D)} \le C\|\tilde{f}\|_{H^{\frac{1}{2}}(\partial D)}$. In particular, we may choose u_0 to be a solution of $\Delta u_0 = 0$ (Example 5.15). Defining the Sobolev space $H_0^1(D, \partial D_D)$ by

$$H_0^1(D, \partial D_D) := \left\{ u \in H^1(D) : u = 0 \text{ on } \partial D_D \right\}$$

equipped with the norm induced by $H^1(D)$, we observe that $w = u - u_0 \in H_0^1(D, \partial D_D)$, where $u \in H^1(D)$ is a solution to (8.10)–(8.12). Furthermore, w satisfies

$$\Delta w + k^2 w = -k^2 u_0 \quad \text{in } D \tag{8.16}$$

and

$$\frac{\partial w}{\partial \nu} + i\lambda w = \tilde{h} \qquad \text{on} \qquad \partial D_I, \tag{8.17}$$

where $\tilde{h} \in H^{-\frac{1}{2}}(\partial D_I)$ is given by

$$\tilde{h} := -\frac{\partial u_0}{\partial \nu} - i\lambda u_0 + h.$$

Multiplying (8.16) by a test function $\overline{\varphi} \in H_0^1(D, \partial D_D)$ and using Green's first identity together with the boundary condition (8.17) we can write (8.10)–(8.12) in the following equivalent variational form: *find $u \in H^1(D)$ such that $w = u - u_0 \in H_0^1(D, \partial D_D)$ and*

$$a(w, \varphi) = L(\varphi) \qquad \text{for all } \varphi \in H_0^1(D, \partial D_D), \tag{8.18}$$

where the sesquilinear form $a(\cdot, \cdot) : H_0^1(D, \partial D_D) \times H_0^1(D, \partial D_D) \to \mathbb{C}$ is defined by

$$a(w, \varphi) := \int_D (\nabla w \cdot \nabla \bar{\varphi} - k^2 w \bar{\varphi}) \, dx + i \int_{\partial D_I} \lambda \, w \, \bar{\varphi} \, ds,$$

and the conjugate linear functional $L : H_0^1(D, \partial D_D) \to \mathbb{C}$ is defined by

$$L(\varphi) = k^2 \int_D u_0 \bar{\varphi} \, dx + \int_{\partial D_I} \tilde{h} \cdot \bar{\varphi} \, dx,$$

where the integral over ∂D_I is interpreted as the duality pairing between $\tilde{h} \in H^{-\frac{1}{2}}(\partial D_I)$ and $\bar{\varphi} \in \tilde{H}^{\frac{1}{2}}(\partial D_I)$ [note that $\bar{\varphi} \in \tilde{H}^{\frac{1}{2}}(\partial D_I)$ since $\tilde{H}^{\frac{1}{2}}(\partial D_I)$ is the trace space of $H_0^1(D, \partial D_D)$].

Next we write $a(\cdot, \cdot)$ as the sum of two terms $a(\cdot, \cdot) = a_1(\cdot, \cdot) + a_2(\cdot, \cdot)$, where

$$a_1(w, \varphi) := \int_D (\nabla w \cdot \nabla \bar{\varphi} + w \, \bar{\varphi}) \, dx + i \int_{\partial D_I} \lambda \, w \, \bar{\varphi} \, ds$$

and

$$a_2(w, \varphi) := -(k^2 + 1) \int_D w \, \bar{\varphi} \, dx.$$

From the Cauchy–Schwarz inequality and the trace Theorem 1.38, since λ is a bounded function on ∂D_I, we have that

$$|a_1(w, \varphi)| \leq C_1 \|w\|_{H^1(D)} \|\varphi\|_{H^1(D)} + C_2 \|w\|_{L^2(\partial D_I)} \|\varphi\|_{L^2(\partial D_I)}$$
$$\leq \tilde{C} \left(\|w\|_{H^1(D)} \|\varphi\|_{H^1(D)} + \|w\|_{H^{\frac{1}{2}}(\partial D)} \|\varphi\|_{H^{\frac{1}{2}}(\partial D)} \right)$$
$$\leq C \|w\|_{H^1(D)} \|\varphi\|_{H^1(D)}$$

and

$$|a_2(w, \varphi)| \leq \tilde{C} \|w\|_{L^2(D)} \|\varphi\|_{L^2(D)} \leq C \|w\|_{H^1(D)} \|\varphi\|_{H^1(D)}.$$

Hence $a_1(\cdot, \cdot)$ and $a_2(\cdot, \cdot)$ are bounded sesquilinear forms. Furthermore, noting that $\varphi = 0$ on ∂D_D, we have that

$$\int_{\partial D_I} \frac{\partial u_0}{\partial \nu} \bar{\varphi} \, ds = \int_{\partial D} \frac{\partial u_0}{\partial \nu} \bar{\varphi} \, ds = \int_D \nabla u_0 \cdot \nabla \bar{\varphi} \, dx.$$

Therefore, from the previous estimates and the trace Theorems 1.38 and 5.7 we have that

$$|L(\varphi)| \leq C_1 \|u_0\|_{H^1(D)} \|\varphi\|_{H^1(D)} + C_2 \|u_0\|_{H^{\frac{1}{2}}(\partial D)} \|\varphi\|_{H^{\frac{1}{2}}(\partial D)}$$
$$+ C_3 \|h\|_{H^{-\frac{1}{2}}(\partial D_I)} \|\varphi\|_{\tilde{H}^{\frac{1}{2}}(\partial D_I)}$$
$$\leq \tilde{C} \left(\|\tilde{f}\|_{H^{\frac{1}{2}}(\partial D)} + \|h\|_{H^{-\frac{1}{2}}(\partial D_I)} \right) \|\varphi\|_{H^1(D)}$$
$$\leq C \left(\|f\|_{H^{\frac{1}{2}}(\partial D_D)} + \|h\|_{H^{-\frac{1}{2}}(\partial D_I)} \right) \|\varphi\|_{H^1(D)}$$

for all $\varphi \in H_0^1(D, \partial D_0)$, which shows that L is a bounded conjugate linear functional and

$$\|L\| \leq C \left(\|f\|_{H^{\frac{1}{2}}(\partial D_D)} + \|h\|_{H^{-\frac{1}{2}}(\partial D_I)} \right), \tag{8.19}$$

with the constant $C > 0$ independent of f and h.

Next, since λ is real, we can write

$$|a_1(w, w)| \geq \|w\|_{H^1(D)}^2,$$

whence $a_1(\cdot, \cdot)$ is strictly coercive.

Therefore, from the Lax–Milgram lemma there exists a bijective bounded linear operator $A : H_0^1(D, \partial D_D) \to H_0^1(D, \partial D_D)$ with bounded inverse such that $(Aw, \varphi) = a_1(w, \varphi)$ for all w and φ in $H_0^1(D, \partial D_D)$. Finally, due to the compact embedding of $H^1(D)$ into $L^2(D)$, there exists a compact bounded linear operator $B : H_0^1(D, \partial D_D) \to H_0^1(D, \partial D_D)$ such that $(Bw, \varphi) = a_2(w, \varphi)$ for all w and φ in $H_0^1(D, \partial D_D)$ (Example 5.17). Therefore, from Theorems 5.16 and 8.2 we obtain the existence of a unique solution to (8.18) and, consequently, to the interior mixed boundary value problem (8.10)–(8.12). The a priori estimate (8.15) follows from (8.19). □

Now let us consider an open disk Ω_R of radius R centered at the origin and containing $\bar{D}$.

Theorem 8.5. *The exterior mixed boundary value problem (8.6)–(8.9) has a solution that satisfies the estimate*

$$\|u\|_{H^1(\Omega_R \setminus \bar{D})} \leq C \left(\|f\|_{H^{\frac{1}{2}}(\partial D_D)} + \|h\|_{H^{-\frac{1}{2}}(\partial D_I)} \right), \tag{8.20}$$

with C a positive constant independent of f and h but depending on R.

Proof. First, exactly in the same way as in Example 5.23, we can show that the exterior mixed boundary value problem (8.6)–(8.9) is equivalent to the following problem:

$$\Delta u + k^2 u = 0 \quad \text{in} \quad \Omega_R \setminus \bar{D}, \tag{8.21}$$

$$u = f \quad \text{on} \quad \partial D_D, \tag{8.22}$$

$$\frac{\partial u}{\partial \nu} + i\lambda u = h \quad \text{on} \quad \partial D_I, \tag{8.23}$$

$$\frac{\partial u}{\partial \nu} = Tu \quad \text{on} \quad \partial \Omega_R, \tag{8.24}$$

where T is the Dirichlet-to-Neumann map. If $\tilde{f} \in H^{\frac{1}{2}}(\partial D)$ is the extension of $f \in H^{\frac{1}{2}}(\partial D_D)$ that satisfies (8.5) with ∂D_0 replaced by ∂D_D, then we construct $u_0 \in H^1(\Omega_R \setminus \bar{D})$ such that $u_0 = \tilde{f}$ on ∂D, $u = 0$ on $\partial \Omega_R$, and $\Delta u_0 = 0$ in $\Omega_R \setminus \bar{D}$ (Example 5.15). Then, for every solution u to (8.21)–(8.24), $w = u - u_0$ is in the Sobolev space $H_0^1(\Omega_R \setminus \bar{D}, \partial D_D)$ defined by

$$H_0^1(\Omega_R \setminus \bar{D}, \partial D_D) := \left\{ u \in H^1(\Omega_R \setminus \bar{D}) : \ u = 0 \text{ on } \partial D_D \right\}$$

and satisfies the variational equation

$$\int_{\Omega_R \setminus \bar{D}} \left(\nabla w \cdot \nabla \bar{\varphi} - k^2 w \bar{\varphi} \right) ds - i \int_{\partial D_I} \lambda w \, \bar{\varphi} \, ds - \int_{\partial \Omega_R} T w \, \bar{\varphi} \, ds$$

$$= k^2 \int_{\Omega_R \setminus \bar{D}} u_0 \bar{\varphi} \, dx - \int_{\partial D_I} \left(\frac{\partial u_0}{\partial \nu} - i \lambda u_0 + h \right) \bar{\varphi} \, ds$$

$$+ \int_{\partial \Omega_R} \left(T u_0 - \frac{\partial u_0}{\partial \nu} \right) \bar{\varphi} \, ds \qquad \text{for all } \varphi \in H_0^1(\Omega_R \setminus \bar{D}, \partial D_D).$$

Making use of Theorem 5.22, the assertion of the theorem can now be proven in the same way as in Theorem 8.4. $\qquad\square$

Remark 8.6. In the case where either $\partial D_I = \emptyset$ (this case corresponds to the Dirichlet boundary value problem) or $\lambda = 0$, the corresponding interior problem may not be uniquely solvable. If nonuniqueness occurs, then k^2 is said to be an eigenvalue of the corresponding boundary value problem. In these cases, Theorem 8.4 holds true under the assumption that k^2 is not an eigenvalue of the corresponding boundary value problem.

Remark 8.7. Due to the change in the boundary conditions, the solution to the mixed boundary value problems (8.6)–(8.9) and (8.10)–(8.12) has a singular behavior near the boundary points in $\overline{\partial D}_D \cup \overline{\partial D}_N$. In particular, even for C^∞ boundary ∂D and analytic incident waves u^i, the solution in general is not in $H^2_{loc}(\mathbb{R}^2 \setminus \bar{D})$. More precisely, the most singular term of the solution behaves like $O(r^{\frac{1}{2}})$, where (r, ϕ) denotes the local polar coordinates centered at the boundary points in $\overline{\partial D}_D \cup \overline{\partial D}_N$ [65]. This is important to take into consideration when finite element methods are used.

8.2 Inverse Scattering Problem for Partially Coated Perfect Conductor

We now consider time-harmonic incident fields given by $u^i(x) = e^{ikx \cdot d}$ with incident direction $d := (\cos \phi, \sin \phi)$ and $x = (x_1, x_2) \in \mathbb{R}^2$. The corresponding scattered field $u^s = u^s(\cdot, \phi)$, which satisfies (8.1)–(8.4), depends also on the incident angle ϕ and has the asymptotic behavior (4.5). The far-field pattern $u_\infty(\theta, \phi)$, $\theta \in [0, 2\pi]$ of the scattered field defines the far-field operator $F : L^2[0, 2\pi] \to L^2[0, 2\pi]$ corresponding to the scattering problem (8.1)–(8.4) by

$$(Fg)(\theta) := \int_0^{2\pi} u_\infty(\theta, \phi) g(\phi) d\phi \qquad g \in L^2[0, 2\pi]. \tag{8.25}$$

The *inverse scattering problem* for a partially coated perfect conductor is given the far-field pattern $u_\infty(\theta, \phi)$ for $\theta \in [0, 2\pi]$ and $\phi \in [0, 2\pi]$ determines *both* D and $\lambda = \lambda(x)$ for $x \in \partial D_I$.

In the same way as in the proof of Theorem 4.3, using Theorem 8.2 we can show the following result.

Theorem 8.8. *Assume that $\partial D_I \neq \emptyset$ and $\lambda \neq 0$. Then the far-field operator corresponding to the scattering problem (8.1)–(8.4) is injective with a dense range.*

Remark 8.9. If $\partial D_I = \emptyset$ or $\lambda = 0$, then all the following results about the far-field operator and the determination of D remain valid assuming the uniqueness for the corresponding interior boundary value problem. Note that the case of $\partial D_I = \emptyset$ corresponds to the scattering problem for a perfect conductor.

Concerning the unique determination of D, the following theorem can be proved in the same way as Theorem 4.5. The only change needed in the proof is that we can always choose the point x^* such that either $\Omega_\epsilon(x^*) \cap \partial D_1 \subset \partial D_{1D}$ or $\Omega_\epsilon(x^*) \cap \partial D_1 \subset \partial D_{1I}$ for some small disk $\Omega_\epsilon(x^*)$ centered at x^* of radius ϵ and satisfying $\Omega_\epsilon(x^*) \cap \bar{D}_2 = \emptyset$, whence one uses either the Dirichlet condition or impedance condition at x^* to arrive at a contradiction.

Theorem 8.10. *Assume that D_1 and D_2 are two partially coated scattering obstacles with corresponding surface impedances λ_1 and λ_2 such that for a fixed wave number the far-field patterns for both scatterers coincide for all incident angles ϕ. Then $D_1 = D_2$.*

Theorem 8.11. *Assume that D_1 and D_2 are two partially coated scattering obstacles with corresponding surface impedances λ_1 and λ_2 such that for a fixed wave number the far-field patterns coincide for all incident angles ϕ. Then $D_1 = D_2$ and $\lambda_1 = \lambda_2$.*

Proof. By Theorem 8.10, we first have that $D_1 = D_2 = D$. Then, following the proof of Theorem 4.7 we can prove that the total fields u_1 and u_2 corresponding to λ_1 and λ_2 coincide in $\mathbb{R}^2 \setminus \bar{D}$, whence $u_1 = u_2$ and $\partial u_1/\partial \nu = \partial u_2/\partial \nu$ on ∂D. From the boundary condition we have

$$u_j = 0 \quad \text{on } \partial D_{D_j}, \qquad \frac{\partial u_j}{\partial \nu} + i\lambda_j u_j = 0 \quad \text{on } \partial D_{I_j}$$

for $j = 1, 2$. First we observe that $\partial D_{D_1} \cap \partial D_{D_2} = \emptyset$, because otherwise $u_1 = \partial u_1/\partial \nu = 0$ on an open arc $\Gamma \subset \partial D$ and a contradiction can be obtained as in the proof of Theorem 4.7. Hence $\partial D_{I_1} = \partial D_{I_2} = \partial D_I$. Next,

$$(\lambda_1 - \lambda_2)u_1 = 0 \qquad \text{on } \partial D_I,$$

and again one can conclude that $\lambda_1 = \lambda_2$, as in Theorem 4.7. $\square$

Having proved the uniqueness results, we now turn our attention to finding an approximation to D and λ. Our reconstruction algorithm is based on solving the far-field equation

$$Fg = \Phi_\infty(\cdot, z) \qquad z \in \mathbb{R}^2,$$

where $\Phi_\infty(\hat{x}, z)$ is the far-field pattern of the fundamental solution (Sect. 4.3). The far-field equation can be written as

$$-(BHg) = \Phi_\infty(\cdot, z) \qquad z \in \mathbb{R}^2,$$

where $B : H^{\frac{1}{2}}(\partial D_D) \times H^{-\frac{1}{2}}(\partial D_I) \to L^2[0, 2\pi]$ maps the boundary data (f, h) to the far-field pattern u_∞ of the radiating solution u to the corresponding exterior mixed boundary value problem (8.6)–(8.9), and $H : L^2[0, 2\pi] \to H^{\frac{1}{2}}(\partial D_D) \times H^{-\frac{1}{2}}(\partial D_I)$ is defined by

$$(Hg)(x) = \begin{cases} v_g(x), & x \in \partial D_D, \\ \dfrac{\partial v_g(x)}{\partial \nu} + i\lambda(x)v_g(x), & x \in \partial D_I, \end{cases}$$

with v_g being the Herglotz wave function with kernel g.

Lemma 8.12. *Any pair $(f, h) \in H^{\frac{1}{2}}(\partial D_D) \times H^{-\frac{1}{2}}(\partial D_I)$ can be approximated in $H^{\frac{1}{2}}(\partial D_D) \times H^{-\frac{1}{2}}(\partial D_I)$ by Hg.*

Proof. Let u be the unique solution to (8.10)–(8.12) with boundary data (f, h). Then the result of this lemma is a consequence of Lemma 6.45 applied to this u and the trace Theorems 1.38 and 5.7. $\qquad\square$

Lemma 8.13. *The bounded linear operator $B : H^{\frac{1}{2}}(\partial D_D) \times H^{-\frac{1}{2}}(\partial D_I) \to L^2[0, 2\pi]$ is compact and injective and has a dense range.*

Proof. The proof proceeds as the proof of Theorem 4.8 making use of Theorems 8.5 and 8.8. $\qquad\square$

Using Lemmas 8.12 and 8.13 we can now prove in a similar way as in Theorem 4.11 the following result.

Theorem 8.14. *Assume that $\partial D_I \neq \emptyset$ and $\lambda \neq 0$. Let u_∞ be the far-field pattern corresponding to the scattering problem (8.1)–(8.4) with associated far-field operator F. Then the following statements hold:*

1. *For $z \in D$ and a given $\epsilon > 0$ there exists a function $g_z^\epsilon \in L^2[0, 2\pi]$ such that*

$$\|Fg_z^\epsilon - \Phi_\infty(\cdot, z)\|_{L^2[0, 2\pi]} < \epsilon$$

   and the Herglotz wave function $v_{g_z^\epsilon}$ with kernel g_z^ϵ converges in $H^1(D)$ as $\epsilon \to 0$.

2. *For $z \notin D$ and a given $\epsilon > 0$ every function $g_z^\epsilon \in L^2[0, 2\pi]$ that satisfies*

$$\|F g_z^\epsilon - \Phi_\infty(\cdot, z)\|_{L^2[0,2\pi]} < \epsilon$$

is such that

$$\lim_{\epsilon \to 0} \|v_{g_z^\epsilon}\|_{H^1(D)} = \infty.$$

An approximation to D can now be obtained as the set of points z, where $\|g_z\|_{L^2[0, 2\pi]}$ becomes large, with g_z the approximate solution to the far-field equation given by Theorem 8.14. Note that the factorization method to characterize D from the range of $(F^*F)^{1/4}$ cannot be established for the scattering problem with mixed boundary conditions. Hence a rigorous justification of the linear sampling method similar to Theorem 7.39 for this case is still an open problem.

Having determined D, in a similar way as in Sect. 4.4, we can now use g_z given by Theorem 8.14 to determine an approximation to the maximum value of λ. In particular, let u_z be the unique solution to

$$\Delta u_z + k^2 u_z = 0 \quad \text{in} \quad D, \tag{8.26}$$

$$u_z = -\Phi(\cdot, z) \quad \text{on} \quad \partial D_D, \tag{8.27}$$

$$\frac{\partial u_z}{\partial \nu} + i\lambda u_z = -\frac{\partial \Phi(\cdot, z)}{\partial \nu} - i\lambda \Phi(\cdot, z) \quad \text{on} \quad \partial D_I, \tag{8.28}$$

where $z \in D$ and $\lambda \in C(\partial D_I)$, $\lambda(x) \geq \lambda_0 > 0$. From the proof of the first part of Theorem 8.14 the following result is valid.

Lemma 8.15. *Assume $\partial D_I \neq \emptyset$ and $\lambda \neq 0$. Let $\epsilon > 0$, $z \in D$, and let u_z be the unique solution of (8.26)–(8.28). Then there exists a Herglotz wave function v_{g_z} with kernel $g_z \in L^2[0, 2\pi]$ such that*

$$\|u_z - v_{g_z}\|_{H^1(D)} \leq \epsilon. \tag{8.29}$$

Moreover, there exists a positive constant $C > 0$ independent of ϵ such that

$$\|F g_z - \Phi_\infty(\cdot, z)\|_{L^2[0, 2\pi]} \leq C\epsilon. \tag{8.30}$$

Now define w_z by

$$w_z := u_z + \Phi(\cdot, z). \tag{8.31}$$

In particular,

$$w_z|_{\partial D_D} = 0 \quad \text{and} \quad \left(\frac{\partial w_z}{\partial \nu} + i\lambda w_z\right)_{|\partial D_I} = 0, \tag{8.32}$$

interpreted in the sense of the trace theorem. Repeating the proof of Theorem 4.12 with minor changes accounting for the boundary conditions (8.32) we have the following result.

Lemma 8.16. *For every $z_1, z_2 \in D$ we have that*

$$2 \int_{\partial D_I} w_{z_1} \lambda \bar{w}_{z_2} \, ds = -4\pi k \, |\gamma|^2 J_0(k \, |z_1 - z_2|)$$

$$- i \left(\overline{u_{z_2}(z_1)} - u_{z_1}(z_2) \right),$$

where $\gamma = e^{i\pi/4}/\sqrt{8\pi k}$ and J_0 is a Bessel function of order zero.

Assuming D is connected, consider a disk $\Omega_r \subset D$ of radius r contained in D (Remark 4.13), and define

$$W := \left\{ f \in L^2(\partial D_I) : \begin{array}{l} f = w_z|_{\partial D_I} \text{ with } w_z = u_z + \Phi(\cdot, z), \\ z \in \Omega_r \text{ and } u_z \text{ the solution of (8.26)–(8.28)} \end{array} \right\}.$$

Lemma 8.17. *W is complete in $L^2(\partial D_I)$.*

Proof. Let φ be a function in $L^2(\partial D_I)$ such that for every $z \in \Omega_r$

$$\int_{\partial D_I} w_z \varphi \, ds = 0.$$

Using Theorem 8.4, let $v \in H^1(D)$ be the unique solution of the interior mixed boundary value problem

$$\begin{aligned}
\Delta v + k^2 v &= 0 && \text{in} && D, \\
v &= 0 && \text{on} && \partial D_D, \\
\frac{\partial v}{\partial \nu} + i\lambda v &= \varphi && \text{on} && \partial D_I.
\end{aligned}$$

Then for every $z \in \Omega_r$, using the boundary conditions and the integral representation formula, we have that

$$\begin{aligned}
0 = \int_{\partial D_I} w_z \varphi \, ds &= \int_{\partial D_I} w_z \left(\frac{\partial v}{\partial \nu} + i\lambda v \right) ds = \int_{\partial D} w_z \left(\frac{\partial v}{\partial \nu} + i\lambda v \right) ds \\
&= \int_{\partial D} \left(u_z \frac{\partial v}{\partial \nu} + i\lambda u_z v + \Phi(\cdot, z) \frac{\partial v}{\partial \nu} + i\lambda \Phi(\cdot, z)v \right) ds \\
&= \int_{\partial D} \left[u_z \frac{\partial v}{\partial \nu} + v \left(-\frac{\partial u_z}{\partial \nu} - \frac{\partial \Phi(\cdot, z)}{\partial \nu} - i\lambda \Phi(\cdot, z) \right) \right] ds \\
&\quad + \int_{\partial D} \left(\Phi(\cdot, z) \frac{\partial v}{\partial \nu} + i\lambda v \Phi(\cdot, z) \right) ds = v(z).
\end{aligned}$$

The unique continuation principle for solutions to the Helmholtz equation now implies that $v(z) = 0$ for all $z \in D$, whence from the trace theorem $\varphi = 0$. $\qquad \square$

Setting $z = z_1 = z_2$ in Lemma 8.16 we arrive at the following integral equation for the determination of λ:

$$2 \int_{\partial D_I} \lambda |u_{z_i} + \Phi(\cdot, z_i)|^2 \, ds = -\frac{1}{4} - \mathrm{Im}(u_z(z))$$

or, noting that $u_z + \Phi(\cdot, z) = 0$ on ∂D_D,

$$2 \int_{\partial D} \lambda |u_{z_i} + \Phi(\cdot, z_i)|^2 \, ds = -\frac{1}{4} - \mathrm{Im}(u_z(z)), \qquad (8.33)$$

where u_z is defined by (8.26)–(8.28). By Lemma 8.17, we see that the left-hand side of this equation is an injective compact integral operator with positive kernel defined on $L^2(\partial D)$. Using the Tikhonov regularization technique (cf. [68]) it is possible to determine λ by finding the regularized solution of (8.33) in $L^2(\partial D)$ (i.e., it is not necessary to know a priori the coated portion ∂D_I). Note that this integral equation has both noisy kernel and noisy right-hand side (recall from Lemma 8.15 that u_z can be approximated by v_{g_z}). For numerical examples using this approach we refer the reader to [27].

In the particular case where the surface impedance is a positive constant $\lambda > 0$, we obtain a simpler formula for λ, namely,

$$\lambda = \frac{-2k\pi|\gamma|^2 - \mathrm{Im}\,(u_z(z))}{\|u_z + \Phi(\cdot, z)\|^2_{L^2(\partial D)}}. \qquad (8.34)$$

Note that expression (8.34) can be used as a target signature to detect whether or not an obstacle is coated. In particular, an object is coated if and only if the denominator is nonzero.

8.3 Numerical Examples

We now present some numerical examples of the preceding reconstruction algorithm when the surface impedance λ is a constant. As explained previously, an approximation for λ in this case is given by

$$\frac{-2k\pi|\gamma|^2 - \mathrm{Im}\,(v_{g_z}(z))}{\|v_{g_z}(\cdot) + \Phi(\cdot, z)\|^2_{L^2(\partial D)}}, \qquad z = (z_1, z_2) \in D, \qquad (8.35)$$

where v_{g_z} is the Herglotz wave function, with kernel g_z the solution of the far-field equation

$$\int_0^{2\pi} u_\infty(\phi, \theta) g_z(\phi) d\phi = \frac{e^{i\pi/4}}{\sqrt{8\pi k}} e^{-ik(z_1 \cos\theta + z_2 \sin\theta)}. \qquad (8.36)$$

We fix the wave number $k = 3$ and select a domain D, boundaries ∂D_D, and ∂D_I (in some examples, $\partial D_D = \emptyset$), and a constant λ. Then, using the incident field $e^{ikx\cdot d}$, where $|d| = 1$, we use the finite-element method to solve the scattering problem (8.1)–(8.4) and compute the far-field pattern. This is obtained as a trigonometric series

$$u_\infty = \sum_{n=-N}^{N} u_{\infty,n} \exp(in\theta).$$

Of course, these coefficients are already in error by the discretization error from using the finite-element method. However, we also add random noise to the Fourier coefficients by setting

$$u_{\infty,a,n} = u_{\infty,n}(1 + \epsilon \chi_n),$$

where ϵ is a parameter and χ_n is given by a random number generator that provides uniformly distributed random numbers in the interval $[-1, 1]$. Thus the input to the inverse solver for computing g is the approximate far-field pattern

$$u_{\infty,a} = \sum_{n=-N}^{N} u_{\infty,a,n} \exp(in\theta).$$

The far-field equation is then solved using Tikhonov regularization and the Morozov discrepancy principle, as described in Chap. 2. In particular, using the preceding expression for $u_{\infty,a}$, the far-field equation (8.36) is rewritten as an ill-conditioned matrix equation for the Fourier coefficients of g, which we write in the form

$$Ag_z = f_z. \tag{8.37}$$

As was already noted, this equation needs to be regularized. We start by computing the singular value decomposition of A,

$$A = U\Lambda V^*,$$

where U and V are unitary and Λ is real diagonal with $\Lambda_{i,i} = \sigma_i$, $1 \le i \le n$. The solution of (8.37) is then equivalent to solving

$$\Lambda V^* g_z = U^* f_z. \tag{8.38}$$

Let

$$\rho_z = (\rho_{z,1}, \rho_{z,2}, \cdots \rho_{z,n})^\top = U^* f_z.$$

Then the Tikhonov regularization of (8.38) leads to solving

$$\min_{g_z \in \mathbb{R}^n} \|\Lambda V^* g_z - f_z\|_{\ell^2}^2 + \alpha \|g\|_{\ell^2}^2,$$

where $\alpha > 0$ is the Tikhonov regularization parameter chosen by using the Morozov discrepancy principle. Defining $u_z = V^* g_z$, we see that the solution to the problem is

$$u_{z,i} = \frac{\sigma_i}{\sigma_i^2 + \alpha} \rho_{z,i}, \qquad 1 \le i \le n,$$

and hence

$$g_z = V u_z \quad \text{and} \quad \|g_z\|_{\ell^2} = \|u_z\|_{\ell^2} = \left(\sum_{i=1}^{n} \frac{\sigma_i^2}{(\sigma_i^2 + \alpha)^2} |\rho_{z,i}|^2 \right)^{\frac{1}{2}}.$$

For the presented examples, we compute the far-field pattern for 100 incident directions and observation directions equally distributed on the unit circle and add random noise of 1 % or 10 % to the Fourier coefficients of the far-field pattern. We choose the sampling points z on a uniform grid of 101×101 points in the square region $[-5,\ 5]^2$ and compute the corresponding g_z. To visualize the obstacle, we plot the level curves of the inverse of the discrete ℓ_2 norm of g_z (note that by the linear sampling method the boundary of the obstacle is characterized as the set of points where the L^2 norm of g starts to become large; see the comments at the end of Sect. 4.3). Then we compute (8.35) at the sampling points in the disk centered at the origin with radius 0.5 (in our examples this circle is always inside D). Although (8.35) is theoretically a constant, because of the ill-posed nature of the far-field equation, we evaluated (8.35) at all the grid points z in the disk and exhibit the maximum, the average, and the median of the computed values of (8.35). In particular, the average, median, and maximum each provide a reasonable approximation to the true impedance.

For our examples we select two scatterers, shown in Fig. 8.1 (the kite and the peanut).

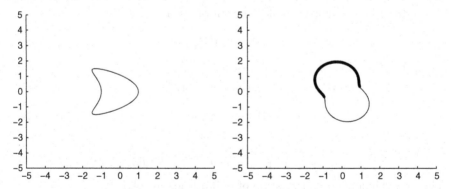

Fig. 8.1. Boundary of scatterers used in this study: kite/peanut. When a mixed condition is used for the peanut, the thicker portion of the boundary is $\partial D_D{}^2$

Kite. We consider the impedance boundary value problem for the kite described by the equation (left curve in Fig. 8.1)

$$x(t) = (1.5 \sin(t), \cos(t) + 0.65 \cos(2t) - 0.65), \qquad 0 \le t \le 2\pi,$$

with impedance $\lambda = 2$, $\lambda = 5$, and $\lambda = 9$. In Fig. 8.2 we show two examples
of the reconstructed kite (the reconstructions for the other tested cases look
similar). In the numerical results for the reconstructed λ shown in Tables 8.1
and 8.2 we use the exact boundary ∂D when we compute the $L^2(\partial D)$ norm
that appears in the denominator of (8.35).

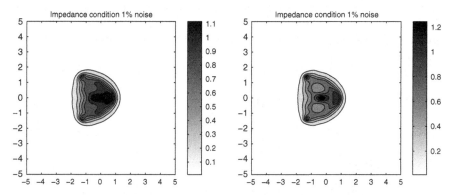

Fig. 8.2. Reconstruction of kite with impedance boundary condition with 1 % noise:
left: with $\lambda = 5$, *right*: with $\lambda = 9^2$

Table 8.1. Reconstruction of surface impedance λ for kite with 1 % noise[2]

	Maximum	Average	Median
$\lambda = 2$	2.050	1.975	1.982
$\lambda = 5$	4.976	4.679	4.787
$\lambda = 9$	8.883	8.342	8.403

Table 8.2. Reconstruction of surface impedance λ for kite with 10 % noise[2]

	Maximum	Average	Median
$\lambda = 2$	2.043	1.960	1.957
$\lambda = 5$	4.858	4.513	4.524
$\lambda = 9$	9.0328	8.013	7.992

Peanut. Next we consider a peanut described by the equation (right curve
in Fig. 8.1)

$$x(t) = \left(\sqrt{\cos^2(t) + 4\sin^2(t)} \, \cos(t), \sqrt{\cos^2(t) + 4\sin^2(t)} \, \sin(t), \ 0 \le t \le 2\pi \right)$$

rotated by $\pi/9$. Here we choose the surface impedance $\lambda = 2$ and $\lambda = 5$ and consider the case of a totally coated peanut (i.e., impedance boundary value problem) as well as of a partially coated peanut (i.e., mixed Dirichlet-impedance boundary value problem, with ∂D_I being the lower half of the peanut, as shown in Fig. 8.1). Two examples of the reconstructed peanut are presented in Fig. 8.3. A natural guess for the boundary of the scatterer is the ellipse shown by a dashed line in Fig. 8.4, and we examine the sensitivity of our formula on the approximation of the boundary using this ellipse to compute $\|v_{g_z} + \Phi(\cdot, z)\|_{L^2(\partial D)}$ in (8.35). The recovered values of λ for our experiments are shown in Tables 8.3 and 8.4.

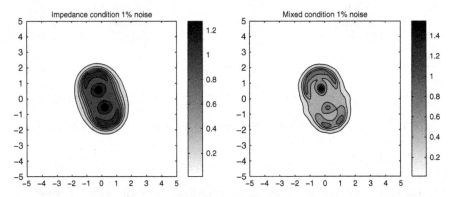

Fig. 8.3. *Left*: reconstruction of peanut with impedance boundary condition with $\lambda = 5$; *right*: reconstruction of peanut with mixed condition with $\lambda = 5$ on impedance part. Both examples are for $k = 3$ with 1 % noise[2]

Table 8.3. Reconstruction of λ for peanut with 1 % noise[2]

	Maximum	Average	Median
$\lambda = 2$ impedance	2.192	1.992	1.979
$\lambda = 2$ imped., approx. bound.	2.395	1.823	1.886
$\lambda = 2$ mixed conditions	2.595	2.207	2.257
$\lambda = 5$ impedance	5.689	4.950	5.181
$\lambda = 5$ imped., approx. bound.	5.534	4.412	4.501
$\lambda = 5$ mixed conditions	5.689	4.950	5.180

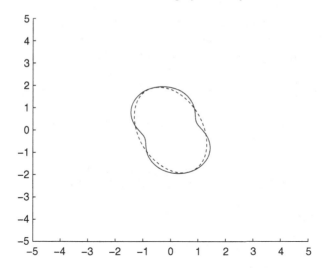

Fig. 8.4. *Dashed line*: approximated boundary used for computing $\|v_{g_z} + \Phi(\cdot; z)\|_{L^2(\partial D)}$ in (8.35) in case of peanut with impedance boundary condition[2]

Table 8.4. Reconstruction of λ for peanut with 10 % noise[2]

	Maximum	Average	Median
$\lambda = 2$ impedance	2.297	1.985	1.978
$\lambda = 2$ imped., approx. bound.	2.301	1.828	1.853
$\lambda = 2$ mixed conditions	2.681	2.335	2.374
$\lambda = 5$ impedance	5.335	4.691	4.731
$\lambda = 5$ imped., approx. bound.	5.806	4.231	4.313
$\lambda = 5$ mixed conditions	5.893	4.649	4.951

8.4 Scattering by Partially Coated Dielectric

We now consider the scattering of time-harmonic electromagnetic waves by an infinitely long, cylindrical, orthotropic dielectric partially coated with a very thin layer of a highly conductive material. Let the bounded domain $D \subset \mathbb{R}^2$ be the cross section of the cylinder, assume that the exterior domain $\mathbb{R}^2 \setminus \bar{D}$ is connected, and let ν be the unit outward normal to the smooth boundary ∂D. The boundary $\partial D = \overline{\partial D_1} \cap \overline{\partial D_2}$ is split into two parts, ∂D_1 and ∂D_2, each an open set relative to ∂D and possibly disconnected. The open arc ∂D_1 corresponds to the uncoated part, and ∂D_2 corresponds to the coated part. We assume that the incident electromagnetic field and the constitutive parameters are as described in Sect. 5.1. In particular, the fields inside D and outside D satisfy (5.5) and (5.6), respectively, and on ∂D_1, the uncoated portion of the boundary, we have the transmission condition (5.7). However, on the coated portion of the cylinder, we have the conductive boundary condition given by

$$\nu \times E^{ext} - \nu \times E^{int} = 0 \quad \text{and} \quad \nu \times H^{ext} - \nu \times H^{int} = \eta(\nu \times E^{ext}) \times \nu, \quad (8.39)$$

where the *surface conductivity* $\eta = \eta(x)$ describes the physical properties of the thin, highly conductive coating [3, 4]. Assuming that η does not depend on the z-coordinate (we recall that the cylinder axis is assumed to be parallel to the z-direction), on ∂D_2 the transmission conditions (8.39) now become

$$v - (u^s + u^i) = -i\eta \frac{\partial}{\partial \nu}(u^s + u^i) \quad \text{and} \quad \frac{\partial v}{\partial \nu_A} - \frac{\partial}{\partial \nu}(u^s + u^i) = 0 \qquad \text{on} \quad \partial D_2,$$

where $\partial v / \partial \nu_A := \nu \cdot A(x) \nabla v$.

The direct scattering problem for a partially coated dielectric can now be formulated as follows: assume that A, n, and D satisfy the assumptions of Sect. 5.1 and $\eta \in C(\overline{\partial D_2})$ satisfies $\eta(x) \geq \eta_0 > 0$ for all $x \in \partial D_2$. Given the incident field u^i satisfying

$$\Delta u^i + k^2 u^i = 0 \qquad \text{in } \mathbb{R}^2,$$

we look for $u^s \in H^1_{loc}(\mathbb{R}^2 \setminus \bar{D})$ and $v \in H^1(D)$ such that

$$\nabla \cdot A \nabla v + k^2 n\, v = 0 \qquad \text{in} \quad D, \tag{8.40}$$

$$\Delta u^s + k^2\, u^s = 0 \qquad \text{in} \quad \mathbb{R}^2 \setminus \bar{D}, \tag{8.41}$$

$$v - u^s = u^i \qquad \text{on} \quad \partial D_1, \tag{8.42}$$

$$v - u^s = -i\eta \frac{\partial(u^s + u^i)}{\partial \nu} + u^i \qquad \text{on} \quad \partial D_2, \tag{8.43}$$

$$\frac{\partial v}{\partial \nu_A} - \frac{\partial u^s}{\partial \nu} = \frac{\partial u^i}{\partial \nu} \qquad \text{on} \quad \partial D, \tag{8.44}$$

$$\lim_{r \to \infty} \sqrt{r}\left(\frac{\partial u^s}{\partial r} - iku^s\right) = 0. \tag{8.45}$$

We start with a brief discussion of the well-posedness of the foregoing scattering problem.

Theorem 8.18. *The problem (8.40)–(8.45) has at most one solution.*

Proof. Let $v \in H^1(D)$ and $u^s \in H^1_{loc}(D_e)$ be the solution of (8.40)–(8.45) corresponding to the incident wave $u^i = 0$. Applying Green's first identity in D and $(\mathbb{R}^2 \setminus \bar{D}) \cap \Omega_R$, where (and in what follows) Ω_R is a disk of radius R centered at the origin and containing $\bar{D}$, and using the transmission conditions we have that

$$\int_D \left(\nabla \bar{v} \cdot A \nabla v - k^2 n |v|^2\right) dy + \int_{\Omega_R \setminus \bar{D}} \left(|\nabla u^s|^2 - k^2 |u^s|^2\right) dy$$

$$= \int_{\partial D} \bar{v} \cdot \frac{\partial v}{\partial \nu_A}\, ds - \int_{\partial D} \overline{u^s} \cdot \frac{\partial u^s}{\partial \nu}\, ds + \int_{\partial \Omega_R} \overline{u^s} \cdot \frac{\partial u^s}{\partial \nu}\, ds$$

$$= i \int_{\partial D_2} \frac{1}{\eta}|v - u^s|^2\, ds + \int_{\partial \Omega_R} \overline{u^s} \cdot \frac{\partial u^s}{\partial \nu}\, ds.$$

Taking the imaginary part of both sides and using the fact that $\mathrm{Im}(A) \leq 0$, $\mathrm{Im}(n) \geq 0$, and $\eta \geq \eta_0 > 0$ we obtain

$$\mathrm{Im} \int_{\partial \Omega_R} u^s \cdot \frac{\partial \overline{u^s}}{\partial \nu} \, ds \geq 0.$$

Finally, an application of Theorem 3.6 and the unique continuation principle yield, as the proof in Lemma 5.25, $u^s = v = 0$. □

We now rewrite the scattering problem in a variational form. Multiplying the equations in (8.40)–(8.45) by a test function φ and using Green's first identity, together with the transmission conditions, we obtain that the total field w defined in Ω_R by $w|_D := v$ and $w|_{\Omega_R \setminus \bar{D}} = u^s + u^i$ satisfies

$$\int_D (\nabla \overline{\varphi} \cdot A \nabla w - k^2 n \overline{\varphi}\, w)\, dy + \int_{\Omega_R \setminus \bar{D}} (\nabla \overline{\varphi} \cdot \nabla w - k^2 \overline{\varphi}\, w)\, dy \qquad (8.46)$$

$$- \int_{\partial D_2} \frac{i}{\eta} [\overline{\varphi}] \cdot [w]\, ds - \int_{\partial \Omega_R} \overline{\varphi} T w \, ds = - \int_{\partial \Omega_R} \overline{\varphi} T u^i \, ds + \int_{\partial \Omega_R} \overline{\varphi} \frac{\partial u^i}{\partial \nu}\, ds,$$

where $T : H^{\frac{1}{2}}(\partial \Omega_R) \to H^{\frac{1}{2}}(\partial \Omega_R)$ is the Dirichlet-to-Neumann operator and $[w] = w^+|_{\partial D} - w^-|_{\partial D}$ denotes the jump of w across ∂D, with w^+ and w^- the traces (in the sense of the trace operator) of $w \in H^1(\Omega_R \setminus \bar{D})$ and $w \in H^1(D)$, respectively. Note that $[w] \in \tilde{H}^{\frac{1}{2}}(\partial D_2)$ since from the transmission conditions $[w]|_{\partial D_1} = 0$.

Hence, the natural variational space for w and φ is $H^1(\Omega_R \setminus \overline{\partial D_2})$. Note that if $u \in H^1(\Omega_R \setminus \overline{\partial D_2})$, then $u \in H^1(D)$, $u \in H^1(\Omega_R \setminus \bar{D})$, $[u]|_{\partial D_1} = 0$, and

$$\|u\|^2_{H^1(\Omega_R \setminus \overline{\partial D_2})} = \|u\|^2_{H^1(D)} + \|u\|^2_{H^1(\Omega_R \setminus \bar{D})}.$$

Now, letting

$$a_1(w, \varphi) := \int_D (\nabla \overline{\varphi} \cdot A \nabla w + \overline{\varphi}\, w)\, dy + \int_{\Omega_R \setminus \bar{D}} (\nabla \overline{\varphi} \cdot \nabla w + \overline{\varphi}\, w)\, dy$$

$$- \int_{\partial D_2} \frac{i}{\eta} [\overline{\varphi}] \cdot [w]\, ds - \int_{\partial \Omega_R} \overline{\varphi} T_0 w\, ds \qquad (8.47)$$

and

$$a_2(w, \varphi) := - \int_{\Omega_R} (n k^2 + 1) \overline{\varphi}\, w\, dy - \int_{\partial \Omega_R} \overline{\varphi}(T_0 - T) w\, ds,$$

where T_0 is the negative definite part of the Dirichlet-to-Neumann mapping defined in Theorem 5.22, the variational formulation of the mixed transmission problem reads: find $w \in H^1(\Omega_R \setminus \overline{\partial D_2})$ such that

$$a_1(w, \varphi) + a_2(w, \varphi) = L(\varphi) \qquad \forall \varphi \in H^1(\Omega_R \setminus \overline{\partial D_2}), \qquad (8.48)$$

where $L(\varphi)$ denotes the bounded conjugate linear functional defined by the right-hand side of (8.46). We leave it as an exercise to the reader to prove that if $w \in H^1(\Omega_R \setminus \overline{\partial D_2})$ solves (8.48), then $v := w|_D$ and $u^s = w|_{\Omega_R \setminus \bar{D}} - u^i$ satisfy (8.40), (8.41) in $\Omega_R \setminus \bar{D}$, the boundary conditions (8.42), (8.43), and (8.44), and $Tu^s = \partial u^s / \partial \nu$ on $\partial \Omega_R$. Exactly in the same way as in Example 5.23 one can show that u^s can be uniquely extended to a solution in $\mathbb{R}^2 \setminus \bar{D}$.

Now using the trace theorem, the Cauchy–Schwarz inequality, the chain of continuous embeddings

$$\tilde{H}^{\frac{1}{2}}(\partial D_2) \subset H^{\frac{1}{2}}(\partial D_2) \subset L^2(\partial D_2) \subset \tilde{H}^{-\frac{1}{2}}(\partial D_2) \subset H^{-\frac{1}{2}}(\partial D_2),$$

and the assumptions on A, n, and η, one can now show in a similar way as in Sect. 5.4 that the sesquilinear form $a_1(\cdot, \cdot)$ is bounded and strictly coercive and the sesquilinear form $a_2(\cdot, \cdot)$ is bounded and gives rise to a compact linear operator due to the compact embedding of $H^1(\Omega_R \setminus \overline{\partial D_2})$ in $L^2(\Omega_R)$. Hence, using the Lax–Milgram lemma and Theorem 5.16, the foregoing analysis, combined with Theorem 8.18, implies the following result.

Theorem 8.19. *The problem (8.40)–(8.45) has exactly one solution $v \in H^1(D)$ and $u^s \in H^1_{loc}(\mathbb{R}^2 \setminus \bar{D})$ that satisfies*

$$\|v\|_{H^1(D)} + \|u^s\|_{H^1(\Omega_R \setminus \overline{D})} \leq C \|u^i\|_{H^1(\Omega_R)},$$

where the positive constant $C > 0$ is independent of u^i but depends on R.

The scattered field u^s again has the asymptotic behavior

$$u^s(x) = \frac{e^{ikr}}{\sqrt{r}} u_\infty(\theta) + O(r^{-3/2}), \qquad r \to \infty,$$

where the corresponding far-field pattern $u_\infty(\cdot)$ depends on the observation direction $\hat{x} := (\cos\theta, \sin\theta)$. In the case of incident plane waves $u^i(x) = e^{ikx \cdot d}$, the interior field v and the scattered field u^s also depend on the incident direction $d := (\cos\phi, \sin\phi)$, as does the corresponding far field pattern $u_\infty(\cdot) := u_\infty(\cdot, \phi)$. The far-field pattern in turn defines the corresponding far-field operator $F : L^2[0, 2\pi] \to L^2[0, 2\pi]$ by (6.7).

As will be seen, the *mixed interior transmission problem* associated with the mixed transmission problem (8.40)–(8.45) plays an important role in studying the far-field operator. Hence, we now proceed to a discussion of this problem. Consider the Sobolev space

$$\mathbb{H}^1(D, \partial D_2) := \left\{ u \in H^1(D) \text{ such that } \frac{\partial u}{\partial \nu} \in L^2(\partial D_2) \right\}$$

equipped with the graph norm

$$\|u\|^2_{\mathbb{H}^1(D,\partial D_2)} := \|u\|^2_{H^1(D)} + \left\|\frac{\partial u}{\partial \nu}\right\|^2_{L^2(\partial D_2)}.$$

Then the *mixed interior transmission problem* corresponding to the mixed transmission problem (8.40)–(8.45) reads: given $f \in H^{\frac{1}{2}}(\partial D)$, $h \in H^{-\frac{1}{2}}(\partial D)$, and $r \in L^2(\partial D_2)$, find $v \in H^1(D)$ and $w \in \mathbb{H}^1(D,\partial D_2)$ such that

$$\nabla \cdot A\nabla v + k^2 n \, v = 0 \qquad\qquad \text{in} \quad D, \qquad\qquad (8.49)$$

$$\Delta w + k^2 \, w = 0 \qquad\qquad \text{in} \quad D, \qquad\qquad (8.50)$$

$$v - w = f|_{\partial D_1} \qquad\qquad \text{on} \quad \partial D_1, \qquad\qquad (8.51)$$

$$v - w = -i\eta \frac{\partial w}{\partial \nu} + f|_{\partial D_2} + r \qquad \text{on} \quad \partial D_2, \qquad\qquad (8.52)$$

$$\frac{\partial v}{\partial \nu_A} - \frac{\partial w}{\partial \nu} = h \qquad\qquad \text{on} \quad \partial D. \qquad\qquad (8.53)$$

Theorem 8.20. *If either $Im(n) > 0$ or $Im\left(\bar{\xi} \cdot A\xi\right) < 0$ at a point $x_0 \in D$, then the mixed interior transmission problem (8.49)–(8.53) has at most one solution.*

Proof. Let v and w be a solution of the homogeneous mixed interior transmission problem (i.e., $f = h = r = 0$). Applying the divergence theorem to $\bar{v}$ and $A\nabla v$ (Corollary 5.8), using the boundary condition, and applying Green's first identity to $\bar{w}$ and w (Remark 6.29) we obtain

$$\int_D \nabla\bar{v} \cdot A\nabla v \, dy - \int_D k^2 n |v|^2 \, dy = \int_D |\nabla w|^2 \, dy - \int_D k^2 |w|^2 \, dy + \int_{\partial D_2} i\eta \left|\frac{\partial w}{\partial \nu}\right|^2 ds.$$

Hence

$$Im\left(\int_D \nabla\bar{v} \cdot A\nabla v \, dy\right) = 0, \quad Im\left(\int_D n|v|^2 \, dy\right) = 0, \quad \text{and} \int_{\partial D_2} \eta \left|\frac{\partial w}{\partial \nu}\right|^2 ds = 0.$$

The last equation implies that $\partial w/\partial \nu = 0$ on ∂D_2, whence w and v satisfy the homogeneous interior transmission problem (6.12)–(6.15). The result of the theorem now follows from Theorem 6.4. $\qquad\square$

The values of k for which the homogeneous mixed interior transmission problem (8.49)–(8.53) has a nontrivial solution are called transmission eigenvalues. From the proof of Theorem 8.20 we have the following result.

Corollary 8.21. *The transmission eigenvalues corresponding to (8.49)–(8.53) form a subset of the transmission eigenvalues corresponding to (6.12)–(6.15) defined in Definition 6.3.*

The preceding corollary justifies the use of the same name for the set of eigenvalues corresponding to both the interior transmission problem and the mixed interior transmission problem. We note that due to the presence of a non-real-valued term in the transmission conditions, the approaches developed in Chap. 6 to prove the existence of transmission eigenvalues cannot be used in the current case. The existence of transmission eigenvalues corresponding to (8.49)–(8.53) is to date an open problem.

From the proof of Theorem 8.20 we also see that if the scatterer is fully coated, i.e., $\partial D_2 = \partial D$, then the solution (v, w) of the homogeneous mixed interior transmission problem satisfies

$$\nabla \cdot A\nabla v + k^2 n v = 0 \text{ in } D, \quad \frac{\partial v}{\partial \nu_A} = 0 \text{ on } \partial D,$$

and

$$\Delta w + k^2 w = 0 \text{ in } D, \quad \frac{\partial w}{\partial \nu} = 0 \text{ on } \partial D.$$

From this it follows that if $\partial D_2 = \partial D$, then the uniqueness of the mixed interior transmission problem is guaranteed if at least one of the foregoing homogeneous Neumann problems has only a trivial solution.

The following important result can be shown in the same way as in Theorem 6.2.

Theorem 8.22. *The far-field operator F corresponding to the scattering problem (8.40)–(8.45) is injective with dense range if and only if there does not exist a Herglotz wave function v_g such that the pair v, v_g is a solution to the homogeneous mixed interior transmission problem (8.49)–(8.53) with $w = v_g$.*

We shall now discuss the solvability of the mixed interior transmission problem (8.49)–(8.53). We will adapt the variational approach used in Sect. 6.2 to solve (6.12)–(6.15). To avoid repetition, we will only sketch the proof, emphasizing the changes due to the boundary terms involving η.

Theorem 8.23. *Assume that k is not a transmission eigenvalue and that there exists a constant $\gamma > 1$ such that*

$$\text{either} \quad \bar{\xi} \cdot Re(A)\,\xi \geq \gamma |\xi|^2 \quad \text{or} \quad \bar{\xi} \cdot Re(A^{-1})\,\xi \geq \gamma |\xi|^2 \quad \forall \xi \in \mathbb{C}^2.$$

Then the mixed interior transmission problem (8.49)–(8.53) has a unique solution (v, w) that satisfies

$$\|v\|^2_{H^1(D)} + \|w\|^2_{\mathbb{H}^1(D,\partial D_2)} \leq C \left(\|f\|_{H^{\frac{1}{2}}(\partial D)} + \|h\|_{H^{-\frac{1}{2}}(\partial D)} + \|r\|_{L^2(\partial D_2)} \right).$$

Proof. We first assume that $\bar{\xi} \cdot Re(A)\,\xi \geq \gamma |\xi|^2$ for some $\gamma > 1$. In the same way as in the proof of Theorem 6.8, we can show that (8.49)–(8.53) is a compact perturbation of the modified mixed interior transmission problem

$$\nabla \cdot A\nabla v - mv = \rho_1 \qquad \text{in} \quad D, \tag{8.54}$$

$$\Delta w - w = \rho_2 \qquad \text{in} \quad D, \tag{8.55}$$

$$v - w = f|_{\partial D_1} \qquad \text{on} \quad \partial D_1, \tag{8.56}$$

$$v - w = -i\eta \frac{\partial w}{\partial \nu} + f|_{\partial D_2} + r \qquad \text{on} \quad \partial D_2, \tag{8.57}$$

$$\frac{\partial v}{\partial \nu_A} - \frac{\partial w}{\partial \nu} = h \qquad \text{on} \quad \partial D, \tag{8.58}$$

where $m \in C(\overline{D})$ such that $m(x) \geq \gamma$. It is now sufficient to study (8.54)–(8.58) since the result of the theorem will then follow by an application of Theorem 5.16 and the fact that k is not a transmission eigenvalue. We first reformulate (8.54)–(8.58) as an equivalent variational problem. To this end, let

$$W(D) := \left\{ \mathbf{w} \in \left(L^2(D) \right)^2 : \nabla \cdot \mathbf{w} \in L^2(D), \ \nabla \times \mathbf{w} = 0, \text{ and } \nu \cdot \mathbf{w} \in L^2(\partial D_2) \right\}$$

equipped with the natural inner product

$$(\mathbf{w}_1, \mathbf{w}_2)_W = (\mathbf{w}_1, \mathbf{w}_2)_{L^2(D)} + (\nabla \cdot \mathbf{w}_1, \nabla \cdot \mathbf{w}_2)_{L^2(D)} + (\nu \cdot \mathbf{w}_1, \nu \cdot \mathbf{w}_2)_{L^2(\partial D_2)}$$

and norm

$$\|\mathbf{w}\|_W^2 = \|\mathbf{w}\|_{L^2(D)}^2 + \|\nabla \cdot \mathbf{w}\|_{L^2(D)}^2 + \|\nu \cdot \mathbf{w}\|_{L^2(\partial D_2)}^2. \tag{8.59}$$

We denote by $\langle \cdot, \cdot \rangle$ the duality pairing between $H^{\frac{1}{2}}(\partial D)$ and $H^{-\frac{1}{2}}(\partial D)$ and recall

$$\langle \varphi, \boldsymbol{\psi} \cdot \nu \rangle = \int_D \varphi \, \nabla \cdot \boldsymbol{\psi} \, dx + \int_D \nabla \varphi \cdot \boldsymbol{\psi} \, dx \tag{8.60}$$

for $(\varphi, \boldsymbol{\psi}) \in H^1(D) \times W(D)$. Then the variational form of (8.54)–(8.58) is as follows: find $U = (v, \mathbf{w}) \in H^1(D) \times W(D)$ such that

$$\mathcal{A}(U, V) = L(V) \qquad \text{for all } V := (\varphi, \boldsymbol{\psi}) \in H^1(D) \times W(D), \tag{8.61}$$

where the sesquilinear form $\mathcal{A}$ defined on $(H^1(D) \times W(D))^2$ is given by

$$\mathcal{A}(U, V) = \int_D A\nabla v \cdot \nabla \bar{\varphi} \, dx + \int_D m \, v \, \bar{\varphi} \, dx + \int_D \nabla \cdot \mathbf{w} \, \nabla \cdot \bar{\boldsymbol{\psi}} \, dx + \int_D \mathbf{w} \cdot \bar{\boldsymbol{\psi}} \, dx$$

$$- i \int_{\partial D_2} \eta \, (\mathbf{w} \cdot \nu) \, (\bar{\boldsymbol{\psi}} \cdot \nu) ds - \langle v, \bar{\boldsymbol{\psi}} \cdot \nu \rangle - \langle \bar{\varphi}, \mathbf{w} \cdot \nu \rangle$$

and the conjugate linear functional L is given by

$$L(V) = \int_D (\rho_1 \, \bar{\varphi} + \rho_2 \, \nabla \cdot \bar{\boldsymbol{\psi}}) \, dx - i \int_{\partial D_2} \eta \, r \, (\bar{\boldsymbol{\psi}} \cdot \nu) \, ds + \langle \bar{\varphi}, h \rangle - \langle f, \bar{\boldsymbol{\psi}} \cdot \nu \rangle.$$

By proceeding exactly as in the proof of Theorem 6.5 we can establish the equivalence between (8.54)–(8.58) and (8.61). In particular, if (v, w) is the unique solution (8.54)–(8.58), then $U = (v, \nabla w)$ is a unique solution to (8.61). Conversely, if U is the unique solution to (8.61), then the unique solution (v, w) to (8.54)–(8.58) is such that $U = (v, \nabla w)$.

Notice that the definitions of $\mathcal{A}$ and L differ from Definitions (6.22) and (6.23) of $\mathcal{A}$ and L corresponding to (6.12)–(6.15) only by an additional $L^2(\partial D_2)$ inner product term, which appears in the W norm given by (8.59). Using the trace theorem and Schwarz's inequality one can show that $\mathcal{A}$ and L are bounded in the respective norms. On the other hand, by taking the real and imaginary parts of $\mathcal{A}(U, U)$, we have from the assumptions on $\mathrm{Re}(A)$, $\mathrm{Im}(A)$, and η that

$$|\mathcal{A}(U,U)| \geq \gamma \|v\|^2_{H^1(D)} + \|\mathbf{w}\|^2_{L^2(D)} + \|\nabla \cdot \mathbf{w}\|^2_{L^2(D)}$$
$$- 2\mathrm{Re}(\langle \bar{v}, \, \nu \cdot \mathbf{w} \rangle) + \eta_0 \|\nu \cdot \mathbf{w}\|^2_{L^2(\partial D_2)}.$$

From the duality pairing (8.60) and Schwarz's inequality we have that

$$2\mathrm{Re}(\langle \bar{v}, \, \nu \cdot \mathbf{w} \rangle) \leq |\langle \bar{v}, \, \mathbf{w} \rangle| \leq \|v\|_{H^1(D)} \left(\|\mathbf{w}\|^2_{L^2(D)} + \|\nabla \cdot \mathbf{w}\|^2_{L^2(D)} \right)^{\frac{1}{2}}.$$

Hence, since $\gamma > 1$, we conclude that

$$|\mathcal{A}(U,U)| \geq \frac{\gamma - 1}{\gamma + 1} \left(\|v\|^2_{H^1(D)} + \|\mathbf{w}\|^2_{L^2(D)} + \|\nabla \cdot \mathbf{w}\|^2_{L^2(D)} \right) + \eta_0 \|\nu \cdot \mathbf{w}\|^2_{L^2(\partial D_2)},$$

which means that $\mathcal{A}$ is coercive, i.e.,

$$|\mathcal{A}(U,U)| \geq C \left(\|v\|^2_{H^1(D)} + \|\mathbf{w}\|^2_{W(D)} \right),$$

where $C = \min((\gamma - 1)/(\gamma + 1), \eta_0)$. Therefore, from the Lax–Milgram lemma we have that the variational problem (8.61) is uniquely solvable, and, hence, so is the modified interior transmission problem (8.54)–(8.58). Finally, the uniqueness of a solution to the mixed interior transmission problem and an application of Theorem 5.16 imply that (8.49)–(8.53) has a unique solution (v, w) that satisfies

$$\|v\|_{H^1(D)} + \|w\|_{\mathbb{H}^1(D, \partial D_2)} \leq C \left(\|f\|_{H^{\frac{1}{2}}(\partial D)} + \|h\|_{H^{-\frac{1}{2}}(\partial D)} + \|r\|_{L^2(\partial D_2)} \right),$$

where $C > 0$ is independent of f, h, r. The case of $\bar{\xi} \cdot \mathcal{R}e(A^{-1}) \xi$ can be treated in a similar way.   □

Another main ingredient that we need to solve the inverse scattering problem for partially coated penetrable obstacles is an approximation property of Herglotz wave functions. In particular, we need to show that if (v, w) is the solution of the mixed interior transmission problem, then w can be approximated by a Herglotz wave function with respect to the $\mathbb{H}^1(D, \partial D_2)$ norm [which is a stronger norm than the $H^1(D)$ used in Lemma 6.45].

Theorem 8.24. *Assume that k is not a transmission eigenvalue, and let (w, v) be the solution of the mixed interior transmission problem (8.49)–(8.53). Then for every $\epsilon > 0$ there exists a Herglotz wave function v_{g_ϵ} with kernel $g_\epsilon \in L^2[0, 2\pi]$ such that*

$$\|w - v_{g_\epsilon}\|_{\mathbb{H}^1(D, \partial D_2)} \leq \epsilon. \tag{8.62}$$

Proof. We proceed in two steps:

1. We first show that the operator $H : L^2[0, 2\pi] \to H^{\frac{1}{2}}(\partial D_1) \times L^2(\partial D_2)$ defined by

$$(Hg)(x) := \begin{cases} v_g(x), & x \in \partial D_1, \\ \dfrac{\partial v_g(x)}{\partial \nu} + i v_g(x), & x \in \partial D_2, \end{cases}$$

has a dense range, where v_g is a Herglotz wave function written in the form

$$v_g(x) = \int_0^{2\pi} e^{-ik(x_1 \cos \theta + x_2 \sin \theta)} g(\theta) ds(\theta), \qquad x = (x_1, x_2).$$

To this end, according to Lemma 6.42, it suffices to show that the corresponding transpose operator $H^\top : \tilde{H}^{-\frac{1}{2}}(\partial D_1) \times L^2(\partial D_2) \to L^2[0, 2\pi]$ defined by

$$\langle Hg, \phi \rangle_{H^{\frac{1}{2}}(\partial D_1), \tilde{H}^{-\frac{1}{2}}(\partial D_1)} + \langle Hg, \psi \rangle_{L^2(\partial D_2), L^2(\partial D_2)}$$
$$= \langle g, H^\top(\phi, \psi) \rangle_{L^2[0, 2\pi], L^2[0, 2\pi]},$$

for $g \in L^2[0, 2\pi]$, $\phi \in \tilde{H}^{-\frac{1}{2}}(\partial D_1)$, $\psi \in L^2(\partial D_2)$, is injective, where $\langle \cdot, \cdot \rangle$ denotes the duality pairing between the denoted spaces. By interchanging the order of integration one can show that

$$H^\top(\phi, \psi)(\hat{x}) = \int_{\partial D} e^{-iky \cdot \hat{x}} \tilde{\phi}(y) \, ds(y) + \int_{\partial D} \frac{\partial e^{-iky \cdot \hat{x}}}{\partial \nu} \tilde{\psi}(y) \, ds(y)$$
$$+ i \int_{\partial D} e^{-iky \cdot \hat{x}} \tilde{\psi}(y) \, ds(y),$$

where $\tilde{\phi} \in H^{-\frac{1}{2}}(\partial D)$ and $\tilde{\psi} \in L^2(\partial D)$ are the extension by zero to the whole boundary ∂D of ϕ and ψ, respectively. Note that from the definition of $\tilde{H}^{-\frac{1}{2}}(\partial D_1)$ in Sect. 8.1 such an extension exists.

Assume now that $H^\top(\phi, \psi) = 0$. Since $H^\top(\phi, \psi)$ is, up to a constant factor, the far-field pattern of the potential

$$P(x) = \int_{\partial D} \Phi(x, y) \tilde{\phi}(y)\, ds(y) + \int_{\partial D} \frac{\partial \Phi(x, y)}{\partial \nu} \tilde{\psi}(y)\, ds(y)$$

$$+\, i \int_{\partial D} \Phi(x, y) \tilde{\psi}(y)\, ds(y),$$

which satisfies the Helmholtz equation in $\mathbb{R}^2 \setminus \bar{D}$, from Rellich's lemma we have that $P(x) = 0$ in $\mathbb{R}^2 \setminus \bar{D}$. As $x \to \partial D$ the following jump relations hold:

$$P^+ - P^-|_{\partial D_1} = 0, \qquad\qquad P^+ - P^-|_{\partial D_2} = \psi$$

$$\frac{\partial P^+}{\partial \nu} - \frac{\partial P^-}{\partial \nu}\bigg|_{\partial D_1} = -\phi, \qquad \frac{\partial P^+}{\partial \nu} - \frac{\partial P^-}{\partial \nu}\bigg|_{\partial D_2} = -i\psi,$$

where by the superscript $+$ and $-$ we distinguish the limit obtained by approaching the boundary ∂D from $\mathbb{R}^2 \setminus \bar{D}$ and D, respectively (see [54], p. 45, for the jump relations of potentials with L^2 densities, and [127] for the jump relations of the single layer potential with $H^{-\frac{1}{2}}$ density). Using the fact that $P^+ = \partial P^+/\partial \nu = 0$ we see that P satisfies the Helmholtz equation and

$$P^-|_{\partial D_1} = 0 \qquad\qquad \frac{\partial P^-}{\partial \nu} + iP^-\bigg|_{\partial D_2} = 0,$$

where the equalities are understood in the L^2 limit sense. Using Green's first identity and a parallel surface argument one can conclude, as in Theorem 8.2, that $P = 0$ in D, whence from the preceding jump relations $\phi = \psi = 0$.

2. Next, we take $w \in \mathbb{H}^1(D, \partial D_2)$, which satisfies the Helmholtz equation in D. By considering w as the solution of (8.10)–(8.12) with $f := w|_{\partial D_1} \in H^{\frac{1}{2}}(\partial D_1)$, $h := \partial w/\partial \nu + iw|_{\partial D_2} \in L^2(\partial D_2) \subset H^{-\frac{1}{2}}(\partial D_2)$, $\lambda = 1$, $\partial D_D = \partial D_1$, and $\partial D_I = \partial D_2$, the a priori estimate (8.15) yields

$$\|w\|_{H^1(D)} + \left\|\frac{\partial w}{\partial \nu}\right\|_{L^2(\partial D_2)} \leq C\|w\|_{H^{\frac{1}{2}}(\partial D_1)} + C\left\|\frac{\partial w}{\partial \nu} + iw\right\|_{L^2(\partial D_2)}.$$

Since v_g also satisfies the Helmholtz equation in D, we can write

$$\|w - v_g\|_{\mathbb{H}^1(D, \partial D_2)} \leq C\|w - v_g\|_{H^{\frac{1}{2}}(\partial D_1)} \tag{8.63}$$

$$+\, C\left\|\frac{\partial(w - v_g)}{\partial \nu} + i(w - v_g)\right\|_{L^2(\partial D_2)}.$$

From the first part of the proof, given ϵ, we can now find $g_\epsilon \in L^2[0, 2\pi]$ that makes the right-hand side of the inequality (8.63) less than ϵ. The theorem is now proved.

$\square$

8.5 Inverse Scattering Problem for Partially Coated Dielectric

The main goal of this section is the solution of the *inverse scattering problem* for partially coated dielectrics, which is formulated as follows: determine *both* D and η from a knowledge of the far-field pattern $u_\infty(\theta, \phi)$ for $\theta, \phi \in [0, 2\pi]$. As shown in Sect. 4.5, it suffices to know the far-field pattern corresponding to $\theta \in [\theta_0, \theta_1] \subset [0, 2\pi]$ and $\phi \in [\phi_0, \phi_1] \subset [0, 2\pi]$. We begin with a uniqueness theorem.

Theorem 8.25. *Let the domains D^1 and D^2 with the boundaries ∂D^1 and ∂D^2, respectively, the matrix-valued functions A_1 and A_2, the functions n_1 and n_2, and the functions η_1 and η_2 determined on the portions $\partial D_2^1 \subseteq \partial D^1$ and $\partial D_2^2 \subseteq \partial D^2$, respectively (either ∂D_2^1 or ∂D_2^2, or both, can be empty sets), satisfy the assumptions of (8.40)–(8.45). Assume that either $\bar{\xi} \cdot Re(A_1)\, \xi \geq \gamma|\xi|^2$ or $\bar{\xi} \cdot Re(A_1^{-1})\, \xi \geq \gamma|\xi|^2$, and either $\bar{\xi} \cdot Re(A_2)\, \xi \geq \gamma|\xi|^2$ or $\bar{\xi} \cdot Re(A_2^{-1})\, \xi \geq \gamma|\xi|^2$ for some $\gamma > 1$. If the far-field patterns $u_\infty^1(\theta, \phi)$ corresponding to D^1, A_1, n_1, η_1 and $u_\infty^2(\theta, \phi)$ corresponding to D^2, A_2, n_2, η_2 coincide for all $\theta, \phi \in [0, 2\pi]$, then $D^1 = D^2$.*

Proof. The proof follows the lines of the uniqueness proof for the inverse scattering problem for an orthotropic medium given in Theorem 6.39. The main two ingredients are the well-posedness of the forward problem established in Theorem 8.19 and the well-posedness of the modified mixed interior transmission problem established in Theorem 8.23. Only minor changes are needed in the proof to account for the space $\mathbb{H}^1(D, \partial D_2) \times H^1(D)$, where the solution of the mixed interior transmission problem exists and replaces $H^1(D) \times H^1(D)$ in the proof of Theorem 6.39. To avoid repetition, we do not present here the technical details. The proof of this theorem for the case of Maxwell's equations in $\mathbb{R}^3$ can be found in [13]. □

The next question to ask concerns the unique determination of the surface conductivity η. From the preceding theorem we can now assume that D is known. Furthermore, we require that for an arbitrary choice of ∂D_2, A, and η there exists at least one incident plane wave such that the corresponding total field u satisfies $\partial u/\partial \nu|_{\partial D_0} \neq 0$, where $\partial D_0 \subset \partial D$ is an arbitrary portion of ∂D. In the context of our application, this is a reasonable assumption since otherwise the portion of the boundary where $\partial u/\partial \nu = 0$ for all incident plane waves would behave like a perfect conductor, contrary to the assumption that the metallic coating is thin enough for the incident field to penetrate into D. We say that k^2 is a Neumann eigenvalue if the homogeneous problem

$$\nabla \cdot A\nabla V + k^2 n\, V = 0 \quad \text{in } D, \qquad \frac{\partial V}{\partial \nu_A} = 0 \quad \text{on } \partial D \qquad (8.64)$$

has a nontrivial solution. In particular, it is easy to show (the reader can try it as an exercise) that if $\text{Im}(A) < 0$ or $\text{Im}(n) > 0$ at a point $x_0 \in D$, then there are no Neumann eigenvalues. The reader can also show as in Example 5.17 that if $\operatorname{Im}(A) = 0$ and $\operatorname{Im}(n) = 0$, then the Neumann eigenvalues exist and form a discrete set.

We can now prove the following uniqueness result for η.

Theorem 8.26. *Assume that k^2 is not a Neumann eigenvalue. Then under the foregoing assumptions and for fixed D and A the surface conductivity η is uniquely determined from the far-field pattern $u_\infty(\theta, \phi)$ for $\theta, \phi \in [0, 2\pi]$.*

Proof. Let D and A be fixed, and suppose there exists $\eta_1 \in C(\overline{\partial D_2^1})$ and $\eta_2 \in C(\overline{\partial D_2^2})$ such that the corresponding scattered fields $u^{s,1}$ and $u^{s,2}$, respectively, have the same far-field patterns $u_\infty^1(\theta, \phi) = u_\infty^2(\theta, \phi)$ for all $\theta, \phi \in [0, 2\pi]$. Then from Rellich's lemma $u^{s,1} = u^{s,2}$ in $\mathbb{R}^2 \setminus \bar{D}$. Hence, from the transmission condition the difference $V = v^1 - v^2$ satisfies

$$\nabla \cdot A \nabla V + k^2 n V = 0 \qquad \text{in} \quad D, \tag{8.65}$$

$$\frac{\partial V}{\partial \nu_A} = 0 \qquad \text{on} \quad \partial D, \tag{8.66}$$

$$V = -i(\tilde{\eta}_1 - \tilde{\eta}_2)\frac{\partial u^1}{\partial \nu} \qquad \text{on} \quad \partial D, \tag{8.67}$$

where $\tilde{\eta}_1$ and $\tilde{\eta}_2$ are the extension by zero of η_1 and η_2, respectively, to the whole of ∂D and $u^1 = u^{s,1} + u^i$. Since k^2 is not a Neumann eigenvalue, (8.65) and (8.66) imply that $V = 0$ in D, and hence (8.67) becomes

$$(\tilde{\eta}_1 - \tilde{\eta}_2)\frac{\partial u^1}{\partial \nu} = 0 \qquad \text{on} \quad \partial D$$

for all incident waves. Since for a given $\partial D_0 \subset \partial D$ there exists at least one incident plane wave such that $\partial u^1/\partial \nu|_{\partial D_0} \neq 0$, the continuity of η_1 and η_2 in $\overline{\partial D_2^1}$ and $\overline{\partial D_2^2}$, respectively, implies that $\tilde{\eta}_1 = \tilde{\eta}_2$. $\qquad \square$

As the reader saw in Chaps. 4 and 6 and Sect. 8.1, our method for solving the inverse problem is based on finding an approximate solution to the far-field equation

$$Fg = \Phi_\infty(\cdot, z), \qquad z \in \mathbb{R}^2,$$

where F is the far-field operator corresponding to the scattering problem (8.54)–(8.58). If we consider the operator $B : \mathbb{H}^1(D, \partial D_2) \to L^2[0, 2\pi]$, which takes the incident field u^i satisfying

$$\Delta u^i + k^2 u^i = 0 \qquad \text{in} \quad D$$

to the far-field pattern u_∞ of the solution to (8.40)–(8.45) corresponding to this incident field, then the far-field equation can be written as

$$(Bv_g)(\hat{x}) = \Phi_\infty(\hat{x}, z), \qquad z \in \mathbb{R}^2,$$

where v_g is the Herglotz wave function with kernel g. Note that the formulation of the scattering problem and Theorem 8.19 remains valid if the incident field u^i is defined as a solution to the Helmhotz equation only in D (or in a neighborhood of ∂D) since the traces of u^i only appear in the boundary conditions. From the well-posedness of (8.40)–(8.45) we see that B is a bounded linear operator. Furthermore, in the same way as in Theorem 6.48, one can show that B is, in addition, a compact operator. Assuming that k^2 is not a transmission eigenvalue, one can now easily see that the range of B is dense in $L^2[0, 2\pi]$ since it contains the range of F, which from Theorem 8.22 is dense in $L^2[0, 2\pi]$. We next observe that

$$\Phi_\infty(\cdot, z) \in \text{Range}(B) \iff z \in D, \qquad (8.68)$$

provided that k is not a transmission eigenvalue. Indeed, if $z \in D$, then the solution u^i of $(Bu^i)(\hat{x}) = \Phi_\infty(\hat{x}, z)$ is $u^i = w_z$, where $w_z \in \mathbb{H}^1(D, \partial D_2)$ and $v_z \in H^1(D)$ is the unique solution of the mixed interior transmission problem

$$\nabla \cdot A \nabla v_z + k^2 n\, v_z = 0 \qquad\qquad \text{in} \quad D, \qquad (8.69)$$

$$\Delta w_z + k^2\, w_z = 0, \qquad\qquad \text{in} \quad D, \qquad (8.70)$$

$$v_z - (w_z + \Phi(\cdot, z)) = 0 \qquad\qquad \text{on} \quad \partial D_1, \qquad (8.71)$$

$$v_z - (w_z + \Phi(\cdot, z)) = -i\eta\, \frac{\partial}{\partial \nu}(w_z + \Phi(\cdot, z)) \qquad \text{on} \quad \partial D_2, \qquad (8.72)$$

$$\frac{\partial v_z}{\partial \nu_A} - \frac{\partial}{\partial \nu}(w_z + \Phi(\cdot, z)) = 0 \qquad\qquad \text{on} \quad \partial D. \qquad (8.73)$$

On the other hand, for $z \in \mathbb{R}^2 \setminus \bar{D}$ the fact that $\Phi(\cdot, z)$ has a singularity at z, together with Rellich's lemma, implies that $\Phi_\infty(\cdot, z)$ is not in the range of B. Notice that since in general the solution w_z of (8.69)–(5.5) is not a Herglotz wave function, the far-field equation in general does not have a solution for any $z \in \mathbb{R}^2$. However, for $z \in D$, from Theorem 8.24 we can approximate w_z by a Herglotz function v_g, and its kernel g is an approximate solution of the far-field equation. Finally, noting that if u^s, v solves (8.40)–(8.45) with $u^i \in \mathbb{H}^1(D, \partial D_2)$, then u^i, v solves the mixed interior transmission problem (8.69)–(8.73) with $\Phi(\cdot, z)$ replaced by u^s and $Bu^i = u_\infty$, where u_∞ is the far-field pattern of u^s, one can easily deduce that B is injective, provided that k is not a transmission eigenvalue. The foregoing discussion now implies, in the same way as in Theorem 6.50, the following result.

Theorem 8.27. *Assume that k is not a transmission eigenvalue and D, A, n, and η satisfy the assumptions in the formulation of the scattering problem (8.40)–(8.45). Then, if F is the far-field operator corresponding to (8.40)–(8.45), we have that*

1. *For $z \in D$ and a given $\epsilon > 0$ there exists a function $g_z^\epsilon \in L^2[0, 2\pi]$ such that*

$$\|Fg_z^\epsilon - \Phi_\infty(\cdot, z)\|_{L^2[0, 2\pi]} < \epsilon,$$

*and the Herglotz wave function $v_{g_z^\epsilon}$ with kernel g_z^ϵ converges in $\mathbb{H}^1(D, \partial D_2)$
to w_z as $\epsilon \to 0$, where (v_z, w_z) is the unique solution of (8.69)–(8.73).*
2. For $z \notin D$ and a given $\epsilon > 0$ every function $g_z^\epsilon \in L^2[0, 2\pi]$ that satisfies

$$\|Fg_z^\epsilon - \Phi_\infty(\cdot, z)\|_{L^2[0, 2\pi]} < \epsilon$$

is such that

$$\lim_{\epsilon \to 0} \|v_{g_z^\epsilon}\|_{\mathbb{H}^1(D, \partial D_2)} = \infty.$$

The approximate solution g of the far-field equation given by Theorem 8.27 (assuming that it can be determined using regularization methods) can be used as in the previous inverse problems considered in Chaps. 4 and 6 and Sect. 8.1 to reconstruct an approximation to D. In particular, the boundary ∂D of D can be visualized as the set of points z where the L^2 norm of g_z becomes large.

Provided that an approximation to D is obtained as was done previously, our next goal is to use the same g to estimate the maximum of the surface conductivity η. To this end, we define W_z by

$$W_z := w_z + \Phi(\cdot, z), \tag{8.74}$$

where (v_z, w_z) satisfy (8.69)–(8.73). In particular, since $w_z \in \mathbb{H}^1(D, \partial D_2)$, $\Delta w_z \in L^2(D)$ and $z \in D$, we have that $W_z|_{\partial D} \in H^{\frac{1}{2}}(\partial D)$, $\partial W_z/\partial \nu|_{\partial D} \in H^{-\frac{1}{2}}(\partial D)$ and $\partial W_z/\partial \nu|_{\partial D_2} \in L^2(\partial D_2)$.

Lemma 8.28. *For every two points z_1 and z_2 in D we have that*

$$-2\int_D \nabla v_{z_1} \cdot \mathrm{Im}(A)\nabla \overline{v}_{z_2}\, dx + 2k^2 \int_D \mathrm{Im}(n)v_{z_1}\overline{v}_{z_2}\, dx + 2\int_{\partial D_2} \eta(x)\frac{\partial W_{z_1}}{\partial \nu}\frac{\partial \overline{W}_{z_2}}{\partial \nu}\, ds$$

$$= -4k\pi|\gamma|^2 J_0(k|z_1 - z_2|) + i\left(w_{z_1}(z_2) - \overline{w}_{z_2}(z_1)\right),$$

where w_{z_1}, W_{z_1} and w_{z_2}, W_{z_2} are defined by (8.69)–(8.73) and (8.74), respectively, and J_0 is a Bessel function of order zero.

Proof. Let z_1 and z_2 be two points in D and $v_{z_1}, w_{z_1}, W_{z_1}$ and $v_{z_2}, w_{z_2}, W_{z_2}$ the corresponding functions defined by (8.69)–(8.73). Applying the divergence theorem (Corollary 5.8) to $v_{z_1}, \overline{v}_{z_2}$ and using (8.69)–(8.73), together with the fact that A is symmetric, we have that

$$\int_{\partial D}\left(v_{z_1}\frac{\partial \overline{v}_{z_2}}{\partial \nu_{\overline{A}}} - \overline{v}_{z_2}\frac{\partial v_{z_1}}{\partial \nu_A}\right)ds = \int_D \left(\nabla v_{z_1} \cdot \overline{A}\nabla \overline{v}_{z_2} - \nabla \overline{v}_{z_2} \cdot A\nabla v_{z_1}\right)dx$$

$$+ \int_D \left(v_{z_1}\nabla \cdot \overline{A}\nabla \overline{v}_{z_2} - \overline{v}_{z_2}\nabla \cdot A\nabla v_{z_1}\right)dx = -2i\int_D \nabla v_{z_1} \cdot \mathrm{Im}(A)\nabla \overline{v}_{z_2}\, dx$$

$$+ 2ik^2 \int_D \mathrm{Im}(n)v_{z_1}\overline{v}_{z_2}\, dx. \tag{8.75}$$

On the other hand, from the boundary conditions we have

$$
\int_{\partial D} \left(v_{z_1} \frac{\partial \overline{v}_{z_2}}{\partial \nu_{\overline{A}}} - \overline{v}_{z_2} \frac{\partial v_{z_1}}{\partial \nu_A} \right) ds
$$

$$
= \int_{\partial D} \left(W_{z_1} \frac{\partial \overline{W}_{z_2}}{\partial \nu} - \overline{W}_{z_2} \frac{\partial W_{z_1}}{\partial \nu} \right) ds - 2i \int_{\partial D_2} \eta(x) \frac{\partial W_{z_1}}{\partial \nu} \frac{\partial \overline{W}_{z_2}}{\partial \nu} ds.
$$

Hence

$$
-2i \int_D \nabla v_{z_1} \cdot \mathrm{Im}(A) \nabla \overline{v}_{z_2} \, dx + 2ik^2 \int_D \mathrm{Im}(n) v_{z_1} \overline{v}_{z_2} \, dx
$$

$$
+2i \int_{\partial D_2} \eta(x) \frac{\partial W_{z_1}}{\partial \nu} \frac{\partial \overline{W}_{z_2}}{\partial \nu} \, ds = \int_{\partial D} \left(W_{z_1} \frac{\partial \overline{W}_{z_2}}{\partial \nu} - \overline{W}_{z_2} \frac{\partial W_{z_1}}{\partial \nu} \right) ds
$$

$$
= \int_{\partial D} \left(\Phi(\cdot, z_1) \frac{\partial \overline{\Phi(\cdot, z_2)}}{\partial \nu} - \overline{\Phi(\cdot, z_2)} \frac{\partial \Phi(\cdot, z_1)}{\partial \nu} \right) ds
$$

$$
+ \int_{\partial D} \left(w_{z_1} \frac{\partial \overline{\Phi(\cdot, z_2)}}{\partial \nu} - \overline{\Phi(\cdot, z_2)} \frac{\partial w_{z_1}}{\partial \nu} \right) ds
$$

$$
+ \int_{\partial D} \left(\Phi(\cdot, z_1) \frac{\partial \overline{w}_{z_2}}{\partial \nu} - \overline{w}_{z_2} \frac{\partial \Phi(\cdot, z_1)}{\partial \nu} \right) ds.
$$

Green's second identity applied to the radiating solution $\Phi(\cdot, z)$ of the Helmholtz equation in D_e implies that

$$
\int_{\partial D} \left(\Phi(\cdot, z_1) \frac{\partial \overline{\Phi(\cdot, z_2)}}{\partial \nu} - \overline{\Phi(\cdot, z_2)} \frac{\partial \Phi(\cdot, z_1)}{\partial \nu} \right) ds = -2ik \int_0^{2\pi} \Phi_\infty(\cdot, z_1) \overline{\Phi_\infty(\cdot, z_2)} ds
$$

$$
= -2ik \int_0^{2\pi} |\gamma|^2 e^{-ik\hat{x} \cdot z_1} e^{ik\hat{x} \cdot z_2} \, ds = -4ik\pi |\gamma|^2 J_0(k|z_1 - z_2|),
$$

and from the representation formula for w_{z_1} and w_{z_2} we now obtain

$$
-2i \int_D \nabla v_{z_1} \cdot \mathrm{Im}(A) \nabla \overline{v}_{z_2} \, dx + 2ik^2 \int_D \mathrm{Im}(n) v_{z_1} \overline{v}_{z_2} \, dx
$$

$$
+2i \int_{\partial D_2} \eta(x) \frac{\partial W_{z_1}}{\partial \nu} \frac{\partial \overline{W}_{z_2}}{\partial \nu} \, ds = -4ik\pi |\gamma|^2 J_0(k|z_1 - z_2|) + \overline{w}_{z_2}(z_1) - w_{z_1}(z_2).
$$

Dividing both sides of the foregoing relation by i we have the result.   □

Assuming D is connected, consider a ball $\Omega_r \subset D$ of radius r contained in D (Remark 4.13), and define a subset of $L^2(\partial D_2)$ by

$$V := \left\{ f \in L^2(\partial D_2) : \quad f = \frac{\partial W_z}{\partial \nu}\bigg|_{\partial D_2} \quad \begin{matrix} \text{with } W_z = w_z + \Phi(\cdot, z), \\ z \in \Omega_r \text{ and } w_z, v_z \text{ the solution of } (8.69)\text{--}(8.73) \end{matrix} \right\}.$$

Lemma 8.29. *Assume that k is not a transmission eigenvalue. Then V is complete in $L^2(\partial D_2)$.*

Proof. Let φ be a function in $L^2(\partial D_2)$ such that for every $z \in \Omega_r$

$$\int_{\partial D_2} \frac{\partial W_z}{\partial \nu} \varphi \, ds = 0.$$

Since k^2 is not a transmission eigenvalue, we can construct $v \in H^1(D)$ and $w \in \mathbb{H}^1(D, \partial D_2)$ as the unique solution of the following mixed interior transmission problem:

$$
\begin{array}{llll}
(i) & \nabla \cdot A \nabla v + k^2 n\, v = 0 & \text{in} & D, \\[2mm]
(ii) & \Delta w + k^2\, w = 0 & \text{in} & D, \\[2mm]
(iii) & v - w = 0 & \text{on} & \partial D_1, \\[2mm]
(iv) & v - w = -i\eta\, \dfrac{\partial w}{\partial \nu} + \varphi & \text{on} & \partial D_2, \\[3mm]
(v) & \dfrac{\partial v}{\partial \nu_A} - \dfrac{\partial w}{\partial \nu} = 0 & \text{on} & \partial D.
\end{array}
$$

Then we have

$$
\begin{aligned}
0 &= \int_{\partial D_2} \frac{\partial W_z}{\partial \nu} \varphi \, ds = \int_{\partial D} \frac{\partial W_z}{\partial \nu}(v - w)\, ds + i \int_{\partial D_2} \eta \frac{\partial W_z}{\partial \nu}\frac{\partial w}{\partial \nu}\, ds \\
&= \int_{\partial D} \frac{\partial W_z}{\partial \nu} v \, ds - \int_{\partial D} \frac{\partial W_z}{\partial \nu} w \, ds + i \int_{\partial D_2} \eta \frac{\partial W_z}{\partial \nu}\frac{\partial w}{\partial \nu}\, ds. \quad (8.76)
\end{aligned}
$$

From the equations for v_z and v, the divergence theorem, and the transmission boundary conditions we have

$$
\begin{aligned}
\int_{\partial D} \frac{\partial W_z}{\partial \nu} v \, ds &= \int_{\partial D} \frac{\partial v_z}{\partial \nu_A} v \, ds = \int_{\partial D} \frac{\partial v}{\partial \nu_A} v_z \, ds \\
&= \int_{\partial D} \frac{\partial w}{\partial \nu} W_z \, ds - i \int_{\partial D_2} \eta \frac{\partial W_z}{\partial \nu}\frac{\partial w}{\partial \nu}\, ds. \quad (8.77)
\end{aligned}
$$

Finally, substituting (8.77) into (8.76) and using the integral representation formula we obtain

$$0 = \int_{\partial D} \left(\frac{\partial w}{\partial \nu} W_z - \frac{\partial W_z}{\partial \nu} w \right) ds = \int_{\partial D} \left(\frac{\partial w}{\partial \nu} w_z - \frac{\partial w_z}{\partial \nu} w \right) ds$$

$$= \int_{\partial D} \left(\frac{\partial w}{\partial \nu} \Phi(\cdot, z) - \frac{\partial \Phi(\cdot, z)}{\partial \nu} w \right) ds = w(z) \qquad \forall z \in \Omega_r. \qquad (8.78)$$

The unique continuation principle for the Helmholtz equation now implies that $w = 0$ in D. Then (cf. the proof of Theorem 8.2) $v = 0$, and therefore $\varphi = 0$, which proves the lemma.   □

We now assume that $\text{Im}(A) = 0$, $\text{Im}(n) = 0$, and that k is not a transmission eigenvalue. Then setting $z = z_1 = z_2$ in Lemma 8.28 we arrive at the following integral equation for η:

$$\int_{\partial D_2} \eta(x) \left| \frac{\partial}{\partial \nu} (w_z(x) + \Phi(x, z)) \right|^2 ds = -\frac{1}{4} - \text{Im} (w_z(z)), \qquad z \in D. \quad (8.79)$$

If we denote by $\tilde{\eta} \in L^2(\partial D)$ the extension by zero to the whole boundary of the surface conductivity η, then we can assume that the region of integration in the integral in (8.79) is ∂D instead of ∂D_2. By Lemma 8.29, we see that the left-hand side of (8.79) is an injective compact integral operator with positive kernel defined in $L^2(\partial D)$ (replacing η by $\tilde{\eta}$). Using Tikhonov regularization techniques (cf. [68]) it is possible to determine $\tilde{\eta}$ (and hence η without knowing a priori the portion ∂D_2) by finding a regularized solution of the integral equation in $L^2(\partial D)$ with noisy kernel and noisy right-hand side (recall from Theorem 8.27 that w_z and its derivatives can be approximated by v_{g_z} and its derivative, respectively). For numerical examples using this approach we refer the reader to [27].

In the particular case where the coating is homogeneous, i.e., the surface conductivity is a positive constant $\eta > 0$, we have that

$$\eta = \frac{-2k\pi|\gamma|^2 - \text{Im} (w_z(z))}{\left\| \frac{\partial}{\partial \nu} (w_z(\cdot) + \Phi(\cdot, z)) \right\|^2_{L^2(\partial D_2)}}. \qquad (8.80)$$

A drawback of (8.80) is that the extent of the coating ∂D_2 is in general not known. Hence, if ∂D_2 is replaced by ∂D, these expressions in practice only provide a lower bound for the maximum of η, unless it is known a priori that D is completely coated.

8.6 Numerical Examples

We now present some numerical tests of the preceding inversion scheme using synthetic data. For our examples, in (8.40)–(8.45) we choose $A = (1/4)I$, $n = 1$, and η equal to a constant. The far-field data are computed using a finite-element method on a domain that is terminated by a rectangular perfectly matched layer (PML), and the far-field equation is solved by the same procedure as described at the end of Sect. 8.1 to compute g [27].

We present some results for an ellipse given by the parametric equations $x = 0.5\cos(s)$ and $y = 0.2\sin(s)$, $s \in [0, 2\pi]$. For the ellipse we consider either a fully coated or partially coated object, shown in Fig 8.5.

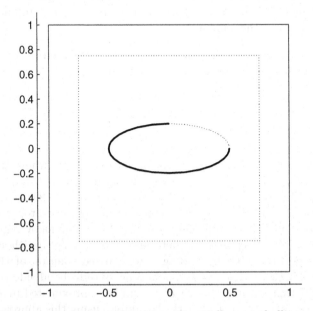

Fig. 8.5. Diagram showing coated portion of partially coated ellipse as *thick line*. *Dotted square*: inner boundary of PML; *solid square*: boundary of finite-element computational domain[3]

We begin by assuming an exact knowledge of the boundary in order to assess the accuracy of (8.80). Having computed g using regularization methods to solve the far-field equation, we approximate (8.80) using the trapezoidal rule with 100 integration points and use $z_0 = (0,0)$. In Fig. 8.6 we show the results of the reconstruction of a range of conductivities η for a fully coated ellipse and partially coated ellipse. Recall that for the partially coated ellipse, (8.80) with ∂D_2 replaced by ∂D provides only a lower bound for η. For each exact η we compute the far-field data, add noise, and compute an approximation to w_z, as discussed previously and in Sect. 8.1.

---

[3]Reprinted from F. Cakoni, D. Colton, and P. Monk, The determination of the surface conductivity of a partially coated dielectric, SIAM J. Appl. Math. 65 (2005), 767–789. Copyright ©2005 Society for Industrial and Applied Mathematics. Reprinted with permission. All rights reserved.

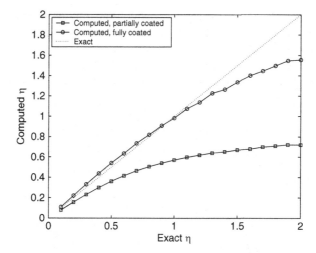

Fig. 8.6. Computation of η using exact boundary for fully coated and partially coated ellipses. Clearly, in all cases the approximation of η deteriorates for large conductivities[3]

We now wish to investigate the solution of the full inverse problem. We start by using the linear sampling method to approximate the boundary of the scatterer, which is based on the behavior of g given by Theorem 8.27. In particular, we compute $1/\|g\|$ for z on a uniform grid in the sampling domain. In the upcoming numerical results we have chosen 61 incident directions equally distributed on the unit circle and we sample on a 101×101 grid on the square $[-1, 1] \times [-1, 1]$.

Having computed g using Tikhonov regularization and the Morozov discrepancy principle to solve the far-field equation, for each sample point we have a discrete level set function $1/\|g\|$. Choosing a contour value C then provides a reconstruction of the support of the given scatterer. We extract the edge of the reconstruction and then fit this using a trigonometric polynomial of degree M assuming that the reconstruction is starlike with respect to the origin (for more advanced applications it would be necessary to employ a more elaborate smoothing procedure). Thus, for an angle θ the radius of the reconstruction is given by

$$r(\theta) = \mathrm{Re}\left(\sum_{n=-M}^{M} r_n \exp(in\theta)\right),$$

where r is measured from the origin (since in all the examples here the origin is within the scatterer). The coefficients r_n are found using a least-squares fit to the boundary identified in the previous step of the algorithm. Once we have a parameterization of the reconstructed boundary, we can compute the normal to the boundary and evaluate (8.80) for some choice of z_0 [in the examples always $z_0 = (0, 0)$] using the trapezoidal rule with 100 points. This provides our reconstruction of η. The results of the experiments for a fully coated ellipse are shown in Figs. 8.7 and 8.8. For more details on the choice of the contour value C that provides a good reconstruction of the boundary of the scatterer we refer the reader to [27].

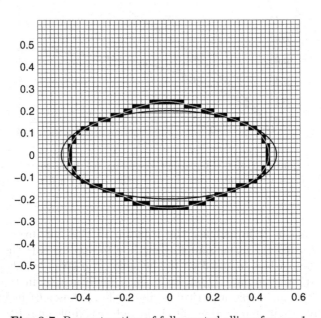

Fig. 8.7. Reconstruction of fully coated ellipse for $\eta = 1$

In the case of a partially coated ellipse (Fig. 8.5), the inversion algorithm is unchanged (both the boundary of the scatterer and η are reconstructed). The result of the reconstruction of D when $\eta = 1$ is shown in Fig. 8.9, and the results for a range of η are shown in Fig. 8.10. We recall again that for a partially coated obstacle (8.80) only provides a lower bound for η (i.e., ∂D_2 is replaced by ∂D).

8.7 Scattering by Cracks

In the last sections of this chapter we will discuss the scattering of a time-harmonic electromagnetic plane wave by an infinite cylinder having an open arc in $\mathbb{R}^2$ as cross section. We assume that the cylinder is a perfect conductor that is (possibly) coated on one side with a material with (constant) surface impedance λ. This leads to a (possibly) mixed boundary value problem for the Helmholtz equation defined in the exterior of an open arc in $\mathbb{R}^2$. Our aim is to establish the existence and uniqueness of a solution to this scattering problem

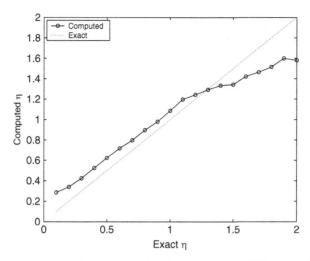

Fig. 8.8. Determination of range of η for (reconstructed) fully coated ellipse. For each exact η we apply the reconstruction algorithm using a range of cutoffs and plot the corresponding reconstruction. An exact reconstruction would lie on the *dotted line*[3]

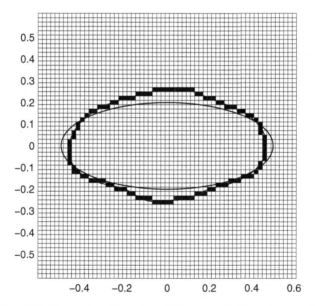

Fig. 8.9. Reconstruction of partially coated ellipse for $\eta = 1$

and to then use this knowledge to study the inverse scattering problem of determining the shape of the open arc (or "crack") from a knowledge of the far-field pattern of the scattered field [15].

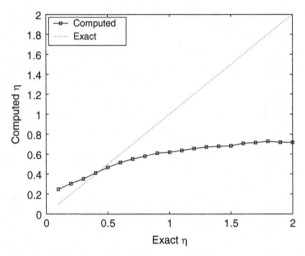

Fig. 8.10. Determination of range of η for (reconstructed) partially coated ellipse[3]

The inverse scattering problem for cracks was initiated by Kress [110] (see also [112, 114, 128]). In particular, Kress considered the inverse scattering problem for a perfectly conducting crack and used Newton's method to reconstruct the shape of the crack from a knowledge of the far-field pattern corresponding to a single incident wave. Kirsch and Ritter [108] used the factorization method (Chap. 7) to reconstruct the shape of the open arc from a knowledge of the far-field pattern assuming a Dirichlet or Neumann boundary condition.

Let $\Gamma \subset \mathbb{R}^2$ be a smooth, open, nonintersecting arc. More precisely, we consider $\Gamma \subset \partial D$ to be a portion of a smooth curve ∂D that encloses a region D in $\mathbb{R}^2$. We choose the unit normal ν on Γ to coincide with the outward normal to ∂D. The scattering of a time-harmonic incident wave u^i by a thin, infinitely long, cylindrical perfect conductor leads to the problem of determining u satisfying

$$\Delta u + k^2 u = 0 \quad \text{in} \quad \mathbb{R}^2 \setminus \bar{\Gamma}, \tag{8.81}$$

$$u^\pm = 0 \quad \text{on} \quad \Gamma, \tag{8.82}$$

where $u^\pm(x) = \lim_{h \to 0+} u(x \pm h\nu)$ for $x \in \Gamma$. The total field u is decomposed as $u = u^s + u^i$, where u^i is an entire solution of the Helmholtz equation, and u^s is the scattered field that is required to satisfy the Sommerfeld radiation condition

$$\lim_{r \to \infty} \sqrt{r} \left(\frac{\partial u^s}{\partial r} - iku^s \right) = 0 \tag{8.83}$$

uniformly in $\hat{x} = x/|x|$ with $r = |x|$. In particular, the incident field can again be a plane wave given by $u^i(x) = e^{ikx \cdot d}$, $|d| = 1$.

In the case where one side of the thin cylindrical obstacle Γ is coated by a material with constant surface impedance $\lambda > 0$, we obtain the following mixed crack problem for the total field $u = u^s + u^i$:

$$\Delta u + k^2 u = 0 \quad \text{in} \quad \mathbb{R}^2 \setminus \bar{\Gamma}, \tag{8.84}$$

$$u^- = 0 \quad \text{on} \quad \Gamma, \tag{8.85}$$

$$\frac{\partial u^+}{\partial \nu} + i\lambda u^+ = 0 \quad \text{on} \quad \Gamma, \tag{8.86}$$

where again $\partial u^{\pm}(x)/\partial \nu = \lim_{h \to 0^+} \nu \cdot \nabla u(x \pm h\nu)$ for $x \in \Gamma$ and u^s satisfies the Sommerfeld radiation condition (8.83).

Recalling the Sobolev spaces $H^1_{loc}(\mathbb{R}^2 \setminus \bar{\Gamma})$, $H^{\frac{1}{2}}(\Gamma)$, and $H^{-\frac{1}{2}}(\Gamma)$ from Sects. 8.1 and 8.4, we observe that the preceding scattering problems are particular cases of the following more general boundary value problems in the exterior of Γ:

Dirichlet crack problem: Given $f \in H^{\frac{1}{2}}(\Gamma)$, find $u \in H^1_{loc}(\mathbb{R}^2 \setminus \bar{\Gamma})$ such that

$$\Delta u + k^2 u = 0 \quad \text{in} \quad \mathbb{R}^2 \setminus \bar{\Gamma}, \tag{8.87}$$

$$u^{\pm} = f \quad \text{on} \quad \Gamma, \tag{8.88}$$

$$\lim_{r \to \infty} \sqrt{r} \left(\frac{\partial u}{\partial r} - iku \right) = 0. \tag{8.89}$$

Mixed crack problem: Given $f \in H^{\frac{1}{2}}(\Gamma)$ and $h \in H^{-\frac{1}{2}}(\Gamma)$, find $u \in H^1_{loc}(\mathbb{R}^2 \setminus \bar{\Gamma})$ such that

$$\Delta u + k^2 u = 0 \quad \text{in} \quad \mathbb{R}^2 \setminus \bar{\Gamma}, \tag{8.90}$$

$$u^- = f \quad \text{on} \quad \Gamma, \tag{8.91}$$

$$\frac{\partial u^+}{\partial \nu} + i\lambda u^+ = h \quad \text{on} \quad \Gamma, \tag{8.92}$$

$$\lim_{r \to \infty} \sqrt{r} \left(\frac{\partial u}{\partial r} - iku \right) = 0. \tag{8.93}$$

Note that the boundary conditions in both problems are assumed in the sense of the trace theorems. In particular, $u^+|_\Gamma$ is the restriction to Γ of the trace $u \in H^{\frac{1}{2}}(\partial D)$ of $u \in H^1_{loc}(\mathbb{R}^2 \setminus \bar{D})$, whereas $u^-|_\Gamma$ is the restriction to Γ of the trace $u \in H^{\frac{1}{2}}(\partial D)$ of $u \in H^1(D)$. Since $\nabla u \in L^2_{loc}(\mathbb{R}^2)$, the same comment is valid for $\partial u^{\pm}/\partial \nu$, where $\partial u/\partial \nu \in H^{-\frac{1}{2}}(\partial D)$ is interpreted in the sense of Theorem 5.7.

It is easy to see that the scattered field u^s in the scattering problem for a perfect conductor and for a partially coated perfect conductor satisfies the Dirichlet crack problem with $f = -u^i|_\Gamma$ and the mixed crack problem with $f = -u^i|_\Gamma$ and $h = -\partial u^i/\partial \nu - i\lambda u^i|_\Gamma$, respectively.

We now define $[u] := u^+ - u^-|_\Gamma$ and $\left[\dfrac{\partial u}{\partial \nu}\right] := \dfrac{\partial u^+}{\partial \nu} - \dfrac{\partial u^-}{\partial \nu}\bigg|_\Gamma$, the jump

of u and $\dfrac{\partial u}{\partial \nu}$, respectively, across the crack Γ.

Lemma 8.30. *If u is a solution to the Dirichlet crack problem (8.87)–(8.89) or the mixed crack problem (8.90)–(8.93), then $[u] \in \tilde{H}^{\frac{1}{2}}(\Gamma)$ and $\left[\dfrac{\partial u}{\partial \nu}\right] \in \tilde{H}^{-\frac{1}{2}}(\Gamma)$.*

Proof. Let $u \in H^1_{loc}(\mathbb{R}^2 \setminus \bar{\Gamma})$ be a solution to (8.87)–(8.89) or (8.90)–(8.93). Then from the trace theorem and Theorem 5.7, $[u] \in H^{\frac{1}{2}}(\partial D)$ and $[\partial u/\partial \nu] \in H^{-\frac{1}{2}}(\partial D)$. But the solution u of the Helmholtz equation is such that $u \in C^\infty$ away from Γ, whence $[u] = [\partial u/\partial \nu] = 0$ on $\partial D \setminus \bar{\Gamma}$. Hence by definition (Sect. 8.1), $[u] \in \tilde{H}^{\frac{1}{2}}(\Gamma)$ and $[\partial u/\partial \nu] \in \tilde{H}^{-\frac{1}{2}}(\Gamma)$.   □

We first establish uniqueness for the problems (8.87)–(8.89) and (8.90)–(8.93).

Theorem 8.31. *The Dirichlet crack problem (8.87)–(8.89) and the mixed crack problem (8.90)–(8.93) have at most one solution.*

Proof. Denote by Ω_R a sufficiently large ball with radius R containing $\bar{D}$. Let u be a solution to the homogeneous Dirichlet or mixed crack problem, i.e., u satisfies (8.87)–(8.89) with $f = 0$ or (8.90)–(8.93) with $f = h = 0$. Obviously, $u \in H^1(\Omega_R \setminus \bar{D}) \cup H^1(D)$ satisfies the Helmholtz equation in $\Omega_R \setminus \bar{D}$, and D and from the preceding lemma u satisfies the following transmission conditions on the complementary part $\partial D \setminus \bar{\Gamma}$ of ∂D:

$$u^+ = u^- \quad \text{and} \quad \frac{\partial u^+}{\partial \nu} = \frac{\partial u^-}{\partial \nu} \quad \text{on } \partial D \setminus \bar{\Gamma}. \tag{8.94}$$

By an application of Green's first identity for u and $\bar{u}$ in D and $\Omega_R \setminus \bar{D}$ and using the transmission conditions (8.94) we see that

$$\int_{\partial \Omega_R} u \frac{\partial \bar{u}}{\partial \nu} ds = \int_{\Omega_R \setminus \bar{D}} |\nabla u|^2 dx + \int_D |\nabla u|^2 dx - k^2 \int_{\Omega_R \setminus \bar{D}} |u|^2 dx - k^2 \int_D |u|^2 dx$$

$$+ \int_\Gamma u^+ \frac{\partial \bar{u}^+}{\partial \nu} ds - \int_\Gamma u^- \frac{\partial \bar{u}^-}{\partial \nu} ds. \tag{8.95}$$

For problem (8.87)–(8.89) the boundary condition (8.88) implies

$$\int_\Gamma u^+ \frac{\partial \bar{u}^+}{\partial \nu} ds = \int_\Gamma u^- \frac{\partial \bar{u}^-}{\partial \nu} ds = 0,$$

while for problem (8.90)–(8.89), since $\lambda > 0$, the boundary conditions (8.92) and (8.91) imply

$$\int_\Gamma u^+ \frac{\partial \overline{u}^+}{\partial \nu} ds - \int_\Gamma u^- \frac{\partial \overline{u}^-}{\partial \nu} ds = i\lambda \int_\Gamma |u^+|^2 ds.$$

Hence for both problems we can conclude that

$$\mathrm{Im} \int_{\partial \Omega_R} u \frac{\partial \overline{u}}{\partial \nu} ds \geq 0,$$

whence from Theorem 3.6 and the unique continuation principle we obtain that $u = 0$ in $\mathbb{R}^2 \setminus \bar{\Gamma}$.   □

To prove the existence of a solution to the foregoing crack problems, we will use an integral equation approach. In Chap. 3 the reader was introduced to the use of integral equations of the second kind to solve boundary value problems. Here we will employ a *first-kind* integral equation approach that is based on applying the Lax–Milgram lemma to boundary integral operators [127]. In this sense the method of first-kind integral equations is similar to variational methods.

We start with the representation formula (Remark 6.29)

$$u(x) = \int_{\partial D} \left(\frac{\partial u(y)}{\partial \nu_y} \Phi(x,y) - u(y) \frac{\partial}{\partial \nu_y} \Phi(x,y) \right) ds_y, \quad x \in D, \tag{8.96}$$

$$u(x) = \int_{\partial D} \left(u(y) \frac{\partial}{\partial \nu_y} \Phi(x,y) - \frac{\partial u(y)}{\partial \nu_y} \Phi(x,y) ds_y \right) ds_y, \quad x \in \mathbb{R}^2 \setminus \bar{D},$$

where $\Phi(\cdot, \cdot)$ is again the fundamental solution to the Helmholtz equation defined by

$$\Phi(x,y) := \frac{i}{4} H_0^{(1)}(k|x-y|), \tag{8.97}$$

with $H_0^{(1)}$ being a Hankel function of the first kind of order zero. Making use of the known jump relations of the single and double layer potentials across the boundary ∂D (Sect. 7.1.1) and by eliminating the integrals over $\partial D \setminus \bar{\Gamma}$, from (8.94) we obtain

$$\frac{1}{2}(u^- + u^+) = -S_\Gamma \left[\frac{\partial u}{\partial \nu} \right] + K_\Gamma[u] \qquad \text{on } \Gamma, \tag{8.98}$$

$$\frac{1}{2}\left(\frac{\partial u^-}{\partial \nu} + \frac{\partial u^+}{\partial \nu} \right) = -K'_\Gamma \left[\frac{\partial u}{\partial \nu} \right] + T_\Gamma[u] \qquad \text{on } \Gamma, \tag{8.99}$$

where S, K, K', T are the boundary integral operators

$$S : H^{-\frac{1}{2}}(\partial D) \longrightarrow H^{\frac{1}{2}}(\partial D), \qquad K : H^{\frac{1}{2}}(\partial D) \longrightarrow H^{\frac{1}{2}}(\partial D),$$

$$K' : H^{-\frac{1}{2}}(\partial D) \longrightarrow H^{-\frac{1}{2}}(\partial D), \qquad T : H^{\frac{1}{2}}(\partial D) \longrightarrow H^{-\frac{1}{2}}(\partial D),$$

defined by (7.3), (7.4), (7.5), and (7.6), respectively, and $S_\Gamma, K_\Gamma, K'_\Gamma, T_\Gamma$ are the corresponding operators restricted to Γ defined by

$$(S_\Gamma\psi)(x) := \int_\Gamma \psi(y)\Phi(x,y)ds_y, \qquad \psi \in \tilde{H}^{-\frac{1}{2}}(\Gamma), \qquad x \in \Gamma,$$

$$(K_\Gamma\psi)(x) := \int_\Gamma \psi(y)\frac{\partial}{\partial\nu_y}\Phi(x,y)ds_y, \qquad \psi \in \tilde{H}^{\frac{1}{2}}(\Gamma), \qquad x \in \Gamma,$$

$$(K'_\Gamma\psi(x)) := \int_\Gamma \psi(y)\frac{\partial}{\partial\nu_x}\Phi(x,y)ds_y, \qquad \psi \in \tilde{H}^{-\frac{1}{2}}(\Gamma), \qquad x \in \Gamma,$$

$$(T_\Gamma\psi)(x) := \frac{\partial}{\partial\nu_x}\int_\Gamma \psi(y)\frac{\partial}{\partial\nu_y}\Phi(x,y)ds_y, \qquad \psi \in \tilde{H}^{-\frac{1}{2}}(\Gamma), \qquad x \in \Gamma.$$

Recalling that functions in $\tilde{H}^{\frac{1}{2}}(\Gamma)$ and $\tilde{H}^{-\frac{1}{2}}(\Gamma)$ can be extended by zero to functions in $H^{\frac{1}{2}}(\partial D)$ and $H^{-\frac{1}{2}}(\partial D)$, respectively, the foregoing restricted operators are well defined. Moreover, they have the following mapping properties:

$$S_\Gamma : \tilde{H}^{-\frac{1}{2}}(\Gamma) \longrightarrow H^{\frac{1}{2}}(\Gamma), \qquad K_\Gamma : \tilde{H}^{\frac{1}{2}}(\Gamma) \longrightarrow H^{\frac{1}{2}}(\Gamma),$$
$$K'_\Gamma : \tilde{H}^{-\frac{1}{2}}(\Gamma) \longrightarrow H^{-\frac{1}{2}}(\Gamma), \qquad T_\Gamma : \tilde{H}^{\frac{1}{2}}(\Gamma) \longrightarrow H^{-\frac{1}{2}}(\Gamma).$$

In the case of the Dirichlet crack problem, since $[u] = 0$ and $u^+ = u^- = f$, the relation (8.98) gives the following first-kind integral equation for the unknown jump of the normal derivative of the solution across Γ:

$$S_\Gamma\left[\frac{\partial u}{\partial \nu}\right] = -f. \tag{8.100}$$

In the case of the mixed crack problem, the unknowns are both $[u] \in \tilde{H}^{\frac{1}{2}}(\Gamma)$ and $\left[\frac{\partial u}{\partial \nu}\right] \in \tilde{H}^{-\frac{1}{2}}(\Gamma)$. Using the boundary conditions (8.91) and (8.92), together with the relations (8.98) and (8.99), we obtain the following integral equation of the first kind for the unknowns $[u]$ and $\left[\frac{\partial u}{\partial \nu}\right]$:

$$\begin{pmatrix} S_\Gamma & -K_\Gamma + I \\ K'_\Gamma - I & -T_\Gamma - i\lambda I \end{pmatrix} \begin{pmatrix} \left[\frac{\partial u}{\partial \nu}\right] \\ [u] \end{pmatrix} = \begin{pmatrix} -f \\ i\lambda f - h \end{pmatrix}. \tag{8.101}$$

We let A_Γ denote the matrix operator in (8.101) and note that A_Γ is a continuous mapping from $\tilde{H}^{-\frac{1}{2}}(\Gamma) \times \tilde{H}^{\frac{1}{2}}(\Gamma)$ to $H^{\frac{1}{2}}(\Gamma) \times H^{-\frac{1}{2}}(\Gamma)$.

Lemma 8.32. *The operator* $S_\Gamma : H^{-\frac{1}{2}}(\Gamma) \to H^{\frac{1}{2}}(\Gamma)$ *is invertible with bounded inverse.*

Proof. From Theorem 7.3 we have that the bounded linear operator $S_i : H^{-\frac{1}{2}}(\partial D) \to H^{\frac{1}{2}}(\partial D)$, defined by (7.3) with k replaced by i in the fundamental solution, satisfies

$$(S_i\psi, \psi) \geq C\|\psi\|^2_{H^{-\frac{1}{2}}(\partial D)} \qquad \text{for } \psi \in H^{-\frac{1}{2}}(\partial D),$$

where $(\cdot, \cdot)$ denotes the conjugated duality pairing between $H^{\frac{1}{2}}(\partial D)$ and $H^{-\frac{1}{2}}(\partial D)$ defined by Definition 7.1. Furthermore, the operator $S_c = S - S_i$ is compact from $H^{-\frac{1}{2}}(\partial D)$ to $H^{\frac{1}{2}}(\partial D)$. Since for any $\psi \in \tilde{H}^{-\frac{1}{2}}(\Gamma)$ its extension by zero $\tilde{\psi}$ is in $H^{-\frac{1}{2}}(\partial D)$, we have that for $\psi \in \tilde{H}^{-\frac{1}{2}}(\Gamma)$

$$(S_{i\Gamma}\psi, \psi) = \left(S_i\tilde{\psi}, \tilde{\psi}\right) \geq C\|\tilde{\psi}\|^2_{H^{-\frac{1}{2}}(\partial D)} = C\|\psi\|^2_{\tilde{H}^{-\frac{1}{2}}(\Gamma)},$$

and $S_{c\Gamma}$ is compact from $\tilde{H}^{-\frac{1}{2}}(\Gamma)$ to $H^{\frac{1}{2}}(\Gamma)$, where $S_{i\Gamma}, S_{c\Gamma} : \tilde{H}^{-\frac{1}{2}}(\Gamma) \to H^{\frac{1}{2}}(\Gamma)$ are the corresponding restrictions of S_i and S_c.

Applying the Lax–Milgram lemma to the bounded and coercive sesquilinear form

$$a(\psi, \phi) := (S_{i\Gamma}\psi, \phi), \qquad \phi, \psi \in \tilde{H}^{-\frac{1}{2}}(\Gamma)$$

we conclude that $S_{i\Gamma}^{-1} : H^{\frac{1}{2}}(\Gamma) \to H^{-\frac{1}{2}}(\Gamma)$ exists and is bounded. Since S_c is compact, an application of Theorem 5.16 to $S_\Gamma = S_{i\Gamma} + S_{c\Gamma} : \tilde{H}^{-\frac{1}{2}}(\Gamma) \to H^{\frac{1}{2}}(\Gamma)$ gives that the injectivity of S_Γ implies that S_Γ is invertible with bounded inverse. Hence it remains to show that S_Γ is injective. To this end, let $\alpha \in \tilde{H}^{-\frac{1}{2}}(\Gamma)$ be such that $S_\Gamma\alpha = 0$. Define the potential

$$u(x) = -\int_\Gamma \alpha(y)\Phi(x, y)\, ds_y = -\int_{\partial D} \tilde{\alpha}(y)\Phi(x, y)\, ds_y \qquad x \in \mathbb{R}^2 \setminus \bar{\Gamma},$$

where $\tilde{\alpha} \in H^{-\frac{1}{2}}(\partial D)$ is the extension by zero of α. This potential satisfies the Helmholtz equation in $\mathbb{R}^2 \setminus \bar{\Gamma}$, the Sommerfeld radiation condition, and, moreover, $u \in H^1_{loc}(\mathbb{R}^2 \setminus \bar{\Gamma})$. Note that from the jump relations for single layer potentials we have that $\tilde{\alpha} = [\partial u/\partial \nu]$ on ∂D. Furthermore, the continuity of S across ∂D and the fact that $S_\Gamma\alpha = S\tilde{\alpha} = 0$ imply that $u^\pm|_\Gamma = -S\tilde{\alpha} = 0$. Hence u satisfies the homogeneous Dirichlet crack problem and from Theorem 8.31 $u = 0$ in $\mathbb{R}^2 \setminus \bar{\Gamma}$, whence $\tilde{\alpha} = [\partial u/\partial \nu] = 0$. This proves that S_Γ is injective. $\qquad \square$

Lemma 8.33. *The operator* $A_\Gamma : \tilde{H}^{-\frac{1}{2}}(\Gamma) \times \tilde{H}^{\frac{1}{2}}(\Gamma) \to H^{\frac{1}{2}}(\Gamma) \times H^{-\frac{1}{2}}(\Gamma)$ *is invertible with bounded inverse.*

Proof. The proof follows that of Lemma 8.32. Let $\tilde{\zeta} = (\tilde{\phi}, \tilde{\psi}) \in H^{-\frac{1}{2}}(\partial D) \times H^{\frac{1}{2}}(\partial D)$ be the extension by zero to ∂D of $\zeta = (\phi, \psi) \in \tilde{H}^{-\frac{1}{2}}(\Gamma) \times \tilde{H}^{\frac{1}{2}}(\Gamma)$. From Theorems 7.3 and 7.5 we have that $S = S_i + S_c$ and $T = T_i + T_c$, where

$$S_c : H^{-\frac{1}{2}}(\partial D) \longrightarrow H^{\frac{1}{2}}(\partial D), \qquad T_c : H^{\frac{1}{2}}(\partial D) \longrightarrow H^{-\frac{1}{2}}(\partial D)$$

are compact and

$$\left(S_i\tilde{\phi}, \tilde{\phi}\right) \geq C\|\tilde{\phi}\|^2_{H^{-\frac{1}{2}}(\partial D)} \qquad \text{for} \qquad \tilde{\phi} \in H^{-\frac{1}{2}}(\partial D), \qquad (8.102)$$

$$\left(-T_i\tilde{\psi}, \tilde{\psi}\right) \geq C\|\tilde{\psi}\|^2_{H^{\frac{1}{2}}(\partial D)} \qquad \text{for} \qquad \tilde{\psi} \in H^{\frac{1}{2}}(\partial D), \qquad (8.103)$$

where $(\cdot, \cdot)$ denotes the conjugated duality pairing between $H^{\frac{1}{2}}(\partial D)$ and $H^{-\frac{1}{2}}(\partial D)$ defined by Definition 7.1. Let K_0 and K_0' be the operators corresponding to the Laplace operator, i.e., defined as K and K' with kernel $\Phi(x, y)$ replaced by $\Phi_0(x, y) = -\frac{1}{2\pi} \ln |x - y|$. Then $K_c = K - K_0$ and $K_c' = K' - K_0'$ are compact since they have continuous kernels [111]. It is easy to show that K_0 and K_0' are adjoint since their kernels are real, i.e.,

$$\left(K_0 \tilde{\psi}, \tilde{\phi}\right) = \left(\tilde{\psi}, K_0' \tilde{\phi}\right) \quad \text{for } \tilde{\phi} \in H^{-\frac{1}{2}}(\partial D) \text{ and } \tilde{\psi} \in H^{\frac{1}{2}}(\partial D). \quad (8.104)$$

Collecting together all the compact terms we can write $A = (A_0 + A_c)$, where

$$A_0 \zeta = \begin{pmatrix} S_i \tilde{\phi} + (-K_0 + I)\tilde{\psi} \\ (K_0' - I)\tilde{\phi} - (T_i + 2i\lambda I)\tilde{\psi} \end{pmatrix} \quad \text{and} \quad A_c \zeta = \begin{pmatrix} S_c \tilde{\phi} - K_c \tilde{\psi} \\ K_c' \tilde{\phi} - T_c \tilde{\psi} \end{pmatrix}.$$

In this decomposition $A_c : H^{-\frac{1}{2}}(\partial D) \times H^{\frac{1}{2}}(\partial D) \to H^{-\frac{1}{2}}(\partial D) \times H^{\frac{1}{2}}(\partial D)$ is compact. Furthermore, we have that

$$\left(A_0 \tilde{\zeta}, \tilde{\zeta}\right) = \left(S_i \tilde{\phi}, \tilde{\phi}\right) + \left(-K_0 \tilde{\psi}, \tilde{\phi}\right) + \left(\tilde{\psi}, \tilde{\phi}\right) + \left(K_0' \tilde{\phi}, \tilde{\psi}\right)$$
$$- \left(\tilde{\phi}, \tilde{\psi}\right) - \left(T_i \tilde{\psi}, \tilde{\psi}\right) - i\lambda \left(\tilde{\psi}, \tilde{\psi}\right). \quad (8.105)$$

Taking the real part of (8.105), from (8.102) and (8.103) we obtain

$$\text{Re} \left[\left(S_i \tilde{\phi}, \tilde{\phi}\right) - \left(T_i \tilde{\psi}, \tilde{\psi}\right)\right] \geq C \left(\|\tilde{\phi}\|^2_{H^{-\frac{1}{2}}(\partial D)} + \|\tilde{\psi}\|^2_{H^{\frac{1}{2}}(\partial D)}\right), \quad (8.106)$$

and (8.104) implies that

$$\text{Re} \left[\left(-K_0 \tilde{\psi}, \tilde{\phi}\right) + \left(K_0' \tilde{\phi}, \tilde{\psi}\right)\right] = \text{Re} \left[-\left(\tilde{\psi}, K_0' \tilde{\phi}\right) + \left(K_0' \tilde{\phi}, \tilde{\psi}\right)\right]$$
$$= \text{Re} \left[-\overline{\left(K_0' \tilde{\phi}, \tilde{\psi}\right)} + \left(K_0' \tilde{\phi}, \tilde{\psi}\right)\right] = 0. \quad (8.107)$$

Finally,
$$\text{Re} \left[\left(\tilde{\psi}, \tilde{\phi}\right) - \left(\tilde{\phi}, \tilde{\psi}\right) - i\lambda \left(\tilde{\psi}, \tilde{\psi}\right)\right] = 0. \quad (8.108)$$

Combining (8.106)–(8.108) we now have that

$$\left|\left(A_0 \tilde{\zeta}, \tilde{\zeta}\right)\right| \geq \text{Re} \left(A_0 \tilde{\zeta}, \tilde{\zeta}\right) \geq C \|\tilde{\zeta}\|^2 \quad \text{for } \tilde{\zeta} \in H^{-\frac{1}{2}}(\partial D) \times \tilde{H}^{\frac{1}{2}}(\partial D). \quad (8.109)$$

Recalling that $\tilde{\zeta}$ is the extension by zero of $\zeta = (\phi, \psi) \in \tilde{H}^{-\frac{1}{2}}(\Gamma) \times \tilde{H}^{\frac{1}{2}}(\Gamma)$, we can rewrite (8.109) as

$$|(A_{0\Gamma}\zeta, \zeta)| \geq C \|\zeta\|^2 \quad \text{for } \zeta \in \tilde{H}^{-\frac{1}{2}}(\Gamma) \times \tilde{H}^{\frac{1}{2}}(\Gamma),$$

where $A_{0,\Gamma}$ is the restriction to Γ of A_0 defined for $\zeta \in \tilde{H}^{-\frac{1}{2}}(\Gamma) \times \tilde{H}^{\frac{1}{2}}(\Gamma)$. The corresponding restriction $A_{c\Gamma} : \tilde{H}^{-\frac{1}{2}}(\Gamma) \times \tilde{H}^{\frac{1}{2}}(\Gamma) \to H^{\frac{1}{2}}(\Gamma) \times H^{-\frac{1}{2}}(\Gamma)$ of A_c clearly remains compact. Hence, the Lax–Milgram lemma, together with Theorem 5.16, implies, in the same way as in Lemma 8.32, that A_Γ is invertible with bounded inverse if and only if A_Γ injective.

We now show that A_Γ is injective. To this end, let $\zeta = (\alpha, \beta) \in \tilde{H}^{-\frac{1}{2}}(\Gamma) \times \tilde{H}^{\frac{1}{2}}(\Gamma)$ be such that $A_\Gamma \zeta = 0$, and let $\tilde{\zeta} = (\tilde{\alpha}, \tilde{\beta}) \in H^{-\frac{1}{2}}(\partial D) \times \tilde{H}^{\frac{1}{2}}(\partial D)$ be its extension by zero. Define the potential

$$u(x) = -\int_\Gamma \alpha(y)\Phi(x,y)ds_y + \int_\Gamma \beta(y)\frac{\partial}{\partial\nu_y}\Phi(x,y)ds_y \quad x \in \mathbb{R}^2 \setminus \bar{\Gamma}. \quad (8.110)$$

This potential is well defined in $\mathbb{R}^2 \setminus \bar{\Gamma}$ since the densities α and β can be extended by zero to functions in $H^{-\frac{1}{2}}(\partial D)$ and $H^{\frac{1}{2}}(\partial D)$, respectively. Moreover, $u \in H^1_{loc}(\mathbb{R}^2 \setminus \bar{\Gamma})$ satisfies the Helmholtz equation in $\mathbb{R}^2 \setminus \bar{\Gamma}$ and the Sommerfeld radiation condition. One can easily show that $\alpha = [\partial u/\partial\nu]$ and $\beta = [u]$. In particular, the jump relations of the single and double layer potentials and the first equation of $A_\Gamma \zeta = 0$ imply

$$u^-|_\Gamma = -S\left[\frac{\partial u}{\partial\nu}\right] + K[u] - [u] = 0. \quad (8.111)$$

We also have that

$$\frac{\partial u^+}{\partial\nu}\bigg|_\Gamma = -K'\left[\frac{\partial u}{\partial\nu}\right] + T[u] + \left[\frac{\partial u}{\partial\nu}\right],$$

and from the fact that $u^+ = [u]$ on Γ (8.111) and the second equation of $A_\Gamma \zeta = 0$ we have that

$$\frac{\partial u^+}{\partial\nu} + i\lambda u^+\bigg|_\Gamma = -K'\left[\frac{\partial u}{\partial\nu}\right] + \left[\frac{\partial u}{\partial\nu}\right] + T[u] + i\lambda[u] = 0. \quad (8.112)$$

Hence u defined by (8.110) is a solution of the mixed crack problem with zero boundary data, and from the uniqueness Theorem 8.31 $u = 0$ in $\mathbb{R}^2 \setminus \bar{\Gamma}$, and hence $\zeta = ([\partial u/\partial\nu], [u]) = 0$.

$\square$

Theorem 8.34. *The Dirichlet crack problem (8.87)–(8.89) has a unique solution. This solution satisfies the a priori estimate*

$$\|u\|_{H^1(\Omega_R \setminus \bar{\Gamma})} \leq C\|f\|_{H^{\frac{1}{2}}(\Gamma)}, \quad (8.113)$$

where Ω_R is a disk of radius R containing $\bar{\Gamma}$, and the positive constant C depends on R but not on f.

Proof. Uniqueness is proved in Theorem 8.31. The solution of (8.87)–(8.89) is given by

$$u(x) = - \int_\Gamma \left[\frac{\partial u(y)}{\partial \nu} \right] \Phi(x,y) ds_y, \quad x \in \mathbb{R}^2 \setminus \bar{\Gamma},$$

where $[\partial u / \partial \nu]$ is the unique solution of (8.100) given by Lemma 8.32. Estimate (8.113) is a consequence of the continuity of S_Γ^{-1} from $H^{\frac{1}{2}}(\Gamma)$ to $\tilde{H}^{-\frac{1}{2}}(\Gamma)$ and the continuity of the single layer potential from $\tilde{H}^{-\frac{1}{2}}(\Gamma)$ to $H^1_{loc}(\mathbb{R}^2 \setminus \bar{\Gamma})$.   □

Theorem 8.35. *The mixed crack problem (8.90)–(8.93) has a unique solution. This solution satisfies the estimate*

$$\|u\|_{H^1(\Omega_R \setminus \bar{\Gamma})} \le C(\|f\|_{H^{\frac{1}{2}}(\Gamma)} + \|h\|_{H^{-\frac{1}{2}}(\Gamma)}), \tag{8.114}$$

where Ω_R is a disk of radius R containing $\bar{\Gamma}$, and the positive constant C depends on R but not on f and h.

Proof. Uniqueness is proved in Theorem 8.31. The solution of (8.90)–(8.93) is given by

$$u(x) = - \int_\Gamma \left[\frac{\partial u(y)}{\partial \nu_y} \right] \Phi(x,y) ds_y + \int_\Gamma [u(y)] \frac{\partial}{\partial \nu_y} \Phi(x,y) ds_y \quad x \in \mathbb{R}^2 \setminus \bar{\Gamma},$$

where $\left(\left[\dfrac{\partial u}{\partial \nu} \right], [u] \right)$ is the unique solution of (8.101) given by Lemma 8.33. Estimate (8.114) is a consequence of the continuity of A_Γ^{-1} from $H^{\frac{1}{2}}(\Gamma) \times H^{-\frac{1}{2}}(\Gamma)$ to $\tilde{H}^{-\frac{1}{2}}(\Gamma) \times \tilde{H}^{\frac{1}{2}}(\Gamma)$, the continuity of the single layer potential from $\tilde{H}^{-\frac{1}{2}}(\Gamma)$ to $H^1_{loc}(\mathbb{R}^2 \setminus \bar{\Gamma})$, and the continuity of the double layer potential from $\tilde{H}^{\frac{1}{2}}(\Gamma)$ to $H^1_{loc}(\mathbb{R}^2 \setminus \bar{\Gamma})$.   □

Remark 8.36. More generally, one can consider the Dirichlet crack problem with boundary data having a jump across Γ, that is, $u^\pm = f^\pm$ on Γ, where both f^+ and f^- are in $H^{\frac{1}{2}}(\Gamma)$. In this case, the right-hand side of the integral equation (8.100) will be replaced by $-(f^+ + f^-)/2$.

We end our discussion on direct scattering problems for cracks with a remark on the regularity of solutions. It is in fact known that the solution of the crack problem with Dirichlet boundary conditions has a singularity near a crack tip no matter how smooth the boundary data are. In particular, the solution does not belong to $H^{\frac{3}{2}}(\mathbb{R}^2 \setminus \bar{\Gamma})$ due to the fact that the solution has a singularity of the form $r^{\frac{1}{2}} \phi(\theta)$, where (r, θ) are the polar coordinates centered at the crack tip. In the case of the crack problem with mixed boundary conditions, one would expect a stronger singular behavior of the solution near the tips. Indeed, for this case the solution of the mixed crack problem with smooth boundary data belongs to $H^{\frac{5}{4}-\epsilon}(\mathbb{R}^2 \setminus \bar{\Gamma})$ for all $\epsilon > 0$ but not to $H^{\frac{5}{4}}(\mathbb{R}^2 \setminus \bar{\Gamma})$ due to the presence of a term of the form $r^{\frac{1}{4}+i\eta} \phi(\theta)$ in the asymptotic expansion of the solution in a neighborhood of the crack tip where η is a real number. A complete investigation of crack singularities can be found in [64].

8.8 Inverse Scattering Problem for Cracks

We now turn our attention to the inverse scattering problem for cracks. To this end, we recall that the approximation properties of Herglotz wave functions are a fundamental ingredient of the linear sampling method for solving the inverse problem. Hence, we first show that traces on Γ of the solution to crack problems can be approximated by the corresponding traces of Herglotz wave functions. More precisely, let v_g be a Herglotz wave function written in the form

$$v_g(x) = \int_0^{2\pi} g(\phi)e^{-ik(x_1 \cos \phi + x_2 \sin \phi)}\,d\phi, \qquad x = (x_1, x_2) \in \mathbb{R}^2,$$

and consider the operator $H : L^2[0, 2\pi] \to H^{\frac{1}{2}}(\Gamma) \times H^{-\frac{1}{2}}(\Gamma)$ defined by

$$(Hg)(x) := \begin{cases} v_g^- & \text{on } \Gamma, \\[2mm] \dfrac{\partial v_g^+}{\partial \nu} + i\lambda v_g^+ & \text{on } \Gamma. \end{cases} \tag{8.115}$$

Theorem 8.37. *The range of $H : L^2[0, 2\pi] \to H^{\frac{1}{2}}(\Gamma) \times H^{-\frac{1}{2}}(\Gamma)$ is dense.*

Proof. From Corollary 6.43, we only need to show that the transpose operator $H^\top : \tilde{H}^{-\frac{1}{2}}(\Gamma) \times \tilde{H}^{\frac{1}{2}}(\Gamma) \to L^2[0, 2\pi]$ is injective. To characterize the transpose operator, recall that $H^\top$ is defined by

$$\langle Hg, (\alpha, \beta)\rangle = \langle g, H^\top(\alpha, \beta)\rangle \tag{8.116}$$

for $g \in L^2[0, 2\pi]$ and $(\alpha, \beta) \in \tilde{H}^{-\frac{1}{2}}(\Gamma) \times \tilde{H}^{\frac{1}{2}}(\Gamma)$. Note that the left-hand side of (8.116) is the duality pairing between $H^{\frac{1}{2}}(\Gamma) \times H^{-\frac{1}{2}}(\Gamma)$ and $\tilde{H}^{-\frac{1}{2}}(\Gamma) \times \tilde{H}^{\frac{1}{2}}(\Gamma)$, while the right-hand side is the $L^2[0, 2\pi]$ inner product without conjugation. One can easily see from (8.116) by changing the order of integration that

$$H^\top(\alpha, \beta)(\phi) := \int_\Gamma \alpha(x)e^{-ikx\cdot d}ds_x + i\lambda \int_\Gamma \beta(x)e^{-ikx\cdot d}ds_x$$
$$+ \int_\Gamma \beta(x)\frac{\partial}{\partial \nu_x}e^{-ikx\cdot d}ds_x, \quad \phi \in [0, 2\pi],$$

where $d = (\cos \phi, \sin \phi)$. Hence $\gamma H^\top(\alpha, \beta)$ coincides with the far-field pattern of the potential

$$\gamma^{-1}V(z) := \int_\Gamma \alpha(x)\Phi(z, x)ds_x + i\lambda \int_\Gamma \beta(x)\Phi(z, x)ds_x$$
$$+ \int_\Gamma \beta(x)\frac{\partial}{\partial \nu_x}\Phi(z, x)ds_x, \quad z \in \mathbb{R}^2 \setminus \bar{\Gamma},$$

where $\gamma = \dfrac{e^{i\pi/4}}{\sqrt{8\pi k}}$. Note that V is well defined in $\mathbb{R}^2 \setminus \bar{\Gamma}$ since the densities α and β can be extended by zero to functions in $H^{-\frac{1}{2}}(\partial D)$ and $H^{\frac{1}{2}}(\partial D)$, respectively. Moreover, $V \in H^1_{loc}(\mathbb{R}^2 \setminus \bar{\Gamma})$ satisfies the Helmholtz equation in $\mathbb{R}^2 \setminus \bar{\Gamma}$ and the Sommerfeld radiation condition. Now assume that $H^\top(\alpha, \beta) = 0$. This means that the far-field pattern of V is zero, and from Rellich's lemma and the unique continuation principle we conclude that $V = 0$ in $\mathbb{R}^2 \setminus \bar{\Gamma}$. Using the jump relations across ∂D for the single and double layer potentials with α and β defined to be zero on $\partial D \setminus \bar{\Gamma}$ we now obtain

$$\beta = [V]_\Gamma,$$

$$\alpha + i\lambda\beta = -\left[\frac{\partial V}{\partial \nu}\right]_\Gamma,$$

and hence $\alpha = \beta = 0$. Thus $H^\top$ is injective and the theorem is proven. $\square$

As a special case of the preceding theorem we obtain the following theorem.

Theorem 8.38. *Every function in $H^{\frac{1}{2}}(\Gamma)$ can be approximated by the trace of a Herglotz wave function $v_g|_\Gamma$ on Γ with respect to the $H^{\frac{1}{2}}(\Gamma)$ norm.*

Assuming the incident field $u^i(x) = e^{ikx\cdot d}$ is a plane wave with incident direction $d = (\cos\phi, \sin\phi)$, the *inverse problem* we now consider is to determine the shape of the crack Γ from a knowledge of the far-field pattern $u_\infty(\cdot, \phi)$, $\phi \in [0, 2\pi]$, of the scattered field $u^s(\cdot, \phi)$. The scattered field is either the solution of the Dirichlet crack problem (8.87)–(8.89) with $f = -e^{ikx\cdot d}|_\Gamma$ or of the mixed crack problem (8.90)–(8.93) with $f = -e^{ikx\cdot d}|_\Gamma$ and $h = -\left(\dfrac{\partial}{\partial\nu} + i\lambda\right) e^{ikx\cdot d}|_\Gamma$. In either case, the far-field pattern is defined by the asymptotic expansion of the scattered field

$$u^s(x, \phi) = \frac{e^{ikr}}{\sqrt{r}} u_\infty(\theta, \phi) + O(r^{-3/2}), \qquad r = |x| \to \infty.$$

Theorem 8.39. *Assume Γ_1 and Γ_2 are two perfectly conducting or partially coated cracks with surface impedance λ_1 and λ_2 such that the far-field patterns $u^1_\infty(\theta, \phi)$ and $u^2_\infty(\theta, \phi)$ coincide for all incidence angles $\phi \in [0, 2\pi]$ and for all observation angles $\theta \in [0, 2\pi]$. Then $\Gamma_1 = \Gamma_2$.*

Proof. Let $G := \mathbb{R}^2 \setminus (\bar{\Gamma}_1 \cup \bar{\Gamma}_2)$ and $x_0 \in G$. Using Lemma 4.4 and the well-posedeness of the forward crack problems one can show, as in Theorem 4.5, that the scattered fields w^s_1 and w^s_2 corresponding to the incident field $u^i = -\Phi(\cdot, x_0)$ [i.e., w^s_j, $j = 1, 2$ satisfy (8.87)–(8.89) with $f = -\Phi(\cdot, x_0)|_{\Gamma_j}$, or (8.90)–(8.93) with $f = -\Phi(\cdot, x_0)|_{\Gamma_j}$ and $h = -\left(\frac{\partial}{\partial\nu} + i\lambda\right) \Phi(\cdot, x_0)|_{\Gamma_j}$] coincide in G.

Now assume that $\Gamma_1 \neq \Gamma_2$. Then, without loss of generality there exists $x^* \in \Gamma_1$ such that $x^* \notin \Gamma_2$. We can choose a sequence $\{x_n\}$ from G such that $x_n \to x^*$ as $n \to \infty$ and $x_n \notin \bar{\Gamma}_2$. Hence we have that $w_{n,1}^s = w_{n,2}^s$ in G, where $w_{n,1}^s$ and $w_{n,2}^s$ are as above, with x_0 replaced by x_n. Consider $w_n^s = w_{n,2}^s$ as the scattered wave corresponding to Γ_2. From the boundary data $(w_n^s)^- = -\Phi(\cdot, x_n)$ on Γ_2 and from (8.113) or (8.114) we have that $\|w_n^s\|_{H^1(\Omega_R \setminus \bar{\Gamma}_2)}$ is uniformly bounded with respect to n, whence from the trace theorem $\|w_n^s\|_{H^{\frac{1}{2}}(\Omega_r(x^*) \cap \Gamma_1)}$ is uniformly bounded with respect to n, where $\Omega_r(x^*)$ is a small neighborhood centered at x^* not intersecting Γ_2. On the other hand, considering $w_n^s = w_{n,1}^s$ as the scattered wave corresponding to Γ_1, from the boundary conditions $(w_n^s)^- = -\Phi(\cdot, x_n)$ on Γ_1 we have $\|w_n^s\|_{H^{\frac{1}{2}}(\Omega_r(x^*)) \cap \Gamma_1)} \to \infty$ as $n \to \infty$ since $\|\Phi(\cdot, x_n)\|_{H^{\frac{1}{2}}(\Omega_r(x^*) \cap \Gamma_1)} \to \infty$ as $n \to \infty$. This is a contradiction. Therefore, $\Gamma_1 = \Gamma_2$. $\square$

To solve the inverse problem, we will use the linear sampling method, which is based on a study of the far-field equation

$$Fg = \Phi_\infty^L, \qquad (8.117)$$

where $F : L^2[0, 2\pi] \to L^2[0, 2\pi]$ is the far-field operator defined by

$$(Fg)(\theta) := \int_0^{2\pi} u_\infty(\theta, \phi) g(\phi) d\phi$$

and Φ_∞^L is a function to be defined shortly. In particular, due to the fact that the scattering object has an empty interior, we need to modify the linear sampling method previously developed for obstacles with nonempty interior. Assume for the moment that the crack is partially coated, and define the operator $B : H^{\frac{1}{2}}(\Gamma) \times H^{-\frac{1}{2}}(\Gamma) \to L^2[0, 2\pi]$, which maps the boundary data (f, h) to the far-field pattern of the solution to the corresponding scattering problem (8.90)–(8.93). By superposition, we have the relation

$$Fg = -BHg,$$

where Hg is defined by (8.115) with the Herglotz wave function v_g now written as

$$v_g(x) = \int_0^{2\pi} g(\phi) e^{ikx \cdot d} \, d\phi.$$

We now define the compact operator $\mathcal{F} : \tilde{H}^{-\frac{1}{2}}(\Gamma) \times \tilde{H}^{\frac{1}{2}}(\Gamma) \longrightarrow L^2[0, 2\pi]$ by

$$\mathcal{F}(\alpha, \beta)(\theta) = \gamma \int_\Gamma \alpha(y) e^{-ik\hat{x} \cdot y} \, ds_y + \gamma \int_\Gamma \beta(y) \frac{\partial}{\partial \nu_y} e^{-ik\hat{x} \cdot y} \, ds_y, \qquad (8.118)$$

where $\hat{x} = (\cos\theta, \sin\theta)$ and $\gamma = e^{i\pi/4}/\sqrt{8\pi k}$, and observe that for a given pair $(\alpha, \beta) \in \tilde{H}^{-\frac{1}{2}}(\Gamma) \times \tilde{H}^{\frac{1}{2}}(\Gamma)$, the function $\mathcal{F}(\alpha, \beta)(\hat{x})$ is the far-field pattern of the radiating solution $P(\alpha, \beta)(x)$ of the Helmholtz equation in $\mathbb{R}^2 \setminus \bar{\Gamma}$, where the potential P is defined by

$$P(\alpha, \beta)(x) := \int_\Gamma \alpha(y)\Phi(x, y)ds_y + \int_\Gamma \beta(y)\frac{\partial}{\partial\nu_y}\Phi(x, y)ds_y. \qquad (8.119)$$

Proceeding as in the proof of Theorem 8.37, using the jump relations across ∂D for the single and double layer potentials with densities extended by zero to ∂D we obtain that $\alpha := -[\partial P/\partial\nu]_\Gamma$ and $\beta := [P]_\Gamma$. Moreover, P satisfies

$$\begin{pmatrix} P^-(\alpha, \beta)|_\Gamma \\ \left(\frac{\partial}{\partial\nu} + i\lambda\right)P^+(\alpha, \beta)|_\Gamma \end{pmatrix} = M \begin{pmatrix} \alpha \\ \beta \end{pmatrix}, \qquad (8.120)$$

where the operator $M : \tilde{H}^{-\frac{1}{2}}(\Gamma) \times \tilde{H}^{\frac{1}{2}}(\Gamma) \to H^{\frac{1}{2}}(\Gamma) \times H^{-\frac{1}{2}}(\Gamma)$ is given by

$$\begin{pmatrix} S_\Gamma & K_\Gamma - I \\ K'_\Gamma - I + i\lambda S_\Gamma & T_\Gamma + i\lambda(I + K_\Gamma) \end{pmatrix}. \qquad (8.121)$$

The operator M is related to the operator A_Γ given in (8.101) by the relation $M = \begin{pmatrix} I & 0 \\ i\lambda kI & I \end{pmatrix} A_\Gamma \begin{pmatrix} I & 0 \\ 0 & -I \end{pmatrix}$, whence $M^{-1} : H^{\frac{1}{2}}(\Gamma) \times H^{-\frac{1}{2}}(\Gamma) \to \tilde{H}^{-\frac{1}{2}}(\Gamma) \times \tilde{H}^{\frac{1}{2}}(\Gamma)$ exists and is bounded. In particular, we have that

$$\mathcal{F}(\alpha, \beta) = BM(\alpha, \beta). \qquad (8.122)$$

In the case of the Dirichlet crack problem (8.87)–(8.89), by proceeding exactly as we did previously, we have $\mathcal{F}_D(\alpha) = BS_\Gamma(\alpha)$, where $\alpha \in \tilde{H}^{-\frac{1}{2}}(\Gamma)$, $B : H^{\frac{1}{2}}(\Gamma) \to L^2[0, 2\pi]$, $\mathcal{F}_D : \tilde{H}^{-\frac{1}{2}}(\Gamma) \to L^2[0, 2\pi]$ is defined by

$$\mathcal{F}_D(\alpha)(\theta) := \gamma \int_\Gamma \alpha(y)e^{-ik\hat{x}\cdot y}\, ds_y \qquad (8.123)$$

and S_Γ is given by (8.100).

Lemma 8.40. *The operator* $\mathcal{F} : \tilde{H}^{-\frac{1}{2}}(\Gamma) \times \tilde{H}^{\frac{1}{2}}(\Gamma) \longrightarrow L^2[0, 2\pi]$ *defined by (8.118) is injective and has a dense range.*

Proof. Injectivity follows from the fact that $\mathcal{F}(\alpha, \beta)$ is the far-field pattern of $P(\alpha, \beta)$ for $(\alpha, \beta) \in \tilde{H}^{-\frac{1}{2}}(\Gamma) \times \tilde{H}^{\frac{1}{2}}(\Gamma)$ given by (8.119). Hence $\mathcal{F}(\alpha, \beta) = 0$ implies $P(\alpha, \beta) = 0$, and so $\alpha := -[\partial P/\partial\nu]_\Gamma = 0$ and $\beta := [P]_\Gamma = 0$. We now note that the transpose operator $\mathcal{F}^\top : L^2[0, 2\pi] \to H^{\frac{1}{2}}(\Gamma) \times H^{-\frac{1}{2}}(\Gamma)$ is given by

$$\gamma^{-1}\mathcal{F}^{\top}g(y) := \begin{cases} v_g^-(y) \\ \dfrac{\partial v_g^+(y)}{\partial \nu_y} \end{cases} \quad y \in \Gamma, \qquad (8.124)$$

where $v_g(y) = \int_0^{2\pi} g(\phi)e^{-ik\hat{x}\cdot y}d\phi$, $\hat{x} = (\cos\phi, \sin\phi)$. From Corollary 6.43, it is enough to show that $\mathcal{F}^{\top}$ is injective. But $\mathcal{F}^{\top}g = 0$ implies that there exists a Herglotz wave function v_g such that $v_g|_\Gamma = 0$ and $\left.\dfrac{\partial v_g}{\partial \nu}\right|_\Gamma = 0$ (note that the limit of v_g and its normal derivative from both sides of the crack is the same). From the representation formula (8.96) and the analyticity of v_g, we now have that $v_g = 0$ in $\mathbb{R}^2$, and therefore $g = 0$. This proves the lemma. □

We obtain a similar result for the operator $\mathcal{F}_D$ corresponding to the Dirichlet crack problem. But in this case $\mathcal{F}_D$ has a dense range only under certain restrictions. More precisely, the following result holds.

Lemma 8.41. *The operator $\mathcal{F}_D : \tilde{H}^{-\frac{1}{2}}(\Gamma) \to L^2[0, 2\pi]$ defined by (8.123) is injective. The range of $\mathcal{F}_D$ is dense in $L^2[0, 2\pi]$ if and only if there does not exist a Herglotz wave function that vanishes on Γ.*

Proof. Injectivity can be proved in the same way as in Lemma 8.40 if one replaces the potential V by the single layer potential.

The dual operator $\mathcal{F}_D^{\top} : L^2[0, 2\pi] \to H^{\frac{1}{2}}(\Gamma)$ in this case coincides with $v_g|_\Gamma$. Hence $\mathcal{F}_D^{\top}$ is injective if and only if there does not exist a Herglotz wave function that vanishes on Γ. □

In polar coordinates $x = (r, \theta)$ the functions

$$u_n(x) = J_n(kr)\cos n\theta, \qquad v_n(x) = J_n(kr)\sin n\theta, \qquad n = 0, 1, \cdots,$$

where J_n denotes a Bessel function of order n, provide examples of Herglotz wave functions. Therefore, by Lemma 8.41, for any straight-line segment the range $\mathcal{F}_D$ (and consequently the range of the far-field operator) is not dense. The same is true for circular arcs with radius R such that kR is a zero of one of the Bessel functions J_n.

From the foregoing analysis we can factorize the far-field operator corresponding to the mixed crack problem as

$$(Fg) = -\mathcal{F}M^{-1}Hg, \qquad g \in L^2[0, 2\pi], \qquad (8.125)$$

and the far-field operator corresponding to the Dirichlet crack problem as

$$(Fg) = -\mathcal{F}_D S_\Gamma^{-1}(v_g|_\Gamma), \qquad g \in L^2[0, 2\pi]. \qquad (8.126)$$

The following lemma will help us to choose an appropriate right-hand side of the far-field equation (8.117).

Lemma 8.42. *For any smooth, nonintersecting arc L and two functions $\alpha_L \in \tilde{H}^{-\frac{1}{2}}(L)$, $\beta_L \in \tilde{H}^{\frac{1}{2}}(L)$ we define $\Phi_\infty^L \in L^2[0, 2\pi]$ by*

$$\Phi_\infty^L(\theta) := \gamma \int_L \alpha_L(y) e^{-ik\hat{x}\cdot y} ds_y + \gamma \int_L \beta_L(y) \frac{\partial}{\partial \nu_y} e^{-ik\hat{x}\cdot y} ds_y \qquad (8.127)$$

$\hat{x} = (\cos\theta, \sin\theta)$. Then, $\Phi_\infty^L \in R(\mathcal{F})$ if and only if $L \subset \Gamma$, where $\mathcal{F}$ is given by (8.118)

Proof. First assume that $L \subset \Gamma$. Then, since $\tilde{H}^{\pm\frac{1}{2}}(L) \subset \tilde{H}^{\pm\frac{1}{2}}(\Gamma)$, it follows directly from the definition of $\mathcal{F}$ that $\Phi_\infty^L \in R(\mathcal{F})$.

Now let $L \not\subset \Gamma$, and assume, on the contrary, that $\Phi_\infty^L \in R(\mathcal{F})$, i.e., there exist $\alpha \in \tilde{H}^{-\frac{1}{2}}(\Gamma)$ and $\beta \in \tilde{H}^{\frac{1}{2}}(\Gamma)$ such that

$$\Phi_\infty^L(\theta) = \gamma \int_\Gamma \alpha(y) e^{-ik\hat{x}\cdot y} ds_y + \gamma \int_\Gamma \beta(y) \frac{\partial}{\partial \nu_y} e^{-ik\hat{x}\cdot y} ds_y.$$

Then, by Rellich's lemma and the unique continuation principle, we have that the potentials

$$\Phi^L(x) = \int_L \alpha_L(y) \Phi(x, y) ds_y + \int_L \beta_L(y) \frac{\partial}{\partial \nu_y} \Phi(x, y) ds_y \qquad x \in \mathbb{R}^2 \setminus \bar{L},$$

$$P(x) = \int_\Gamma \alpha(y) \Phi(x, y) ds_y + \int_\Gamma \beta(y) \frac{\partial}{\partial \nu_y} \Phi(x, y) ds_y \qquad x \in \mathbb{R}^2 \setminus \bar{\Gamma}$$

coincide in $\mathbb{R}^2 \setminus (\bar{\Gamma} \cup \bar{L})$. Now let $x_0 \in L$, $x_0 \notin \Gamma$, and let $\Omega_\epsilon(x_0)$ be a small ball with center at x_0 such that $\Omega_\epsilon(x_0) \cap \Gamma = \emptyset$. Hence P is analytic in $\Omega_\epsilon(x_0)$, while Φ^L has a singularity at x_0, which is a contradiction. Hence $\Phi_\infty^L \notin R(\mathcal{F})$. □

Remark 8.43. The statement and proof of Lemma 8.42 remain valid for the operator $\mathcal{F}_D$ given by (8.123) if we set $\beta_L = 0$ in (8.127).

Now let us denote by $\mathcal{L}$ the set of open, nonintersecting, smooth arcs and look for a solution $g \in L^2[0, 2\pi]$ of the far-field equation

$$-Fg = \mathcal{F}M^{-1}Hg = \Phi_\infty^L \qquad \text{for } L \in \mathcal{L}, \qquad (8.128)$$

where Φ_∞^L is given by (8.127) and F is the far-field operator corresponding to the mixed crack problem. If $L \subset \Gamma$, then the corresponding (α_L, β_L) is in $\tilde{H}^{-\frac{1}{2}}(\Gamma) \times \tilde{H}^{\frac{1}{2}}(\Gamma)$. Since $M(\alpha_L, \beta_L) \in H^{\frac{1}{2}}(\Gamma) \times H^{-\frac{1}{2}}(\Gamma)$, then from Theorem 8.37 for every $\epsilon > 0$ there exists a $g_L^\epsilon \in L^2[0, 2\pi]$ such that

$$\|M(\alpha_L, \beta_L) - Hg_L^\epsilon\|_{H^{\frac{1}{2}}(\Gamma) \times H^{-\frac{1}{2}}(\Gamma)} < \epsilon,$$

whence from the continuity of M^{-1}

$$\|(\alpha_L, \beta_L) - M^{-1} H g_L^\epsilon\|_{\tilde{H}^{-\frac{1}{2}}(\Gamma) \times \tilde{H}^{\frac{1}{2}}(\Gamma)} < C\epsilon, \qquad (8.129)$$

with a positive constant C. Finally (8.125), the continuity of $\mathcal{F}$ and the fact that $\mathcal{F}(\alpha_L, \beta_L) = \Phi_\infty^L$ imply that

$$\|F g_L^\epsilon + \Phi_\infty^L\|_{L^2[0, 2\pi]} < \tilde{C}\epsilon. \qquad (8.130)$$

For some constant $\tilde{C} > 0$ independent of ϵ.

Next, we assume that $L \not\subset \Gamma$. Let $g_n := g_L^{\epsilon_n}$ be such that

$$\|F g_n + \Phi_\infty^L\|_{L^2[0, 2\pi]} < \epsilon_n \qquad (8.131)$$

for some null sequence ϵ_n, and assume that $H g_n$ is bounded in $H^{\frac{1}{2}}(\Gamma) \times H^{-\frac{1}{2}}(\Gamma)$. Thus, without loss of generality we may assume that $H g_n \rightharpoonup (\phi, \psi)$ converge weakly to some $(\phi, \psi) \in H^{\frac{1}{2}}(\Gamma) \times H^{-\frac{1}{2}}(\Gamma)$. The boundedness of M^{-1} implies that $M^{-1} H g_n$ converges weakly to some $(\alpha, \beta) \in \tilde{H}^{-\frac{1}{2}}(\Gamma) \times \tilde{H}^{\frac{1}{2}}(\Gamma)$, and the boundedness of $\mathcal{F}$ implies that $\mathcal{F} M^{-1} H g_n$ converges weakly to $(\mathcal{F}(\alpha, \beta)$ in $L^2[0, 2\pi]$. But from (8.131) we have that $\mathcal{F} M^{-1} H g_n$ converges strongly to $\Phi_\infty^L := (\mathcal{F}(\alpha_L, \beta_L)$, and hence $\Phi_\infty^L = \mathcal{F}(\alpha, \beta)$, which contradicts Lemma 8.42.

We summarize these results in the following theorem, noting that for $L \in \mathcal{L}$ we have that $\rho \to 0$ as $\delta \to 0$.

Theorem 8.44. *Assume that Γ is a nonintersecting, smooth, open arc. For a given nonintersecting smooth arc L, consider Φ_∞^L given in Lemma 8.41 for some $(\alpha_L, \beta_L) \in \tilde{H}^{-\frac{1}{2}}(\Gamma) \times \tilde{H}^{\frac{1}{2}}(\Gamma)$. If F is the far-field operator corresponding to the scattering problem (8.84)–(8.86) and (8.83), then the following is true:*

1. *For $L \subset \Gamma$ and a given $\epsilon > 0$ there exists a function $g_L^\epsilon \in L^2[0, 2\pi]$ satisfying*

$$\|F g_L^\epsilon + \Phi_\infty^L\|_{L^2[0, 2\pi]} < \epsilon$$

   such that $\|v_{g^\epsilon L}\|_{H^1(\Omega_R)}$ is bounded, $v_{g_L^\epsilon}$ is the Herglotz wave function with kernel g_L, and Ω_R is a large enough disk of radius R. Furthermore, the corresponding $H g_L^\epsilon$ given by (8.115) converges to $M(\alpha_L, \beta_L)$ in $H^{\frac{1}{2}}(\Gamma) \times H^{-\frac{1}{2}}(\Gamma)$, where M is given by (8.121).

2. *For $L \not\subset \Gamma$ and a given $\epsilon > 0$ every function $g_L^\epsilon \in L^2[0, 2\pi]$ that satisfies*

$$\|F g_L + \Phi_\infty^L\|_{L^2[0, 2\pi]} < \epsilon$$

   is such that $\lim_{\epsilon \to 0} \|v_{g_L}\|_{H^1(\Omega_R)} = \infty$.

Remark 8.45. The statement and proof of Theorem 8.44 remain valid in the case where F is the far-field operator corresponding to the Dirichlet crack if we set $\beta_L = 0$ in the definition of Φ_∞^L and assume that there does not exist a Herglotz wave function that vanishes on Γ.

In particular, if $L \subset \Gamma$, then we can find a bounded solution to the far-field equation (8.128) with discrepancy ϵ, whereas if $L \not\subset \Gamma$, then there exist solutions to the far-field equation with discrepancy $\epsilon + \delta$ with an arbitrarily large norm in the limit as $\delta \to 0$. For numerical purposes we need to replace Φ_∞^L in the far-field equation (8.128) by an expression independent of L. To this end, assuming that there does not exist a Herglotz wave function that vanishes on L, we can conclude from Lemma 8.41 that the class of potentials of the form

$$\int_L \alpha(y) e^{-ik\hat{x}\cdot y}\, ds_y, \qquad \alpha \in \tilde{H}^{-\frac{1}{2}}(L) \tag{8.132}$$

is dense in $L^2[0, 2\pi]$, and hence for numerical purposes we can replace Φ_∞^L in (8.128) by an expression of the form (8.132). Finally, we note that as L degenerates to a point z, with α_L an appropriate delta sequence, we have that the integral in (8.132) approaches $-\gamma e^{-ik\hat{x}\cdot z}$. Hence, it is reasonable to replace Φ_∞^L by $-\Phi_\infty$, where $\Phi_\infty(\hat{x}, z) := \gamma e^{-ik\hat{x}\cdot z}$ when numerically solving the far-field equation (8.128).

8.9 Numerical Examples

As we explained in the last paragraph of the previous section, to determine the shape of a crack, we compute a regularized solution to the far-field equation

$$\int_0^{2\pi} u_\infty(\theta, \phi) g(\phi)\, d\phi = \gamma e^{-ik\hat{x}\cdot z} \qquad \hat{x} = (\cos\phi, \sin\phi), \ z \in \mathbb{R}^2,$$

where u_∞ is the far-field data of the scattering problem. This is the same far-field equation we used in all the inverse problems presented in this chapter, which emphasizes one of the advantages of the linear sampling method, namely, it does not make use of any a priori information on the geometry of the scattering object.

To solve the far-field equation, we apply the same procedure as in Sect. 8.3. In all our examples, we use synthetic data corrupted with random noise. We show reconstruction examples for four different cracks, all of which are subject to the Dirichlet boundary condition.

1. The curve given by the parametric equation (Fig. 8.11, top left)

$$\Gamma := \left\{ \varrho(s) = \left(2\sin\frac{s}{2}, \sin s \right) : \frac{\pi}{4} \le s \le \frac{7\pi}{4} \right\}.$$

2. The line given by the parametric equation (Fig. 8.11, top right)

$$\Gamma := \{ \varrho(s) = (-2 + s, 2s) : -1 \le s \le 1 \}.$$

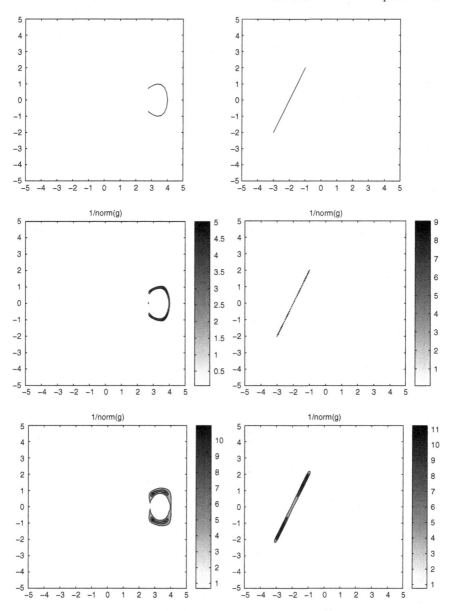

Fig. 8.11. The true object (*top*), reconstruction with 0.5 % noise (*middle*), and with 5 % noise (*bottom*). The wave number is $k = 3^4$

---

[4]Reprinted from F. Cakoni and D. Colton, The linear sampling method for cracks, Inverse Problems 19 (2003), 279–295.

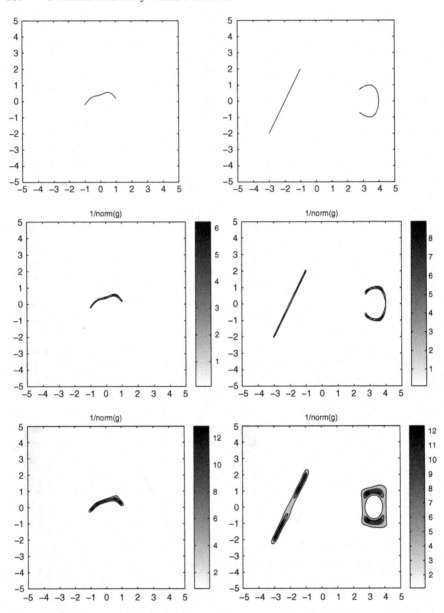

Fig. 8.12. The true object (*top*), reconstruction with 0.5 % noise (*middle*), and with 5 % noise (*bottom*). The wave number is $k = 3^4$

3. The curve given by the parametric equation (Fig. 8.12, top left)

$$\Gamma := \left\{ \varrho(s) = \left(s, 0.5 \cos \frac{\pi s}{2} + 0.2 \sin \frac{\pi s}{2} - 0.1 \cos \frac{3\pi s}{2} \right) : -1 \leq s \leq 1 \right\}.$$

4. Two disconnected curves described as in curves 1 and 2 above (Fig. 8.12, top right).

In all our examples, $k = 3$, and the far-field data are given for 32 incident directions and 32 observation directions equally distributed on the unit circle.

9

Inverse Spectral Problems for Transmission Eigenvalues

We previously encountered transmission eigenvalues and their role in inverse scattering theory in Chap. 6. We now return to this topic and consider the inverse spectral problem for transmission eigenvalues in the simplest possible case, i.e., when the inhomogeneous medium is an isotropic spherically stratified medium in $\mathbb{R}^3$ and the eigenfunctions corresponding to the transmission eigenvalues are spherically symmetric. In this case the inverse spectral problem for transmission eigenvalues reduces to a problem in ordinary differential equations analogous to the inverse Sturm–Liouville problem (cf. [98, 142]) with the important distinction that the spectral problem under consideration is now no longer self-adjoint. Nevertheless, using tools from analytic function theory, we will be able to obtain a partial answer to the question of when a knowledge of the transmission eigenvalues corresponding to spherically symmetric eigenfunctions uniquely determines the (spherically symmetric) index of refraction.

We begin our investigation by recalling the basic results in analytic function theory that will be needed in our analysis, in particular the Hadamard factorization theorem, Laguerre's theorem on critical points of entire functions, and the Paley–Wiener theorem. We then introduce the concept of transformation operators for ordinary differential equations [45,98]. This chapter is then concluded by showing the existence of complex transmission eigenvalues and a proof of an inverse spectral theorem for transmission eigenvalues.

9.1 Entire Functions

We first collect the results from analytic function theory that will be needed for our investigation of the inverse spectral problem for transmission eigenvalues. Only a few proofs will be given. For full proofs and further details we refer the reader to [8] and [123].

F. Cakoni and D. Colton, *A Qualitative Approach to Inverse Scattering Theory*, 263
Applied Mathematical Sciences 188, DOI 10.1007/978-1-4614-8827-9_9,
© Springer Science+Business Media New York 2014

An entire function of a complex variable z is a function f that is analytic in the entire complex plane, i.e.,

$$f(z) = \sum_{n=0}^{\infty} a_n z^n,$$

where the series converges for all finite values of z. We define the maximum modulus $M(r)$ by

$$M(r) := \max_{0 \le \theta \le 2\pi} \left| f(re^{i\theta}) \right|. \tag{9.1}$$

The *order* ρ of $f(z)$ is now defined as

$$\rho := \varlimsup_{r \to \infty} \frac{\log \log M(r)}{\log r}, \tag{9.2}$$

and if $f(z)$ is of order ρ, then the *type* τ is defined as

$$\tau := \varlimsup_{r \to \infty} \frac{\log M(r)}{r^\rho}. \tag{9.3}$$

For example, $f(z) = e^{az}$, $a > 0$, is easily seen to be of order one and type a. Entire functions of order one and finite type are said to be of *exponential type*.

A basic result on entire functions of finite order is the Hadamard factorization theorem. To state this theorem, we first need to introduce the Weierstrass prime factors $E(z, p)$ defined by

$$E(z, 0) := 1 - z,$$
$$E(x, p) := (1 - z) \exp \left\{ z + \frac{1}{2} z + \cdots + \frac{1}{p} z^p \right\}, \, p \ge 1. \tag{9.4}$$

Note that if $|z| \le 1/2$, then $1 - z = e^{\log{(1-z)}}$, and hence

$$E(z, p) = \exp \left\{ \log{(1 - z)} + z + \frac{z^2}{2} + \cdots + \frac{z^p}{p} \right\}$$
$$= e^w,$$

where

$$w = -\sum_{k=p+1}^{\infty} \frac{z^k}{k}.$$

Since $|z| \le 1/2$, we have that

$$|w| \le |z|^{p+1} \sum_{k=p+1}^{\infty} \frac{|z|^{k-p-1}}{k}$$
$$\le |z|^{p+1} \sum_{j=0}^{\infty} 2^{-j}$$
$$\le 2 |z|^{p+1},$$

and hence

$$|1 - E(z, p)| = |1 - e^w|$$
$$\leq e\,|w|$$
$$\leq 2e\,|z|^{p+1}, \qquad |z| \leq \frac{1}{2}. \tag{9.5}$$

Now let $\{z_n\}$ be a set in the complex plane having no finite point of accumulation, and let σ be the *exponent of convergence* of the series

$$\sum_{n=1}^{\infty} \frac{1}{|z_n|^\alpha}, \tag{9.6}$$

i.e., the smallest number σ such that

$$\sum_{n=1}^{\infty} \frac{1}{|z_n|^\alpha} < \infty$$

for all $\alpha > \sigma$, where by hypothesis σ is supposed to be finite. Given σ, we define the integer p as follows:

1. If σ is not an integer, then $p = [\sigma]$, where $[\sigma]$ is the greatest integer less than or equal to σ.
2. If σ is an integer and (9.6) converges for $\alpha = \sigma$, then $p = \sigma - 1$.
3. If σ is an integer and (9.6) does not converge for $\alpha = \sigma$, then $p = \sigma$.

Recall that if $\sum_{n=1}^{\infty} |a_n| < \infty$, then the infinite product $\prod_{n-1}^{\infty} (1 - a_n)$ converges. Hence from the estimate (9.5) we can conclude from the definition of p that the infinite product

$$P(z) := \prod_{n-1}^{\infty} E\left(\frac{z}{z_n}, p\right)$$

converges for all finite z and defines an entire function of z. It can be shown that the entire function $P(z)$ is of order σ. We are now in a position to state the Hadamard factorization theorem.

Theorem 9.1 (Hadamard Factorization Theorem). *Suppose $f(z)$ is an entire function (not identically zero) of finite order ρ and $z_1, z_2, \ldots$ are its zeros (other than the origin) repeated as often as their multiplicities. Then there exists a polynomial $q(z)$ of degree less than or equal to ρ and an integer $p \leq \rho$ chosen as earlier such that $f(z)$ can be expressed as*

$$f(z) = z^k e^{q(z)} \prod_{n=1}^{\infty} E\left(\frac{z}{z_n}, p\right),$$

where k is the multiplicity of the zero at the origin ($k = 0$ if $f(0) \neq 0$).

The following theorem is an easy consequence of the Hadamard factorization theorem, and the theorem and its corollaries will be needed in the sequel.

Theorem 9.2 (Laguerre). *Let $f(z)$ be an entire function of order less than two that is real for real z and has only real zeros. Then the zeros of $f'(z)$ are also all real and are separated from each other by the zeros of $f(z)$.*

Proof. By Hadamard's factorization theorem we have that

$$f(z) = cz^k e^{az} \prod_{n=1}^{\infty} \left(1 - \frac{z}{z_n}\right) e^{z/z_n}, \tag{9.7}$$

where k is zero or a positive integer and c, a, and $z_1, z_2, \ldots$ are all real. Then, taking the logarithm of (9.7) and differentiating gives

$$\frac{f'(z)}{f(z)} = \frac{k}{z} + a + \sum_{n=1}^{\infty}\left(\frac{1}{z - z_n} + \frac{1}{z_n}\right). \tag{9.8}$$

Hence for $z = x + iy$ we have that

$$\mathrm{Im}\left\{\frac{f'(z)}{f(z)}\right\} = -iy\left\{\frac{k}{x^2 + y^2} + \sum_{n=1}^{\infty}\frac{1}{(x - z_n)^2 + y^2}\right\},$$

which is only equal to zero if $y = 0$. Hence $f'(z)$ cannot be zero except on the real axis.

Now differentiate (9.8) again to arrive at

$$\frac{d}{dz}\left(\frac{f'(z)}{f(z)}\right) = -\frac{k}{z^2} - \sum_{n=1}^{\infty}\frac{1}{(z - z_n)^2},$$

which is negative if z is real. Hence $f'(z)/f(z)$ is monotonically decreasing as z goes through real values from z_n to z_{n+1} and hence cannot vanish more than once between z_n and z_{n+1}. However, from (9.8) we see that $f'(z)/f(z)$ changes sign as z goes through real values from z_n to z_{n+1}. Hence $f'(z)/f(z)$ vanishes just once in this interval, and the proof of the theorem is complete. $\square$

Corollary 9.3. *Let $f(z)$ satisfy the assumptions of Theorem 9.2. Then $f'(z)$ cannot vanish more than once inside an interval where it does not change sign.*

A slight modification of the proof of Laguerre's theorem can be used to obtain the following extension of Corollary 9.3 [122].

Corollary 9.4. *Let $f(z)$ be an entire function of order less than two that is real for real z. Suppose that $f(z)$ has infinitely many real zeros and only a finite number of complex ones. Then $f'(z)$ vanishes only once on each interval (z_n, z_{n+1}) formed by two consecutive real zeros of $f(z)$ when the interval is sufficiently far from the origin.*

Laguerre's theorem is not true in general for entire functions of order two. For example, if $f(z) = ze^{z^2}$, then $f'(z) = (2z^2 + 1)e^{z^2}$, and the zeros of $f'(z)$ are complex. On the other hand, if $f(z) = (z^2 - 4)e^{z^2/3}$, then $f'(z) = \frac{2}{3}z(z^2 - 1)e^{z^2/3}$, and the zeros of $f'(z)$ are real but are not separated by those of $f(z)$.

The final topic of this section is the celebrated Paley–Wiener theorem. To motivate this theorem, let $\varphi \in L^2[-A, A]$ and define $f(z)$ by

$$f(z) = \int_{-A}^{A} \varphi(t)e^{izt}\, dt. \tag{9.9}$$

Then it is easily verified that $f(z)$ is an entire function of exponential type at most A. Furthermore, by Plancherel's theorem we have that

$$\int_{-\infty}^{\infty} |f(x)|^2\, dx = 2\pi \int_{-A}^{A} |\varphi(t)|^2\, dt < \infty.$$

The Paley–Wiener theorem says that any entire function of exponential type that is square integrable on the real axis is of the form (9.9).

Theorem 9.5 (Paley–Wiener). *Let $f(z)$ be an entire function of exponential type at most A such that*

$$\int_{-\infty}^{\infty} |f(x)|^2\, dx < \infty.$$

Then there exists $\varphi \in L^2[-A, A]$ such that

$$f(z) = \int_{-A}^{A} \varphi(t)e^{izt}\, dt.$$

9.2 Transformation Operators

To investigate the inverse spectral problem for transmission eigenvalues, we will need to introduce the tool of transformation operators (cf. [45, 98]). These operators map solutions of ordinary differential equations with constant coefficients onto solutions of ordinary differential equations with variable coefficients and have historically played an important role in the inverse scattering problem in quantum mechanics. Here we will only consider the simplest case in which the solution $\frac{1}{k}\sin kx$ of $y'' + k^2 y = 0$ is mapped onto the solution of

$$y'' + (k^2 - p(x))y = 0, \quad 0 \le x \le a, \tag{9.10a}$$
$$y(0) = 0, \quad y'(0) = 1, \tag{9.10b}$$

where $p \in C^1[0, a]$. In particular, we look for a solution of (9.10a), (9.10b) in the form

$$y(x) = \frac{\sin kx}{k} + \int_0^x K(x, t) \frac{\sin kt}{k} \, dt \quad , 0 \le x \le a. \tag{9.11}$$

Substituting (9.11) into (9.10a), (9.10b) and integrating by parts shows that (9.11) will be a solution of (9.10a), (9.10b) provided $K(x, t)$ satisfies

$$K_{xx} - K_{tt} - p(x)K = 0, \quad 0 < t < x < a, \tag{9.12a}$$
$$K(x, 0) = 0, \quad 0 \le x \le a, \tag{9.12b}$$
$$K(x, x) = \frac{1}{2} \int_0^x p(s) \, ds, \quad 0 \le x \le a. \tag{9.12c}$$

To establish the existence of the transformation operator (9.11), we need to show the existence of a function $K(x, t)$ that satisfies (9.12a)–(9.12c) for $\Delta_0 := \{(x, t) : 0 < t < x < a\}$ and is twice continuously differentiable in Δ_0. Following Kirsch [98] we will now use the method of successive approximations to show that $K(x, t)$ exists.

We first extend (9.12a)–(9.12c) to the region $\Delta := \{(x, t) : |t| < x < a\}$ by considering the problem

$$K_{xx} - K_{tt} - p(x)K = 0 \quad \text{in } \Delta, \tag{9.13a}$$
$$K(x, x) = \frac{1}{2} \int_0^x p(s) \, ds, \tag{9.13b}$$
$$K(x, -x) = -\frac{1}{2} \int_0^x p(s) \, ds. \tag{9.13c}$$

Setting $x = \xi + \eta$, $t = \xi - \eta$, and $k(\xi, \eta) = K(\xi + \eta, \xi - \eta)$ we can rewrite (9.13a)–(9.13c) for $(\xi, \eta) \in D := \{(\xi, \eta) : 0 < \xi < a, 0 < \eta < a, 0 < \eta + \xi < a\}$ as

$$\frac{\partial^2 k}{\partial \xi \partial \eta} - p(\xi + \eta)k = 0 \quad \text{in } D, \tag{9.14a}$$
$$k(\xi, 0) = \frac{1}{2} \int_0^\xi p(s) \, ds, \tag{9.14b}$$
$$k(0, \eta) = -\frac{1}{2} \int_0^\eta p(s) \, ds. \tag{9.14c}$$

This initial-value problem is equivalent to the Volterra integral equation

$$k(\xi, \eta) = \int_0^\eta \int_0^\xi p(s + t)k(s, t) \, ds \, dt$$
$$- \frac{1}{2} \int_0^\eta p(s) \, ds + \frac{1}{2} \int_0^\xi p(t) \, dt \tag{9.15}$$

for $(\xi, \eta) \in D$. The integral equation (9.15) can now be solved in a straightforward manner in $C(\overline{D})$ by the method of successive approximations (cf. [87]), and we have thus established the existence of a unique solution $K(x,t) \in C^2(\overline{D})$ of (9.13a)–(9.13c). By uniqueness, the solution $K(x,t)$ is odd with respect to t, i.e., $K(x,0) = 0$, and hence $K(x,t) \in C^2(\overline{D}_0)$ is a solution of (9.12a)–(9.12c).

An integration by parts in (9.11) now shows that for $k > 0$ we have the asymptotic relation

$$y(x) = \frac{\sin kx}{k} + O\left(\frac{1}{k^2}\right)$$

uniformly for $x \in [0, a]$, whereas differentiating (9.11) with respect to x shows that

$$y'(x) = \cos kx + O\left(\frac{1}{k}\right)$$

uniformly for $x \in [0, a]$ as $k \to +\infty$. For subsequent use will also need the following result due to Rundell and Sacks [147] (see also [98], p. 162).

Theorem 9.6. Let $K(x,t) \in C^2(\overline{\Delta}_0)$ satisfy (9.12a)–(9.12c). Then $p \in C^1[0, a]$ is uniquely determined by the Cauchy data $K(a,t)$, $K_x(a,t)$.

9.3 Transmission Eigenvalues

The transmission eigenvalue problem for an isotropic, spherically stratified medium in $\mathbb{R}^3$ is to find a nontrivial solution $v, w \in L^2(B_a)$, $v - w \in H_0^2(B_a)$, to

$$\Delta w + k^2 n(r) w = 0 \quad \text{in } B_a, \tag{9.16a}$$

$$\Delta v + k^2 v = 0 \quad \text{in } B_a, \tag{9.16b}$$

$$v - w = 0 \quad \text{on } \partial B_a, \tag{9.16c}$$

$$\frac{\partial v}{\partial r} - \frac{\partial w}{\partial r} = 0 \quad \text{on } \partial B_a, \tag{9.16d}$$

where B_a is the ball $B_a := \{x : |x| < a\}$ and $n \in C^3[0, a]$. Values of k such that a nontrivial solution to (9.16a)–(9.16d) exists are called *transmission eigenvalues*. If we look for axially symmetric eigenfunctions

$$w(x) = a_0 \frac{y(r)}{r},$$

$$v(x) = b_0 \frac{\sin kr}{kr},$$

with $r = |x|$ and constants a_0, b_0 such that $y(r)$ satisfies

$$y'' + k^2 n(r) y = 0, \tag{9.17a}$$

$$y(0) = 0, y'(0) = 1, \tag{9.17b}$$

then k is a transmission eigenvalue if and only if

$$d(k) := \det \begin{vmatrix} y(a) & -\dfrac{\sin ka}{k} \\ y'(a) & -\cos ka \end{vmatrix} = 0. \tag{9.18}$$

To reduce the initial-value problem (9.17a), (9.17b) to the form discussed in Sect. 9.2, we use the Liouville transformation

$$\xi := \int_0^r \sqrt{n(\rho)}\, d\rho,$$

$$z(\xi) := [n(r)]^{1/4}\, y(r)$$

to arrive at the initial-value problem

$$z'' + [k^2 - p(\xi)]\, z = 0,$$
$$z(0) = 0,\ z'(0) = [n(0)]^{-1/4},$$

where

$$p(\xi) := \frac{n''(r)}{4\,[n(r)]^2} - \frac{5}{16}\frac{[n'(r)]^2}{[n(r)]^3}. \tag{9.19}$$

Assuming for the time being that $k > 0$, we see from Sect. 9.2 that

$$z(\xi) = \frac{\sin k\xi}{k\,[n(0)]^{1/4}} + O\left(\frac{1}{k^2}\right)$$

and

$$z'(\xi) = \frac{\cos k\xi}{[n(0)]^{1/4}} + O\left(\frac{1}{k}\right),$$

and hence the solution $y(r)$ of (9.17a), (9.17b) has the asymptotic behavior

$$y(r) = \frac{1}{k\,[n(0)n(r)]^{1/4}} \sin\left(k\int_0^r \sqrt{n(\rho)}\, d\rho\right) + O\left(\frac{1}{k^2}\right) \tag{9.20a}$$

and

$$y'(r) = \left[\frac{n(r)}{n(0)}\right]^{1/4} \cos\left(k\int_0^r \sqrt{n(\rho)}\, d\rho\right) + O\left(\frac{1}{k}\right) \tag{9.20b}$$

as $k > 0$ tends to infinity. If we assume that $n(a) = 1$, then from (9.18), (9.20b), (9.20b) we have that

$$d(k) = \frac{1}{k\,[n(0)]^{1/4}} \left\{\sin k\left(a - \int_0^a \sqrt{n(\rho)}\, d\rho\right) + O\left(\frac{1}{k}\right)\right\}. \tag{9.21}$$

Hence if $k > 0$ and $n(a) = 1$, then we see that if

$$a - \int_0^a \sqrt{n(\rho)} \, d\rho \neq 0,$$

then there exists an infinite number of positive transmission eigenvalues having spherically symmetric eigenfunctions.

In the case where $n(a) \neq 1$, we have that

$$d(k) = \frac{1}{k} \left(B \sin k\delta \cos ka - C \cos k\delta \sin ka \right) + O\left(\frac{1}{k^2}\right), \tag{9.22}$$

where

$$B = \frac{1}{(n(0)n(a))^{1/4}}, \qquad C = \left(\frac{n(a)}{n(0)}\right)^{1/4}$$

and

$$\delta = \int_0^a \sqrt{n(\rho)} \, d\rho, \tag{9.23}$$

and in this case it can also be shown that there exists an infinite number of positive transmission eigenvalues [60].

However, in general, there can also exist complex transmission eigenvalues. To see this, we consider the special case where $n(r) = n_0^2$ where $n_0 > 0$, $n_0 \neq 1$, is a constant [122].

Theorem 9.7. *Let $n(r) = n_0^2$, where n_0 is a positive constant not equal to one. Then, if n_0 is an integer or the reciprocal of an integer, all the transmission eigenvalues are real. If n_0 is not an integer or the reciprocal of an integer, then there are infinitely many real and complex transmission eigenvalues.*

Proof. If $n(r) = n_0^2$, then

$$y(r) = \frac{1}{kn_0} \sin kn_0 r,$$

and hence from (9.18) we need to consider the zeros of $F(k) := n_0 k d(k)$, where

$$F(k) = \sin(n_0 ka) \cos ka - n_0 \cos(n_0 ka) \sin ka,$$

i.e., k is a transmission eigenvalue if and only if $F(k) = 0$. $F(k)$ is clearly an entire function of k of exponential type, and we can rewrite $F(k)$ as

$$F(k) = (n_0^2 - 1) \int_0^{ka} \sin n_0 x \sin x \, dx. \tag{9.24}$$

It suffices to only consider the case $n_0 > 1$ since $F(k) = 0$ if and only if

$$\int_0^{n_0 ka} \sin(t/n_0) \sin t \, dt = 0.$$

When $n_0 > 1$ is an integer, the roots of $F(k) = 0$ are simply the critical points (i.e., the points where the derivative vanishes) of the entire function

$$\frac{\sin n_0 ka}{\sin ka}.$$

Hence, by Laguerre's theorem, all the zeros of $F(k)$ are real. We have already noted that if $n_0 \neq 1$, then $F(k) = n_0 k d(k)$ has an infinite number of positive zeros. To show that $F(k)$ also has an infinite number of complex zeros when $n_0 > 1$ is not an integer, we note from (9.24) that

$$F'(k) = a\left(n_0^2 - 1\right)\sin\left(n_0 ka\right)\sin ka, \tag{9.25}$$

which has zeros at

$$\{\pi/an_0, 2\pi/an_0, \cdots, j\pi/an_0, \cdots\}$$

and at

$$\{\pi/a, 2\pi/a, \cdots, j\pi/a, \cdots\}.$$

Our goal is to argue that there are infinitely many real intervals where $F(k)$ does not change sign and has at least two consecutive critical points inside. The desired result will then follow from Corollary 9.4 of Laguerre's theorem.

Let m be a fixed positive integer. Then, in the case where the interval $(mn_0 - 1, mn_0)$ contains an integer j, we have that $mn_0\pi = j\pi + \epsilon\pi$ for some ϵ in $(0, 1)$, where ϵ may vary with m and j. This case will certainly be true when n_0 is irrational, and we consider this case first. Then from (9.25) we have that $j\pi/an_0$ and $m\pi/a$ are two consecutive critical points of $F(k)$ and

$$F\left(\frac{m\pi}{a}\right) = (-1)^m \sin\left(n_0 m\pi\right) = (-1)^m \sin\left(j\pi + \epsilon\pi\right) = (-1)^{j+m}\sin\epsilon\pi,$$

$$F\left(\frac{j\pi}{an_0}\right) = (-1)^{j+1}n_0\sin\left(\frac{j\pi}{n_0}\right) = (-1)^{j+1}n_0\sin\left(m\pi - \frac{\epsilon\pi}{n_0}\right)$$

$$= (-1)^{j+m}n_0\sin\left(\frac{\epsilon\pi}{n_0}\right).$$

Hence the signs of $F\left(\frac{m\pi}{a}\right)$ and $F\left(\frac{j\pi}{an_0}\right)$ are identical. Furthermore, $F(k)$ cannot change sign in $\left(\frac{j\pi}{an_0}, \frac{m\pi}{a}\right)$ since otherwise there would be another critical point created inside this interval.

Now assume that n_0 is rational but not an integer. Then $F(\pi/a) = -\sin n_0\pi \neq 0$. We will show that its sign agrees with that of $F(k)$ at the critical point just before π/a. Pick the integer j such that $j < n_0 < j+1$, i.e., $j\pi/an_0 < \pi/a < (j+1)\pi/an_0$. Then, as previously noted, $j\pi/an_0$ is a critical point and

$$F(j\pi/an_0) = (-1)^{j+1}n_0\sin\left(\frac{j\pi}{n_0}\right).$$

We now note that $-\sin n_0\pi$ is positive if j is odd and negative if j is even, with the same conclusion also applying to the sign of $(-1)^{j+1}n_0\sin\left(\frac{j\pi}{n_0}\right)$. Since $F(k)$ is periodic if n_0 is rational, we again can conclude that there are infinitely many real intervals where $F(k)$ does not change sign and has at least two consecutive critical points inside.   □

Example 9.8 ([1]). Let $n(r) = n_0^2$, and again let $F(k) := n_0 kd(k)$. When $n_0 = 1/2$, we have that

$$F(k) = -2n_0\sin^3\left(\frac{ka}{2}\right),$$

and hence $F(k)$ has an infinite set of real zeros at $k = 2\pi j/a, j = 0, 1, 2, \cdots$, each having multiplicity three, and has no complex zeros. On the other hand, when $n_0 = 2/3$, we have that

$$F(k) = -n_0\sin^3\left(\frac{ka}{3}\right)\left[3 + 2\cos\left(\frac{2ka}{3}\right)\right],$$

and hence $F(k)$ has an infinite set of real zeros at $k = 3\pi j/a, j = 0, 1, 2, \cdots$, and an infinite set of simple complex zeros at the roots of $\cos(2ka/3) = -3/2$.

We now consider the case where $n(r)$ is no longer constant and determine where the eigenvalues lie in the complex plane ([56]; see also [85]).

Theorem 9.9. *Assume that $n(a) \neq 1$. Then if complex eigenvalues exist, they all lie in a strip parallel to the real axis.*

Proof. From (9.22) we need to investigate the zeros of

$$d(k) = \frac{1}{k}\left(B\sin k\delta\cos ka - C\cos k\delta\sin ka\right) + O\left(\frac{1}{k^2}\right),$$

where B, C, and δ are defined in (9.23). In particular, since $n(a) \neq 1$, $B/C = 1/\sqrt{n(a)} \neq 1$. We first consider

$$T(k) := B\sin k\delta\cos ka - C\cos k\delta\sin ka$$

and note that $T(k)$ is an entire function of exponential type $\tau := \delta + a$. For future convenience we rewrite $T(k)$ as

$$T(k) = \frac{B-C}{2}\sin k(\delta + a) + \frac{B+C}{2}\sin k(\delta - a). \tag{9.26}$$

Returning to the determinant $d(k)$, it is easily verified that $d(k)$ is an entire function of exponential type, and since $B \neq C$, $d(k)$ is of type τ. We now write

$$kd(k) = T(k) + E(k).$$

Then from (9.26) and the preceding discussion we have that $E(k)$ is an entire function of exponential type at most τ, and from (9.22) for k real, we have that

$$E(k) = O\left(\frac{1}{k}\right).$$

This implies that $E(k)$ is a square integrable function on the real axis. Hence by the Paley–Wiener theorem (Theorem 9.5) there exists $\varphi \in L^2[-\tau, \tau]$ such that

$$E(k) = \int_{-\tau}^{\tau} \varphi(t)e^{ikt}\,dt.$$

Setting $y = \text{Im}(k)$ and using the Cauchy–Schwarz inequality we now have that

$$|E(k)| \leq \|\varphi\|\sqrt{\frac{\sin h\,(2\tau\,|y|)}{|y|}},$$

which implies that $|E(k)|\,e^{-\tau|y|}$ goes to zero as $y = \text{Im}(k)$ goes to infinity.

Now suppose that $d(k)$ has a sequence of zeros k_j with $y_j = \text{Im}(k_j)$ going to infinity. Then $T(k_j) + E(k_j) = 0$, and $E(k_j)e^{-\tau|y_j|}$ goes to zero as y_j goes to infinity. However, from (9.26) the modulus of $T(k_j)e^{-\tau|y_j|}$ goes to $|B - C|/4$ as y_j goes to infinity. Hence $d(k)$ cannot have an infinite number of zeros whose imaginary part goes to infinity. The proof of the theorem is now complete. □

9.4 An Inverse Spectral Theorem

We now turn our attention to deriving an inverse spectral theorem for transmission eigenvalues. In particular, we consider the transmission eigenvalue problem (9.16a)–(9.16d) and ask if the transmission eigenvalues (real and complex) corresponding to axially symmetric eigenfunctions $v(r)$ and $w(r)$ uniquely determine the spherically stratified index of refraction $n(r)$. As shown in Sect. 9.3, the transmission eigenvalue problem for axially symmetric solutions to (9.16a)–(9.16d) can be reduced to the one-dimensional problem of determining eigenvalues k such that there exists a nontrivial solution to the coupled set of ordinary differential equations

$$\begin{aligned} y'' + k^2 n(r)y &= 0, \quad 0 < r < a, \\ y_0'' + k^2 y_0 &= 0, \quad 0 < r < a, \end{aligned} \tag{9.27a}$$

$$\begin{aligned} y(a) &= y_0(a), \\ y'(a) &= y_0'(a), \end{aligned} \tag{9.27b}$$

where $y(0) = y_0(0) = 0$. As we saw in Sect. 9.3, there exist in general an infinite number of real and complex eigenvalues $\{k_j\}$, and the inverse spectral problem is to determine $n(r)$ from a knowledge of $\{k_j\}$ (including multiplicities). The earliest results on this problem were given by McLaughlan and Polyakov [126], with further contributions being given by Aktosun et al. [1] and Aktosun and Papanicolaou [2]. Here we will follow the analysis of Colton and Leung [56].

We assume that $n \in C^3[0, a]$ such that $n(a) = 1$ and $n'(a) = 0$. Using the transformation operator introduced in Sect. 9.2 and Liouville transformation given in Sect. 9.3 we can represent the solution $y(r)$ of (9.27a) satisfying $y'(0) = 1$ in the form

$$y(r) = \frac{1}{(n(0)n(r))^{1/4}} \left[\frac{\sin\left(k \int_0^r \sqrt{n(\rho)}\, d\rho\right)}{k} \right.$$
$$\left. + \int_0^{\int_0^r \sqrt{n(\rho)}\, d\rho} K\left(\int_0^r \sqrt{n(\rho)}\, d\rho, t\right) \frac{\sin kt}{k}\, dt \right], \qquad (9.28)$$

where $K(x, t)$ is the unique solution of (9.13a)–(9.13c). Recall further that k is a transmission eigenvalue if and only if k is a zero of the determinant $d(k)$ defined by (9.18). From (9.11) we see that $d(k)$ is an even entire function of k of order (at most) one. It can furthermore be shown [19] that if

$$\int_0^a \rho^2 \left[1 - n(\rho)\right] d\rho \neq 0,$$

then $d(k)$ has a zero of order two at the origin. Since $d\left(\sqrt{k}\right)$ is an entire function of order (at most) one-half, we can now conclude by the Hadamard factorization theorem (Theorem 9.1) that

$$d(k) = ck^2 \prod_{j=1}^{\infty} \left(1 - k^2/k_j^2\right), \qquad (9.29)$$

where $\{k_j\}$ are the zeros of $d(k)$ (including multiplicities) and c is a constant. Since as $k \to \infty$ along the positive real axis we have the asymptotic behavior (9.21), the constant c is determined if $n(0)$ is known. Hence the transmission eigenvalues, together with $n(0)$, uniquely determine $d(k)$. We are now in a position to prove the following theorem.

Theorem 9.10. *Assume that $n \in C^3[0, a]$, $n(a) = 1$, and $n'(a) = 0$, and $n(0)$ is given. Then, if $0 < n(r) < 1$ for $0 < r < a$, the transmission eigenvalues (including multiplicity) uniquely determine $n(r)$.*

Proof. From (9.28) and the fact that $n(a) = 1$ and $n'(a) = 0$ we have that

$$y(a) = \frac{1}{(n(0))^{1/4}} \left[\frac{\sin k\delta}{k} + \int_0^\delta K(\delta, t) \frac{\sin kt}{k}\, dt \right], \qquad (9.30)$$

$$y'(a) = \frac{1}{(n(0))^{1/4}} \left[\cos k\delta + \frac{\sin k\delta}{2k} \int_0^\delta p(s)\, ds + \int_0^\delta K_\xi(\delta, t) \frac{\sin kt}{k}\, dt \right],$$

where $p(\xi)$ is defined by (9.19) and, as in (9.23),

$$\delta := \int_0^a \sqrt{n(\rho)}\,d\rho. \qquad (9.31)$$

From the discussion preceding Theorem 9.10 and the assumptions of the theorem we can assume that $d(k)$ is known, and hence from (9.21) we can determine δ. Evaluating $d(k)$ at $k = \ell\pi/a$, $\ell = 1, 2, \cdots$ now gives

$$\frac{\ell\pi}{a}d\left(\frac{\ell\pi}{a}\right) = \frac{(-1)^{\ell+1}}{(n(0))^{1/4}}\left[\sin\frac{\ell\pi\delta}{a} + \int_0^\delta K(\delta,t)\sin\frac{\ell\pi t}{a}\,dt\right], \qquad (9.32)$$

and, since $\{\sin\frac{\ell\pi t}{a}\}$ is complete in $L^2[0.\delta]$ if $\delta < a$ [161, p. 97], from (9.32) and the assumptions of the theorem we can conclude that $K(\delta,t)$ is uniquely determined. Now set $k = \ell\pi/\delta$. Then from (9.30) we have that

$$\begin{aligned}
\frac{\ell\pi}{\delta}d\left(\frac{\ell\pi}{\delta}\right) &= -y(a)\frac{\ell\pi}{\delta}\cos\frac{\ell\pi a}{\delta} \\
&+ \frac{\sin\frac{\ell\pi a}{\delta}}{(n(0))^{1/4}}\left[(-1)^\ell + \frac{\delta}{\ell\pi}\int_0^\delta K_\xi(\delta,t)\sin\frac{\ell\pi t}{\delta}\,dt\right],
\end{aligned} \qquad (9.33)$$

and since $K(\delta,t)$ is known by the preceding discussion and $n(0)$ is known by assumption, we have that $y(a)$ is known. Hence we can conclude from (9.33) and the completeness of $\{\sin\frac{\ell\pi t}{\delta}\}$ in $L^2[0.\delta]$ that $K_\xi(\delta,t)$ is uniquely determined. From Theorem 9.6 we can now conclude that, under the assumptions of the theorem, $p(\xi)$ is uniquely determined for $0 \le \xi \le \delta$.

Now suppose that there were two refractive indices $n_1(r)$ and $n_2(r)$ satisfying the assumptions of the theorem such that the transmission eigenvalue problems corresponding to $n_1(r)$ and $n_2(r)$ have the same eigenvalues (including multiplicity). Then the preceding discussion implies that $p(\xi_i)$ is uniquely determined, where

$$\xi_i := \int_0^r \sqrt{n_i(\rho)}\,d\rho \quad, i = 1, 2.$$

Then from (9.19) and the fact that $n_i(a) = 1$ and $n_i'(a) = 0$ we have that $n_i(r(\xi_i))$ satisfies

$$\begin{aligned}
(n_i^{1/4})'' - p(\xi_i)n_i^{1/4} &= 0, \quad 0 < \xi < \delta, \\
n_i^{1/4}(r(\delta)) &= 1, \qquad\qquad\qquad (9.34) \\
\left(n_i^{1/4}\right)'(r(\delta)) &= 0
\end{aligned}$$

for $i = 1, 2$, where the derivatives are with respect to ξ_i. Hence by the uniqueness theorem for the initial-value problem for linear ordinary differential equations we have that $n_1(r(\cdot)) = n_2(r(\cdot))$. But $r_i = r(\xi_i)$ satisfies

$$\frac{dr_i}{d\xi_i} = \frac{1}{\sqrt{n_i(r(\xi_i))}},$$ (9.35)

$$r_i(0) = 0,$$

and this initial-value problem also has a unique solution, i.e., $r_1(\cdot) = r_2(\cdot)$. This now implies that $\xi_1 = \xi_2$, and hence $n_1(r) = n_2(r)$.   □

Theorem 9.10 has also been proved by Aktosun et al. [1] and Aktosun and Papanicolaou [2] without the assumption that $n(0)$ is known. At the time of this writing, a uniqueness theorem analogous to Theorem 9.10 for the case where $n(r) > 1$ for $0 < r < a$ is unknown. However, if all the transmission eigenvalues for (9.16a)–(9.16d) are known (including the case where the eigenfunctions are not necessarily axially symmetric) and $n(r) > 1$ for $0 < r < 1$ with $n(0)$ known, then the unique determination of $n(r)$ from the transmission eigenvalues is established in [19].

10

A Glimpse at Maxwell's Equations

In the preceding chapters, we used the scattering of electromagnetic waves by an infinite cylinder as our model, thereby reducing the three-dimensional Maxwell system to a two-dimensional scalar equation. In this last chapter, we want to briefly indicate the modifications needed to treat three-dimensional electromagnetic scattering problems. In view of the introductory nature of our book, our presentation will be brief, and for details we will refer the reader to Chap. 14 of [129] and the monograph [26].

There are two basic problems that arise in treating three-dimensional electromagnetic scattering problems. The first of these problems is that the formulation of the direct scattering problem must be done in function spaces that are more complicated than those used for two-dimensional problems. The second problem follows from the first in that, due to more complicated function spaces, the mathematical techniques used to study both the direct and inverse problems become rather sophisticated. Nevertheless, the logical scheme one must follow to obtain the desired theorems is basically the same as that followed in the two-dimensional case.

We first consider the scattering of electromagnetic waves by a (possibly) partially coated obstacle D in $\mathbb{R}^3$. We assume that D is a bounded region with smooth boundary ∂D such that $D_e := \mathbb{R}^3 \setminus \bar{D}$ is connected. We assume that the boundary ∂D is split into two disjoint parts, ∂D_D and ∂D_I, where ∂D_D and ∂D_I are disjoint, relatively open subsets (possibly disconnected) of ∂D, and let ν denote the unit outward normal to ∂D. We allow the possibility that either ∂D_D or ∂D_I is the empty set. The direct scattering problem we are interested in is to determine an electromagnetic field E, H such that

$$\operatorname{curl} E - ikH = 0,$$
$$\operatorname{curl} H + ikE = 0 \tag{10.1}$$

for $x \in D_e$ and

$$\nu \times E = 0 \quad \text{on } \partial D_D, \tag{10.2}$$
$$\nu \times \operatorname{curl} E - i\lambda(\nu \times E) \times \nu = 0 \quad \text{on } \partial D_I, \tag{10.3}$$

F. Cakoni and D. Colton, *A Qualitative Approach to Inverse Scattering Theory*, Applied Mathematical Sciences 188, DOI 10.1007/978-1-4614-8827-9_10, © Springer Science+Business Media New York 2014

where $\lambda > 0$ is the surface impedance, which, for the sake of simplicity, is assumed to be a (possibly different) constant on each connected subset of ∂D_I. Note that the case of a perfect conductor corresponds to the case where $\partial D_I = \emptyset$, and the case of an imperfect conductor corresponds to the case where $\partial D_D = \emptyset$. We introduce the incident fields

$$
\begin{aligned}
E^i(x) : &= \frac{i}{k}\,\mathrm{curl}\,\mathrm{curl}\,pe^{ikx\cdot d} \\
&= ik(d \times p) \times de^{ikx\cdot d},
\end{aligned}
\tag{10.4}
$$

$$
\begin{aligned}
H^i(x) : &= \mathrm{curl}\,pe^{ikx\cdot d} \\
&= ikd \times pe^{ikx\cdot d},
\end{aligned}
\tag{10.5}
$$

where $k > 0$ is the wave number, $d \in \mathbb{R}^3$ is a unit vector giving the direction of propagation, and $p \in \mathbb{R}^3$ is the polarization vector. Finally, the scattered field E^s, H^s defined by

$$
\begin{aligned}
E &= E^i + E^s, \\
H &= H^i + H^s
\end{aligned}
\tag{10.6}
$$

is required to satisfy the *Silver–Müller radiation condition*

$$
\lim_{r\to\infty} (H^s \times x - rE^s) = 0
\tag{10.7}
$$

uniformly in $\hat{x} = x/|x|$, where $r = |x|$.

The scattering problem (10.1)–(10.7) is a special case of the exterior mixed boundary value problem

$$
\begin{aligned}
\mathrm{curl}\,\mathrm{curl}\,E - k^2 E &= 0 \quad \text{in } D_e, & (10.8)\\
\nu \times E &= f \quad \text{on } \partial D_D, & (10.9)\\
\nu \times \mathrm{curl}\,E - i\lambda(\nu \times E) \times \nu &= h \quad \text{on } \partial D_I, & (10.10)\\
\lim_{r\to\infty} (H \times x - rE) &= 0 & (10.11)
\end{aligned}
$$

for prescribed functions of f and h, with $H = \frac{1}{ik}\,\mathrm{curl}\,E$. The first problem that needs to be addressed concerns the conditions on f and h under which there exists a unique solution to (10.8)–(10.11). To this end, we define

$$
X(D, \partial D_I) := \left\{ u \in H(\mathrm{curl}, D) : \nu \times u|_{\partial D_I} \in L_t^2(\partial D_I) \right\}
$$

equipped with the norm

$$
\|u\|_{X(D,\partial D)}^2 := \|u\|_{H(\mathrm{curl},D)}^2 + \|\nu \times u\|_{L^2(\partial D_I)}^2,
$$

where

$$
H(\mathrm{curl}, D) := \left\{ u \in \left(L^2(D)\right)^3 : \mathrm{curl}\,u \in \left(L^2(D)\right)^3 \right\},
$$

$$
L_t^2(\partial D_I) := \left\{ u \in \left(L^2(\partial D_I)\right)^3 : \nu \times u = 0 \quad \text{on } \partial D_I \right\},
$$

with norms

$$\|u\|_{H(\mathrm{curl},D)}^2 := \|u\|_{(L^2(D))^3} + \|\mathrm{curl}\,u\|_{(L^2(D))^3},$$

$$\|u\|_{L_t^2(\partial D_I)} = \|u\|_{(L^2(\partial D_I))^3},$$

respectively. As in Chap. 3, we can also define the spaces $X_{loc}(D_e, \partial D_I)$ and $H_{loc}(\mathrm{curl}, D_e)$. Finally, we introduce the trace space of $X(D, \partial D_I)$ on the complementary part ∂D_D by

$$Y(\partial D_D) := \left\{ f \in \left(H^{-1/2}(\partial D_D) \right)^3 : \text{there exists } u \in H_0(\mathrm{curl}, \Omega_R) \right.$$

$$\left. \text{such that } \nu \times u|_{\partial D_I} \in L_t^2(\partial D_I) \text{ and } f = \nu \times u|_{\partial D_D} \right\},$$

where $D \subset \Omega_R = \{x : |x| < R\}$ and

$$H_0(\mathrm{curl}, \Omega_R) := \left\{ u \in H(\mathrm{curl}, \Omega_R) : \nu \times u|_{\partial \Omega_R} = 0 \right\}.$$

The trace space is equipped with the norm

$$\|f\|_{Y(\partial D_D)}^2 := \inf \left\{ \|u\|_{H(\mathrm{curl},\Omega_R)}^2 + \|\nu \times u\|_{L^2(\partial D_I)}^2 \right\},$$

where the minimum is taken over all functions $u \in H_0(\mathrm{curl}, \Omega_R)$ such that $\nu \times u|_{\partial D_I} \in L_t^2(\partial D_I)$ and $f = \nu \times u|_{\partial D_D}$ (for details see [129]). We now have the following theorem [24].

Theorem 10.1. *Given $f \in Y(\partial D_D)$ and $h \in L_t^2(\partial D_I)$, there exists a unique solution $E \in X_{loc}(D_e, \partial D_I)$ to (10.8)–(10.11) such that*

$$\|E\|_{X(D_e \cap \Omega_R, \partial D_I)} \leq C(\|f\|_{Y(\partial D_D)} + \|h\|_{L^2(\partial D_I)})$$

for some positive constant C depending on R but not on f and h.

We now turn our attention to the inverse problem of determining D and λ from a knowledge of the far-field data of the electric field. In particular, from [54] it is known that the solution E^s, H^s to (10.1)–(10.7) has the asymptotic behavior

$$E^s(x) = \frac{e^{ik|x|}}{|x|} \left\{ E_\infty(\hat{x}, d, p) + O\left(\frac{1}{|x|}\right) \right\},$$

$$H^s(x) = \frac{e^{ik|x|}}{|x|} \left\{ H_\infty(\hat{x}, d, p) + O\left(\frac{1}{|x|}\right) \right\} \tag{10.12}$$

as $|x| \to \infty$, where $E_\infty(\cdot, d, p)$ and $H_\infty(\cdot, d, p)$ are tangential vector fields defined on the unit sphere S^2 and are known as the electric and magnetic far-field patterns, respectively. Our aim is to determine λ and D from $E_\infty(\hat{x}, d, p)$

with no a priori assumption or knowledge of Γ_D, Γ_I, and λ. The solution of this inverse scattering problem is unique, and this can be proved following the approach described in Theorem 7.1 of [54] (where only the well-posedness of the direct scattering problem is required).

The derivation of the linear sampling method for the vector case now under consideration follows the same approach as the scalar case discussed in Sect. 8.2. In particular, we begin by defining the *far-field operator* F : $L_t^2(S^2) \to L_t^2(S^2)$ by

$$(Fg)(\hat{x}) := \int_{S^2} E_\infty\left(\hat{x}, d, g(d)\right) ds(d) \tag{10.13}$$

and define the *far-field equation* by

$$Fg = E_{e,\infty}(\hat{x}, z, q), \tag{10.14}$$

where $E_{e,\infty}$ is the electric far-field pattern of the *electric dipole*

$$E_e(x, z, q) := \frac{i}{k} \operatorname{curl}_x \operatorname{curl}_x q\, \Phi(x, z),$$
$$H_e(x, z, q) := \operatorname{curl}_x q\, \Phi(x, z), \tag{10.15}$$

where $q \in \mathbb{R}^3$ is a constant vector and Φ is the fundamental solution of the Helmholtz equation given by

$$\Phi(x, z) := \frac{e^{ik|x-z|}}{4\pi |x - z|}. \tag{10.16}$$

We can explicitly compute $E_{e,\infty}$, arriving at

$$E_{e,\infty}(\hat{x}, z, q) = \frac{ik}{4\pi}(\hat{x} \times q) \times \hat{x} e^{-ik\hat{x}\cdot z}. \tag{10.17}$$

Note that the far-field operator given by (10.13) is linear since $E_\infty(\hat{x}, d, p)$ depends linearly on the polarization p.

We now return to the exterior mixed boundary value problem (10.8)–(10.11) and introduce the linear operator $B : Y(\partial D_D) \times L_t^2(\partial D_I) \to L_t^2(S^2)$ mapping the boundary data (f, h) onto the electric far-field pattern E_∞. In [24], it is shown that this operator is injective and compact and has a dense range in $L_t^2(S^2)$. Using B it is now possible to write the far-field equation as

$$- (B \Lambda E_g)(\hat{x}) = \frac{1}{ik} E_{e,\infty}(\hat{x}, z, q), \tag{10.18}$$

where Λ is the trace operator corresponding to the mixed boundary condition, i.e., $\Lambda u := \nu \times u|_{\partial D_D}$ on ∂D_D and $\Lambda u := \nu \times \operatorname{curl} u - i\lambda(\nu \times u) \times \nu|_{\partial D_I}$ on ∂D_I, and E_g is the electric field of the electromagnetic Herglotz pair with kernel $g \in L_t^2(S^2)$ defined by

$$E_g(x) := \int_{S^2} e^{ikx \cdot d} g(d) \, ds(d),$$

$$H_g(x) := \frac{1}{ik} \operatorname{curl} E_g(x).$$

(10.19)

We note that $E_{e,\infty}(\hat{x}, z, q)$ is in the range of B if and only if $z \in D$ [24].
Finally, we consider the interior mixed boundary value problem

$$\operatorname{curl} \operatorname{curl} E - k^2 E = 0 \quad \text{in } D,$$ (10.20)

$$\nu \times E = f \quad \text{on } \partial D_D,$$ (10.21)

$$\nu \times \operatorname{curl} E - i\lambda(\nu \times E) \times \nu = h \quad \text{on } \partial D_I,$$ (10.22)

where $f \in Y(\partial D_D)$, $h \in L^2(\partial D_I)$. It is shown in [24] that if $\partial D_I \neq \emptyset$, then there exists a unique solution to (10.20)–(10.22) in $X(D, \partial D_I)$ and that the following theorem is valid.

Theorem 10.2. *Assume that $\partial D_I \neq \emptyset$. Then the solution E of the interior mixed boundary value problem* (10.20)–(10.22) *can be approximated in $X(D, \partial D_I)$ by the electric field of an electromagnetic Herglotz pair.*

The factorization (10.18), together with Theorem 10.2, now allows us to prove the following theorem [24].

Theorem 10.3. *Assume that $\partial D_I \neq \emptyset$. Then if F is the far-field operator corresponding to the scattering problem* (10.1)–(10.7), *then we have that*

1. *For $z \in D$ and a given $\epsilon > 0$ there is a function $g_z^\epsilon \in L_t^2(S^2)$ satisfying the inequality*

$$\|F g_z^\epsilon - E_{e,\infty}(\cdot, z, q)\|_{L_t^2(S^2)} < \epsilon,$$

   and the electric field of the electromagnetic Herglotz pair $E_{g_z^\epsilon}$ with kernel g_z^ϵ converges to the unique solution of (10.20)–(10.22) *with $f := \nu \times E_e(\cdot, z, q)$ and $h := \nu \times \operatorname{curl} E_e(\cdot, z, q) - i\lambda(\nu \times E_e(\cdot, z, q)) \times \nu$;*
2. *For $z \in D_e$ and a given $\epsilon > 0$ every function $g_z^\epsilon \in L_t^2(S^2)$ that satisfies*

$$\|F g_z^\epsilon - E_{e,\infty}(\cdot, z, q)\|_{L_t^2(S^2)} < \epsilon$$

   is such that

$$\lim_{\epsilon \to 0} \|E_{g_z^\epsilon}\|_{X(D, \partial D_I)} = \infty.$$

Theorem 10.3 is also valid for the case of a perfect conductor (i.e., $\partial D_I = \emptyset$) provided we modify the far-field operator F in an appropriate manner [16]. For numerical examples demonstrating the use of Theorem 10.3 in reconstructing D, see [24, 42, 48]. By a method analogous to that of Sect. 4.4 for the scalar case, the function g_z can also be used to determine the surface impedance λ [17]. The case of mixed boundary value problems for screens was examined in [29]. We also remark that the factorization method is not established for obstacle scattering for Maxwell's equations.

We next examine the case of Maxwell's equations in an inhomogeneous anisotropic medium (which, of course, includes isotropic media as a special case). We again assume that $D \subset \mathbb{R}^3$ is a bounded domain with connected complement such that its boundary ∂D is in class C^2 with unit outward normal ν. Let N be a 3×3 symmetric matrix whose entries are piecewise continuous, complex-valued functions in $\mathbb{R}^3$ such that N is the identity matrix outside D. We further assume that there exists a positive constant $\gamma > 0$ such that

$$\mathrm{Re} \left(\bar{\xi} \cdot N(x)\xi \right) \geq \gamma |\xi|^2$$

for every $\xi \in \mathbb{C}^3$ where N is continuous and

$$\mathrm{Im} \left(\bar{\xi} \cdot N(x)\xi \right) > 0$$

for every $\xi \in \mathbb{C}^3 \setminus \{\emptyset\}$ and points $x \in D$ where N is continuous. Finally, we assume that $N - I$ is invertible and $\mathrm{Re}(N - I)^{-1}$ is uniformly positive definite in D (partial results for the case where this is not true can be found in [139]).

Now consider the scattering of the time-harmonic incident field (10.4), (10.5) by an anisotropic inhomogeneous medium D with refractive index N satisfying the preceding assumptions. Then the mathematical formulation of the scattering of a time-harmonic plane wave by an anisotropic medium is to find $E \in H_{loc}(\mathrm{curl}, \mathbb{R}^3)$ such that

$$\mathrm{curl}\, \mathrm{curl}\, E - k^2 N E = 0, \tag{10.23}$$

$$E = E^s + E^i, \tag{10.24}$$

$$\lim_{r \to \infty} (\mathrm{curl}\, E^s \times x - ikr E^s) = 0. \tag{10.25}$$

A proof of the existence of a unique solution to (10.23)–(10.25) can be found in [129]. It can again be shown that E^s has the asymptotic behavior given in (10.12). Unfortunately, in general, the electric far-field pattern E_∞ does not uniquely determine N (although it does in the case where the medium is isotropic, i.e., $N(x) = n(x)I$, where n is a scalar [59, 82]). However E_∞ does uniquely determine D [13], and a derivation of the linear sampling method for determining D from E_∞ can be found in [78]. Numerical examples using this approach for determining D when the medium is isotropic can be found in [80]. Finally, a treatment of the factorization method for the case of electromagnetic waves in an isotropic medium is given in [102].

In the analysis of the uniqueness of D and the linear sampling method, the *interior transmission problem* corresponding to (10.23)–(10.25) plays an important role. In particular, the interior transmission eigenvalue problem is to find $E, E_0 \in L^2(D)$ and $E - E_0 \in \mathcal{U}_0(D)$ such that

$$\nabla \times \nabla \times E - k^2 N(x)E = 0 \qquad \text{in} \quad D,$$
$$\nabla \times \nabla \times E_0 - k^2 E_0 = 0 \qquad \text{in} \quad D,$$
$$\nu \times E = \nu \times E_0 \qquad \text{on} \quad \partial D,$$
$$\nu \times \nabla \times E = \nu \times \nabla \times E_0 \qquad \text{on} \quad \partial D,$$

where

$$\mathcal{U}_0(D) := \{u, \nabla \times u \in H(\mathrm{curl}, D) \text{ such that } \nu \times u|_{\partial D} = 0, \nu \times \nabla \times u|_{\partial D} = 0\}$$

equipped with the inner product

$$(u,v)_{\mathcal{U}_0} = (u,v)_{L^2(D)} + (\nabla \times u, \nabla \times v)_{L^2(D)} + (\nabla \times \nabla \times u, \nabla \times \nabla \times v)_{L^2(D)}.$$

The values of k for which the preceding problem has nontrivial solutions are called *transmission eigenvalues*. In [31], it is proven that if $N > I$ or $N < I$, then the set of transmission eigenvalues is discrete and there exists an infinite set of real transmission eigenvalues accumulating at $+\infty$. Estimates similar to those in Theorem 6.22 are also obtained in terms of the first transmission eigenvalue and the matrix refractive index N. Other work in the study of transmission eigenvalue problem for Maxwell's equations are [35, 40, 66, 105].

References

1. Aktosun T, Gintides D, Papanicolaou V (2011) The uniqueness in the inverse problem for transmission eigenvalues for the spherically symmetric variable-speed wave equation. Inverse Problems 27:115004.
2. Aktosun T, Papanicolaou V (2013) Reconstruction of the wave speed from transmission eigenvalues for the spherically symmetric variable-speed wave equation. Inverse Problems 29:065007.
3. Angell T, Kirsch A (1992) The conductive boundary condition for Maxwell's equations. SIAM J. Appl. Math. 52:1597–1610.
4. Angell T, Kirsch A (2004) *Optimization Methods in Electromagnetic Radiation.* Springer, New York.
5. Arens T (2001) Linear sampling methods for 2D inverse elastic wave scattering. Inverse Problems 17:1445–1464.
6. Arens T (2004) Why linear sampling works. Inverse Problems 20:163–173.
7. Arens T, Lechleiter A(2009) The linear sampling method revisited. J. Integral Equations Appl. 21:179–203.
8. Boas Jr, Ralph P (1954) *Entire Functions.* Academic, New York.
9. Bonnet-BenDhia AS, Chesnel L, Haddar H (2011) On the use of t-coercivity to study the interior transmission eigenvalue problem. C. R. Acad. Sci., Ser. I 340:647–651.
10. Bonnet-BenDhia AS, Ciarlet P, Maria Zwölf C (2010) Time harmonic wave diffraction problems in materials with sign-shifting coefficients. J. Comput. Appl. Math 234:1912–1919.
11. Bressan A (2013) *Lecture Notes on Functional Analysis with Applications to Linear Partial Differential Equations.* American Mathematical Society, Providence, RI.
12. Buchanan JL, Gilbert RP, Wirgin A, Xu Y (2004) *Marine Acoustics. Direct and Inverse Problems.* SIAM, Philadelphia.
13. Cakoni F, Colton D (2003) A uniqueness theorem for an inverse electomagnetic scattering problem in inhomogeneous anisotropic media. Proc. Edinb. Math. Soc. 46:293–314.
14. Cakoni F, Colton D (2003) On the mathematical basis of the linear sampling method. Georgian Math. J. 10/3:411–425.
15. Cakoni F, Colton D (2003) The linear sampling method for cracks. Inverse Problems 19:279–295.

F. Cakoni and D. Colton, *A Qualitative Approach to Inverse Scattering Theory,*   287
Applied Mathematical Sciences 188, DOI 10.1007/978-1-4614-8827-9,
© Springer Science+Business Media New York 2014

16. Cakoni F, Colton D (2003) Combined far field operators in electromagnetic inverse scattering theory. Math. Methods Appl. Sci. 26:413–429.
17. Cakoni F, Colton D (2004) The determination of the surface impedance of a partially coated obstacle from far field data. SIAM J. Appl. Math. 64:709–723.
18. Cakoni F, Colton D (2005) Open problems in the qualitative approach to inverse electromagnetic scattering theory. Eur. J. Appl. Math. to appear.
19. Cakoni F, Colton D, Gintides D (2010) The interior transmission eigenvalue problem. SIAM J. Math. Anal. 42:2912–2921.
20. Cakoni F, Colton D, Haddar H (2002) The linear sampling method for anisotropic media. J. Comp. Appl. Math. 146:285–299.
21. Cakoni F, Colton D, Haddar H (2009) The computation of lower bounds for the norm of the index of refraction in an anisotropic media. J. Integral Equations Appl. 21(2):203–227.
22. Cakoni F, Colton D, Haddar H (2010) On the determination of Dirichlet or transmission eigenvalues from far field data. C. R. Math. Acad. Sci. Paris, Ser I 348(7–8):379–383.
23. Cakoni F, Colton D, Monk P (2001) The direct and inverse scattering problems for partially coated obstacles. Inverse Problems 17:1997–2015.
24. Cakoni F, Colton D, Monk P (2004) The electromagnetic inverse scattering problem for partly coated Lipschitz domains. Proc. R. Soc. Edinb. 134A:661–682.
25. Cakoni F, Colton D, Monk P (2010) The determination of boundary coefficients from far field measurements. J. Int. Equations Appl. 42(2):167–191.
26. Cakoni F, Colton D, Monk P (2011) *The Linear Sampling Method in Inverse Electromagnetic Scattering.* CBMS-NSF Regional Conference Series in Applied Mathematics 80, SIAM, Philadelphia.
27. Cakoni F, Colton D, Monk P (2005) The determination of the surface conductivity of a partially coated dielectric. SIAM J. Appl. Math. 65:767–789.
28. Cakoni F, Colton D, Monk P, Sun J (2010) The inverse electromagnetic scattering problem for anisotropic media. Inverse Problems 26:074004.
29. Cakoni F, Darrigrand E (2005) The inverse electromagnetic scattering problem for a mixed boundary value problem for screens. J. Comp. Appl. Math. 174:251–269.
30. Cakoni F, Fares M, Haddar H (2006) Anals of two linear sampling methods applied to electromagnetic imaging of buried objects. Inverse Problems 42:237–255.
31. Cakoni F, Gintides D, Haddar H (2010) The existence of an infinite discrete set of transmission eigenvalues. SIAM J. Math. Anal. 42:237–255.
32. Cakoni F, Haddar H (2013) Transmission eigenvalues in inverse scattering theory *Inverse Problems and Applications, Inside Out* 60, MSRI Publications, Berkeley, CA.
33. Cakoni F, Haddar H (2008), On the existence of transmission eigenvalues in an inhomogeneous medium. Applicable Anal. 88(4):475–493.
34. Cakoni F, Haddar H (2003) Interior transmission problem for anisotropic media. *Mathematical and Numerical Aspects of Wave Propagation* (Cohen et al., eds.), Springer, 613–618.
35. Cakoni F, Kirsch A (2010) On the interior transmission eigenvalue problem (2010) Int. J. Comp. Sci. Math. 3:142–167.
36. Chanillo S, Helffer B, Laptev A (2004) Nonlinear eigenvalues and analytic hypoellipticity. J. Functional Analysis 209:425–443.

37. Charalambopoulos A, Gintides D, Kiriaki K (2002) The linear sampling method for the transmission problem in three-dimensional linear elasticity. Inverse Problems 18:547–558.

38. Charalambopoulos A, Gintides D, Kiriaki K (2003) The linear sampling method for non-absorbing penetrable elastic bodies. Inverse Problems 19:549–561.

39. Chesnel L (2012) Étude de quelques problémes de transmission avec changement de signe. Application aux métamatériaux. Ph.D. thesis. École Doctorale de l'École Polytechnique, France.

40. Chesnel L (2012) Interior transmission eigenvalue problem for Maxwell's equations: the T-coercivity as an alternative approach. Inverse Problems 28:065005.

41. Cheng J, Yamamoto M (2003) Uniqueness in an inverse scattering problem within non-trapping polygonal obstacles with at most two incoming waves. Inverse Problems 19:1361–1384.

42. Collino F, Fares M, Haddar H (2003) Numerical and analytical studies of the linear sampling method in electromagnetic inverse scattering problems. Inverse Problems 19:1279–1298.

43. Colton D (2004) *Partial Differential Equations: An Introduction*. Dover, New York.

44. Colton D, Coyle J, Monk P (2000) Recent developments in inverse acoustic scattering theory. SIAM Rev. 42:369–414.

45. Colton D (1980) *Analytic Theory of Partial Differential Equations*. Pitman Advanced Publishing Program, Boston.

46. Colton D, Erbe C (1996) Spectral theory of the magnetic far field operator in an orthotropic medium, in *Nonlinear Problems in Applied Mathematics*, SIAM, Philadelphia.

47. Colton D, Haddar H (2005) An application of the reciprocity gap functional to inverse scattering theory. Inverse Problems 21:383–398.

48. Colton D, Haddar H, Monk P (2002) The linear sampling method for solving the electromagnetic inverse scattering problem. SIAM J. Sci. Comput. 24:719–731.

49. Colton D, Haddar H, Piana P (2003) The linear sampling method in inverse electromagnetic scattering theory. Inverse Problems 19:S105–S137.

50. Colton D, Kirsch A (1996) A simple method for solving inverse scattering problems in the resonance region. Inverse Problems 12:383–393.

51. Colton D, Kress R (1983) *Integral Equation Methods in Scattering Theory*. Wiley, New York.

52. Colton D, Kress R (1995) Eigenvalues of the far field operator and inverse scattering theory. SIAM J. Math. Anal. 26:601–615.

53. Colton D, Kress R (1995) Eigenvalues of the far field operator for the Helmholtz equation in an absorbing medium. SIAM J. Appl. Math. 55:1724–35.

54. Colton D, Kress R (2013) *Inverse Acoustic and Electromagnetic Scattering Theory*, 3rd edn. Springer, New York.

55. Colton D, Kress R (2001) On the denseness of Herglotz wave functions and electromagnetic Herglotz pairs in Sobolev spaces. Math. Methods Appl. Sci. 24:1289–1303.

56. Colton D, Leung YJ (2013) Complex eigenvalues and the inverse spectral problem for transmission eigenvalues. Inverse Problems 29:104008.

57. Colton D, Kress R, Monk P. (1997) Inverse scattering from an orthotropic medium. J. Comp. Appl. Math. 81:269–298.

58. Colton D, Monk P. (1999) A linear sampling method for the detection of leukemia using microwaves. II. SIAM J. Appl. Math. 69, 241–255.
59. Colton D, Päivarinta L (1992) The uniqueness of a solution to an inverse scattering problem for electromagnetic wave. Arch. Rational Mech. Anal. 119:59–70.
60. Colton D, Päivärinta L, Sylvester J (2007) The interior transmission problem. Inverse Problems Imag. 1:13–28.
61. Colton D, Piana M, Potthast R (1997) A simple method using Morozov's discrepancy principle for solving inverse scattering problems. Inverse Problems 13:1477–1493.
62. Colton D, Sleeman BD (1983) Uniqueness theorems for the inverse problem of acoustic scattering. IMA J. Appl. Math. 31:253–59.
63. Colton D, Sleeman BD (2001) An approximation property of importance in inverse scattering theory. Proc. Edinb. Math. Soc. 44:449–454.
64. Costabel M, Dauge M (2002) Crack singularities for general elliptic systems. Math. Nachr. 235:29–49.
65. Costabel M, Dauge M (1996) A singularly perturbed mixed boundary value problem. Comm. Partial Differential Equations 21:1919–1949.
66. Cossonnière A, Haddar H (2011) The electromagnetic interior transmission problem for regions with cavities. SIAM J. Math. Anal. 43:1698–1715.
67. Coyle J (2000) An inverse electromagnetic scattering problem in a two-layered background. Inverse Problems 16:275–292.
68. Engl HW, Hanke M, Neubauer A (1996) Regularization of Inverse Problems. Kluwer, Dordrecht.
69. Fredholm I (1903) Sur une classe d'équations fonctionelles. Acta Math. 27:365–390.
70. Friedman A (1969) Partial Differential Equations. Holt, Rinehart and Winston, New York.
71. Ghosh Roy DN, Couchman LS (2002) Inverse Problems and Inverse Scattering of Plane Waves. Academic, London.
72. Gilbarg D, Trudinger NS (1983) Elliptic Partial Differential Equations of Second Order, 2nd edn. Springer, Berlin.
73. Gintides D, Kiriaki K (2001) The far-field equations in linear elasticity – an inversion scheme. Z. Angew. Math. Mech. 81:305–316.
74. Griesmaier R, Hanke M, Sylvester J (to appear) Far field splitting for the Helmholtz equation.
75. Grinberg NI, Kirsch A (2002) The linear sampling method in inverse obstacle scattering for impedance boundary conditions. J. Inv. Ill-Posed Problems 10:171–185.
76. Grinberg NI, Kirsch A (2004) The factorization method for obstacles with a-priori separated sound-soft and sound-hard parts. Math. Comput. Simulation 66:267–279
77. Gylys-Colwell F (1996) An inverse problem for the Helmholtz equation. Inverse Problems 12:139–156.
78. Haddar H (2004) The interior transmission problem for anisotropic Maxwell's equations and its applications to the inverse problem. Math. Methods Appl. Sci. 27:2111–2129.
79. Haddar H, Joly P (2002)Stability of thin layer approximation of electromagnetic waves scattering by linear and nonlinear coatings. J. Comp. Appl. Math. 143:201–236.

80. Haddar H, Monk P (2002) The linear sampling method for solving the electromagnetic inverse medium problem. Inverse Problems 18:891–906.

81. Hähner P (2000) On the uniqueness of the shape of a penetrable, anisotropic obstacle. J. Comp. Appl. Math. 116:167–180.

82. Hähner P (2002) Electromagnetic wave scattering: theory. in *Scattering* (Pike and Sabatier, eds.) Academic, New York.

83. Hartman P, Wilcox C (1961) On solutions of the Helmholtz equation in exterior domains. Math. Zeit. 75:228–255.

84. Hitrik M, Krupchyk K, Ola P, Päivärinta L (2010) Transmission eigenvalues for operators with constant coefficients. SIAM J. Math. Anal. 42:2965–2986.

85. Hitrik M, Krupchyk K, Ola P and Päivärinta L (2011) The interior transmission problem and bounds on transmission eigenvalues. Math Res. Lett. 18:279–293.

86. Hitrik M, Krupchyk K, Ola P, Päivärinta L (2011) Transmission eigenvalues for elliptic operators. SIAM J. Math. Anal. 43:2630–2639.

87. Hochstadt H (1973) *Integral Equations.* Wiley, New York.

88. Hooper AE, Hambric HN (1999) Unexploded ordinance (UXO): The problem. *Detection and Identification of Visually Obscured Targets* (Baum, ed.), Taylor and Francis, Philadelphia.

89. Hörmander L (1985) *The Analysis of Linear Partial Differential Operators III.* Springer, Berlin.

90. Hsiao G, Wendland WL (2008) *Boundary Integral Equations.* Springer, Berlin.

91. Ikehata M (1998) Reconstruction of the shape of an obstacle from scattering amplitude at a fixed frequency. Inverse Problems 14:949–954.

92. Ikehata M (1999) Reconstructions of obstacle from boundary measurements. Waves Motion 30:205–223.

93. Isakov V (1988) On the uniqueness in the inverse transmission scattering problem. Comm. Partial Differential Equations 15:1565–1587.

94. Isakov V (1998) *Inverse Problems for Partial Differential Equations.* Springer, New York.

95. John F (1982) *Partial Differential Equations, 4th ed.* Springer Verlag, New York.

96. Jones DS (1974) Integral equations for the exterior acoustic problem. Q. J. Mech. Appl. Math. 27:129–142.

97. Y. Katznelson (9168) *An Introduction to Harmonic Analysis.* Wiley, New York.

98. Kirsch A (2011) *An Introduction to the Mathematical Theory of Inverse Problems,* 2nd edn. Springer, New York.

99. Kirsch A (1998) Characterization of the shape of a scattering obstacle using the spectral data of the far field operator. Inverse Problems 14:1489–1512.

100. Kirsch A (1999) Factorization of the far field operator for the inhomogeneous medium case and an application in inverse scattering theory. Inverse Problems 15:413–29.

101. Kirsch A (2002) The MUSIC-algorithm and the factorization method in inverse scattering theory for inhomogeneous media. Inverse Problems 18:1025–1040.

102. Kirsch A (2004) The factorization method for Maxwell's equations. Inverse Problems 20:S117-S134.

103. Kirsch A (2005) The factorization method for a class of inverse elliptic problems. Math. Nachr. 278:258–277.

104. Kirsch A (2008) An integral equation for the scattering problem for an anisotropic medium and the factorization method. *Advanced Topics in Scattering and Biomedical Engineering, Proceedings of the 8th International Workshop on Mathematical Methods in Scattering Theory and Biomedical Engineering.* World Scientific, New Jersey.

105. Kirsch A (2009) On the existence of transmission eigenvalues. Inverse Problems Imag. 3:155–172.

106. Kirsch A, Kress R (1993) Uniqueness in inverse obstacle scattering. Inverse Problems 9:81–96.

107. Kirsch A, Grinberg N (2008) *The Factorization Method for Inverse Problems.* Oxford University Press, Oxford.

108. Kirsch A, Ritter S (2000) A linear sampling method for inverse scattering from an open arc. Inverse Problems 16:89–105.

109. Kleinman RE, Roach GF (1982) On modified Green's functions in exterior problems for the Helmholtz equation. Proc. R. Soc. Lond. A383:313–332.

110. Kress R (1995) Inverse scattering from an open arc. Math. Methods Appl. Sci. 18:267–293.

111. Kress R (1999) *Linear Integral Equations*, 2nd edn. Springer, New York.

112. Kress R, Lee KM (2003) Integral equation methods for scattering from an impedance crack. J. Comp. Appl. Math. 161:161–177.

113. Kress R, Rundell W (2001) Inverse scattering for shape and impedance. Inverse Problems 17:1075–1085.

114. Kress R, Serranho P (2005) A hybrid method for two-dimensional crack reconstruction. Inverse Problems 21:773–784.

115. Kreyszig E (1978) *Introductory Functional Analysis with Applications.* Wiley, New York.

116. Kusiak S, Sylvester J (2003) The scattering support. Comm. Pure Appl. Math. 56:1525–1548.

117. Kusiak S, Sylvester J (2005) The convex scattering support in a background medium. SIAM J. Math. Anal. 36:1142–1158.

118. Lakshtanov E, Vainberg B (2012) Bounds on positive interior transmission eigenvalues. Inverse Problems 28:105005.

119. Lakshtanov E, Vainberg B (2012) Remarks on interior transmission eigenvalues, Weyl formula and branching billiards. J. Phys. A 25 12:125202.

120. Lakshtanov E, Vainberg B (2012) Ellipticity in the interior transmission problem in anisotropic media. SIAM J. Math. Anal. 44 2:1165–1174.

121. Lebedev NN (1965) *Special Functions and Their Applications.* Prentice-Hall, Englewood Cliffs, NJ.

122. Leung YJ, Colton D (2012) Complex transmission eigenvalues for spherically stratified media. Inverse Problems 28:2944956.

123. Levin B Y (1996) *Lectures on Entire Functions.* American Mathematical Society. Providence, RI.

124. Lions J, Magenes E (1972) *Non-homogeneous Boundary Value Problems and Applications.* Springer, New York.

125. Magnus W (1949) Fragen der Eindeutigkeit und des Verhattens im Unendlichen für Lösungen von $\Delta u + k^2 u = 0$. Abh. Math. Sem. Hamburg 16:77–94.

126. McLaughlin JR, Polyakov PL (1994) On the uniqueness of a spherically symmetric speed of sound from transmission eigenvalues. J. Differential Equations 107:351–382.

127. McLean W (2000) *Strongly Elliptic Systems and Boundary Integral Equations.* Cambridge University Press, Cambridge.
128. Mönch L (1997) On the inverse acoustic scattering problem by an open arc: the sound-hard case. Inverse Problems 13:1379–1392
129. Monk P (2003) *Finite Element Methods for Maxwell's Equations.* Oxford University Press, Oxford.
130. Morozov VA (1984) *Methods for Solving Incorrectly Posed Problems.* Springer, New York.
131. Müller C (1952) Über die ganzen Lösungen der Wellengleichung. Math. Annalen 124:235–264
132. Nintcheu Fata S, Guzina BB (2004) A linear sampling method for near-field inverse problems in elastodynamics. Inverse Problems 20:713–736.
133. Norris AN (1998) A direct inverse scattering method for imaging obstacles with unknown surface conditions. IMA J. Applied Math. 61:267–290.
134. Päivärinta L, Sylvester J. (2008) Transmission eigenvalues. SIAM J. Math. Anal. **40** 738–753.
135. Pelekanos G, Sevroglou V (2003) Inverse scattering by penetrable objects in two-dimensional elastodynamics. J. Comp. Appl. Math. 151:129–140.
136. Piana M (1998) On uniqueness for anisotropic inhomogeneous inverse scattering problems. Inverse Problems 14:1565–1579.
137. Potthast R (1999) Electromagnetic scattering from an orthotropic medium. J. Integral Equations Appl. 11:197–215.
138. Potthast R (2000) Stability estimates and reconstructions in inverse acoustic scattering using singular sources. J. Comp. Appl. Math. 114:247–274.
139. Potthast R (2001) *Point Sourse and Multipoles in Inverse Scattering Theory.* Research Notes in Mathematics, Vol 427, Chapman and Hall/CRC, Boca Raton, FL.
140. Potthast R (2004) A new non-iterative singular sources method for the reconstruction of piecewise constant media. Numer. Math. 98:703–730.
141. Potthast R, Sylvester J, Kusiak S (2003) A 'range test' for determining scatterers with unknown physical properties. Inverse Problems 19:533–47.
142. Pöschel J, Trubowitz E (1987) *Inverse Spectral Theory.* Academic, Boston.
143. Rellich F (1943) Über das asymptotische Verhalten der Lösungen von $\triangle u + \lambda u = 0$ im unendlichen Gebieten. Jber. Deutsch. Math. Verein. 53:57–65.
144. Riesz F (1918) Über lineare Funktionalgleichungen. Acta Math. 41:71–98.
145. Robert D (2004) Non-linear eigenvalue problems. Mat. Contemp. 26:109–127.
146. Rondi L (2003) Unique determination of non-smooth sound-soft scatteres by finitely many far field measurements. Indiana University Math. J. 52:1631–62.
147. Rundell W, Sacks P (1992) Reconstruction techniques for classical inverse Sturm-Liouville problems. Math. Comput. 58:161–183.
148. Rynne BP, Sleeman BD (1991) The interior transmission problem and inverse scattering from inhomogeneous media. SIAM J. Math. Anal. 22:1755–1762.
149. Schechter M (2002) *Principles of Functional Analysis*, 2nd edn. American Mathematical Society, Providence, RI.
150. Sevroglou V (2005) The far-field operator for penetrable and absorbing obstacles in 2D inverse elastic scattering. Inverse Problems 21:717–738.
151. Stefanov P, Uhlmann G (2004) Local uniqueness for the fixed energy fixed angle inverse problem in obstacle scattering. Proc. Am. Math. Soc. 132:1351–54.
152. Stephan EP (1987) Boundary integral equations for screen problems in $\mathbb{R}^3$. Integral Equations Operator Theory 10:236–257.

153. Stephan EP, Wendland W (1984) An augmented Galerkin procedure for the boundary integral method applied to two-dimensional screen and crack problems. Appl. Anal. 18:183–219.
154. Sylvester J (2012) Discreteness of transmission eigenvalues via upper triangular compact operator. SIAM J. Math. Anal. 44:341–354.
155. Tacchino A, Coyle J, Piana M (2002) Numerical validation of the linear sampling method. Inverse Problems 18:511–527.
156. Ursell F (1978) On the exterior problems of acoustics II. Proc. Cambridge Phil. Soc. 84:545–548.
157. Vekua IN (1943) Metaharmonic functions. Trudy Tbilisskogo Matematichesgo Instituta 12:105–174.
158. Xu Y, Mawata C, Lin W (2000) Generalized dual space indicator method for underwater imaging. Inverse Problems 16:1761–1776.
159. You YX, Miao GP (2002) An indicator sampling method for solving the inverse acoustic scattering problem from penetrable obstacles. Inverse Problems 18:859–880.
160. You YX, Miao GP, Liu YZ (2000) A fast method for acoustic imaging of multiple three-dimensional objects. J. Acoust. Soc. Am. 108:31–37.
161. Young RM (2001) An Introduction to Nonharmonic Fourier Series. Academic, San Diego.

Index

addition formula, 51
adjoint operator, 15, 167
analytic Fredholm theorem, 14
anisotropic medium, 85, 284
annihilator, 152

Banach space, 3
basis
  Schauder, 170
  equivalent, 170
  Riesz, 170
Bessel function, 48
Bessel's equation, 48
best approximation, 4
bounded operator, 7

Cauchy-Schwarz inequality, 3
compact operator, 8
compact set, 8
complete set, 5
conjugate linear functional, 94
conormal derivative, 88
crack problem
  Dirichlet, 243
  mixed problem, 243
cutoff function, 91

Dini's theorem, 24
direct sum, 5
Dirichlet eigenvalues, 99
Dirichlet problem
  crack problems, 243
  exterior domain, 102, 176
  Helmholtz equation, 98, 102

interior domain, 98
  Poisson equation, 96
Dirichlet-to-Neumann map, 100, 101
discrepancy principle, 30, 35, 38
double layer potential, 166
dual space, 7
duality pairing, 22

eigenelement, 12
eigenvalue, 12
electric dipole, 282
electric far field pattern, 281
electromagnetic Herglotz pair, 282
entire function, 264
  exponential type, 264
  order, 264
  type, 264

factorization method, 180, 193
far field pattern, 281
far-field equation, 76, 156, 213, 232,
  253, 282
far-field operator, 66, 177, 178, 211,
  232, 253, 282
far-field pattern, 64
fundamental solution, 51

Hadamard factorization theorem, 265
Hankel functions, 49
Helmholtz equation, 51, 208
  crack problems, 243
  exterior Dirichlet problem, 102, 176
  exterior impedance problem, 51, 207
  interior Dirichlet problem, 98

F. Cakoni and D. Colton, *A Qualitative Approach to Inverse Scattering Theory*,   295
Applied Mathematical Sciences 188, DOI 10.1007/978-1-4614-8827-9,
© Springer Science+Business Media New York 2014

interior impedance problem, 207
mixed problems, 207
Herglotz wave function
  approximation properties, 74, 154,
    213, 228, 251, 252, 283
  definition, 50, 113, 282
Hilbert space, 4
Hilbert-Schmidt theorem, 16

ill-posed, 27
impedance boundary value problem
  exterior, 51, 207, 280
  for Maxwell's equations, 279, 280, 283
  interior, 207, 283
imperfect conductor, 46
improperly posed, 27
inner product, 3
inner product space, 4
interior transmission problem, 114–122,
    136, 225, 226, 228, 284
inverse spectral theorem, 275

Jacobi–Anger expansion, 50

Laguerre's theorem, 266
Lax-Milgram lemma, 94
limited aperture data, 81
linear functional, 7
linear sampling method, 72, 74, 76–78,
    151, 152, 154, 156, 157, 159,
    199–201, 213, 233, 257, 283

magnetic far field pattern, 281
Maxwell's equations, 46, 86, 279
  anisotropic medium, 86, 284
  imperfect conductor, 46, 280, 283
  mixed problems, 280, 283
  perfect conductor, 176, 283
mildly ill-posed, 32
minimum norm solution, 38
mixed boundary value problems
  interior impedance, 283
  coated dielectrics, 222
  cracks, 243
  exterior impedance, 207, 280
  interior impedance, 207
  Maxwell's equations, 279, 280, 283
mixed interior transmission problem,
    225

modified interior transmission problem,
    115, 226
modified single layer potential, 58
monotonicity property, 133–135, 147

near-field data, 83
Neumann function, 49
normed space, 1
null space, 11

operator
  adjoint, 15, 167
  boundary, 72, 151, 213, 232, 253, 282
  bounded, 7
  compact, 8
  far field, 232, 282
  far-field, 66, 113, 177, 178, 211, 253
  non negative, 124
  normal, 177
  projection, 8
  resolvent, 12
  self-adjoint, 16
  strictly coercive, 124
  transpose, 152
orthogonal complement, 4
orthogonal projection, 8
orthogonal system, 4
orthonormal basis, 6, 170
orthotropic medium
  definition, 86
  scattering problem, 88, 93

Paley–Wiener theorem, 267
Parseval's equality, 6
Picard's theorem, 31
Poincaré's inequality, 90

quasi-solutions, 38, 41

radiating solution, 100
range, 12, 16
reciprocity relation, 65, 177, 191
regularization parameter, 29
regularization scheme, 29
relatively compact, 8
Rellich's lemma, 55
Rellich's theorem, 20, 89
representation formula, 53
representation theorem, 52

Riesz representation theorem, 14
Riesz's lemma, 10
Riesz's theorem, 11

scattering operator, 193
Schauder basis, 170
sesquilinear form, 94
severely ill-posed, 32
Silver–Müller radiation condition, 46,
    280
single layer potential, 56, 166
singular system, 31
singular value decomposition, 30
singular values, 30
Sobolev embedding theorem, 20
Sobolev spaces, 17
  $H(\mathrm{curl}, D)$, 280
  $H^1(D, \Delta_A)$, 91
  $H_0^1(D)$, 90
  $H_0^1(D, \partial D \setminus \overline{\partial D_0})$, 205
  $H_0^1(D, \partial D_D)$, 208
  $H^p(\partial D)$, 23
  $H^p[0, 2\pi]$, 17
  $H^{-\frac{1}{2}}(\partial D_0)$, 206
  $H^{-p}(\partial D)$, 59
  $H^{-p}[0, 2\pi]$, 21
  $H^{\frac{1}{2}}(\partial D_0)$, 204
  $H_0(\mathrm{curl}, \Omega_R)$, 281
  $H_{00}^{\frac{1}{2}}(\partial D_0)$, 206
  $H_{com}^1(\mathbb{R}^2 \setminus \bar{D})$, 59
  $H_{loc}^1(\mathbb{R}^2 \setminus \bar{D})$, 59
  $X(D, \partial D_I)$, 280
  $Y(\partial D_D)$, 281

$\tilde{H}^{-\frac{1}{2}}(\partial D_0)$, 206
$\tilde{H}^{\frac{1}{2}}(\partial D_0)$, 205
$\mathbb{H}^1(D, \partial D_2)$, 224
$H^1(D)$, 23
Sommerfeld radiation condition, 47
spectral cutoff method, 35
strategy, 30
strictly coercive, 96
surface conductivity, 222
surface impedance, 46

T-coercivity, 137
Tikhonov functional, 37
Tikhonov regularization, 36, 41, 185
trace theorem, 24, 60, 91
transformation operators, 267
transmission eigenfunction, 128
transmission eigenvalue problem,
    122–129, 131, 133, 134, 136,
    138–140, 142–144, 146, 147
transmission eigenvalues, 114, 128, 160,
    225, 269, 285
triangle inequality, 2

unique continuation principle, 54, 108

variational form, 97

wave number, 46
weak convergence, 38
weak solution, 59, 97, 98
well-posed, 27
Wronskian, 49

Printed in the United States
By Bookmasters